T0073530

MODELS IN BIOLOGY

MODELS IN BIOLOGY

History, Philosophy, and Practical Concerns

GEORG F. STRIEDTER

The MIT Press
Cambridge, Massachusetts
London, England

The MIT Press would like to thank the anonymous peer reviewers who provided comments on drafts of this book. The generous work of academic experts is essential for establishing the authority and quality of our publications. We acknowledge with gratitude the contributions of these otherwise uncredited readers.

This book was set in Adobe Garamond Pro and Berthold Akzidenz Grotesk by Westchester Publishing Services. Printed and bound in the United States of America.

Library of Congress Cataloging-in-Publication Data

Names: Striedter, Georg F., 1962– author.
Title: Model systems in biology : history, philosophy, and practical concerns / Georg Striedter.
Description: Cambridge, Massachusetts : The MIT Press, [2022] | Includes bibliographical references and index.
Identifiers: LCCN 2021033979 | ISBN 9780262046947 (hardcover)
Subjects: LCSH: Animal models in research. | Animal experimentation.
Classification: LCC R853.A53 S77 2022 | DDC 616.02/7—dc23
LC record available at https://lccn.loc.gov/2021033979

10 9 8 7 6 5 4 3 2 1

CONTENTS

PREFACE

Comparative medicine is functionally "peppered" throughout the scientific enterprise and is value added to biomedical research.

—MACY AND HORVATH (2017), P. 497

How did this book materialize? The answer is personal: I have been fascinated by animal behavior as long as I can remember, and in college I came to view it through an evolutionary lens. In graduate school, I added neurobiology as a research dimension and, consequently, explored both proximate and ultimate causes of behavior. Through all this work I became increasingly intrigued by species differences. At the same time, it appeared to me that many of my fellow neurobiologists became less and less interested in species differences and sought mainly to find the most convenient and popular "models" for their research. At one point, one of those colleagues asked me why I didn't promote my work on vocal learning in parakeets as a model for the understanding and treatment of human stuttering. I ignored this well-meaning advice and continued to investigate not just how brains control behavior, but also how and why different brains work differently. Of course, I was hardly alone in being fascinated by evolution and keen to study its diverse outcomes. However, it seemed to me that, even among comparative biologists, far more attention was given to commonalities than differences, to features retained from common ancestors rather than, in Darwin's words, descent *with modification*.

Compounding my unease was that, about a decade ago, I began to hear more and more complaints about how much biological research could not be replicated and how many treatments that worked in mice or other "model organisms" did not translate into effective treatments for humans. Still, most of the biologists I knew felt pressure to make their research translationally relevant (i.e., relevant to human health). Feeling this pressure myself, it occurred to me in early 2019 that I might try to make my own knowledge of comparative biology translationally relevant. I did not have a proper name for my pursuit at the time, but I was essentially contemplating the kind of comparative medicine that is mentioned in the Preface's opening quote. Specifically, I thought that I might try to analyze the pros and cons of the various model systems that are so prominent in biological research, asking which models are "best" for which specific purposes and how one might account for the translational failures as well as the undoubted successes.

This book is the result. I tried to find simple answers, but they are few and far between. The subject matter is complex, and good people may disagree about which model systems are ideal for which kinds of research. Therefore, I settled for a critical analysis that largely eschews taking sides. As I wrote the book, I often hoped that my efforts might help young scientists decide which model systems they should choose for their research. In addition, I hoped that I might influence more senior scientists who can still shape the future of their field. I knew from the outset that this project was extremely ambitious, and I still fear that it will be dismissed (or worse). However, I did enjoy surveying so many different areas of biology and trying to make sense of what I learned. To use Sydney Brenner's phrase (cited in Friedberg 2008, 9), I delighted in the challenge to "convert the vast amount of information that we are accumulating into knowledge." Hopefully, I have achieved some measure of success, by which I mean that you will find the book useful. It would also make me happy if my work helps to make comparative medicine a more coherent discipline, less peppered across publications and lines of inquiry.

Many people helped improve this book along the way. Jessica Bolker and Todd Preuss provided key initial triggers for my thoughts about model systems, and they provided helpful feedback on my draft. Christopher Platt, who worked for many years at the National Science Foundation as well as the National Institutes of Health, read the entire book and provided incredibly detailed, useful input. Rachel Ankeny, Jorge Busciglio, Hung Fan, Susan Fitzpatrick, Claudia Kawas, Cheryl Logan, Steve Mahler, and Tom Schilling also provided constructive feedback on specific chapters or sections. I am indebted to all of them. Of course, any remaining errors and odd opinions are my own. Finally, I am grateful to Robert Prior and Kate Gibson from MIT Press for handling the book so expertly, and to my wife, Anna, for her perennial support.

1 CRISES IN BIOMEDICAL RESEARCH

It has been said, indeed, that experiments performed on a dog or a frog may be conclusive in their application to dogs and frogs, but never to man, because man has a physiological and pathological nature proper to himself and different from all other animals. It has been further stated that to be really conclusive for man, experiments would have to be made on man or animals as near to him as possible. It was surely with this idea that Galen chose a monkey for his experiments, and Vesalius a pig, as subjects more closely resembling man in his omnivorous capacity. Even today, many people choose dogs for experiments, not only because it is easier to procure this animal, but also because they think that experiments performed on dogs can more properly be applied to man than those performed on frogs. How well founded are these opinions?

—BERNARD (1927/1957), PP. 122–123

Which animals, if any, should biologists study if they want to learn about human biology? As Claude Bernard, who is often regarded as a founding father of experimental physiology, discussed in the chapter's opening quotation, different people answer this question differently. Bernard himself performed experiments on a variety of different animals—including rabbits, frogs, and dogs—and made some important discoveries, notably the principle of physiological homeostasis (Jørgensen, 2001). However, those same experiments, in conjunction with Bernard's apparently rather cavalier attitude toward animal suffering, also contributed to the development of an organized anti-vivisection movement in Victorian England, which in turn led to the first (and over-due) legislation to regulate animal experiments (French, 1975; Guerrini, 2003).

Ever since those early days of experimental biology, societies have sought some sort of compromise between the desire to minimize animal suffering and the quest for knowledge and specific benefits to human health. The balance of these conflict-ing motivations has shifted over the years and varied across societies (e.g., Francione,

1996; Caruana, 2020). However, a fairly broad consensus has formed around the notion that scientists should minimize the suffering of animals and use "lower" animals as much as possible. As Russell and Burch (1959) discussed in their influential book on the "3Rs," scientists should *refine* experiments, *reduce* animal numbers, and *replace* sentient animals (i.e., animals presumed to have feelings) whenever this is possible without preventing scientists from answering important questions. According to Russell and Burch, the third of these recommendations—replacement—means that experimentalists should work on non-mammals whenever possible or avoid animals entirely by studying cultured cells or computer models. Similarly, the *Guide for the Care and Use of Laboratory Animals* states that researchers should avoid using animals or replace vertebrate animals with "animals that are lower on the phylogenetic scale" whenever possible (National Research Council, 2011b, p. 5).

These guidelines are well-intentioned but leave many questions unanswered. Given that the notion of a "phylogenetic scale" has long been disavowed by evolutionary biologists (Hodos & Campbell, 1969), which animals are lower than others? Is it justifiable to kill numerous lower animals when the findings do not extrapolate to humans? When is it sufficient to work with cultured cells or computer simulations rather than animals? Is it better to work on intact lower animals or cultured human cells? Should the choice of animal species depend on the study's purpose? Does it matter whether scientists are using the animals to test a novel cosmetic or a life-saving COVID-19 vaccine? And is it okay to kill any animals for research that does not promise direct, immediate benefits to human health? Again, different people tend to answer these questions differently. Among biologists, individuals must find their own comfort zone, balancing the perceived importance of their research with their own animal welfare concerns as well as the relevant regulations (Kwon et al., 2010; Franco et al., 2018). Nonscientists as well balance these factors in diverse ways and, therefore, vary in their attitudes toward animal research (Joffe et al., 2016; Bradley et al., 2020).

Importantly, the compromises involved in selecting research subjects have not received as much attention as they deserve, at least among biologists (Orzack, 2012). Anti-vivisectionists do discuss these issues, but their positions tend to be so fervently against all animal research that biologists find them unworkable. Historians and philosophers of science have scrutinized how biologists select their models and use them (e.g., Burian, 1993; Bolker, 2019; Ankeny & Leonelli, 2020; Dietrich et al., 2020), but most biologists pay little heed. Biologists do occasionally pen thoughtful commentaries or reviews on the pros and cons of various model systems (see chapter 2), but the vast majority of these papers advocate primarily for the model that the authors themselves are working on. Few are willing to critique the choices of other biologists, perhaps because they worry about being associated with animal rights activists or exposing their colleagues to extremist attacks (Endersby, 2007, p. 408).

Meanwhile, the balance of which models receive the lion's share of research attention shifts slowly over time (see chapters 3 and 4). One hears occasional outcries from those whose favored models are being phased out, but otherwise the changes receive only fleeting attention. This situation would be acceptable if the research consistently achieved its stated aims. However, biologists over the last two decades have realized that much of the supposedly "translational" research on model systems fails to generalize to humans, at least as judged by the alarmingly high failure rate of clinical trials.

1.1 THE TRANSLATABILITY CRISIS

Louis Pasteur, who developed some of the earliest vaccines (see chapter 5), once wrote that "there are no such things as applied sciences, only applications of science . . . which are related to one another as the fruit is related to the tree that has borne it" (see Wellems, 2010). This is true, I think, but distinctions between "basic" and "applied" science continue to be popular. One reason for this persistence is that, over the last twenty years or so, societies around the globe have increasingly pushed scientists to demonstrate that their research has tangible benefits (Zerhouni, 2003; Maienschein et al., 2008; van der Laan & Boenink, 2015). Within biology, those benefits include the development of novel biofuels, more productive crops, and bacteria that can digest plastic (Ru et al., 2020). However, the vast majority of applied research in biology focuses on the development of novel diagnostics, drugs, and other therapies. Such efforts are often called *translational research* because their aim is to "translate" laboratory discoveries into tools that physicians can use. Although one can distinguish various types or phases of translational research (Sung et al., 2003; Fort et al., 2017), the research covered in this book involves mainly efforts to apply knowledge obtained from animals or in vitro preparations (i.e., preclinical research) to the enhancement of human health (i.e., clinical research).

Before new drugs or vaccines can be used in humans, they must be approved by regulatory agencies such as the Food and Drug Administration (FDA) in the United States. This regulatory approval usually requires a series of human clinical trials to show that the compounds are safe and effective, unless such trials would be unethical or simply not feasible (e.g., if they involve exposure to chemical weapons). The clinical trials, in turn, are usually based on preclinical research with cultured cells and studies on one or more animal species. Sadly, each step along this process tends to be more expensive than the previous, with late-phase clinical trials often costing hundreds of millions of dollars, and the entire process typically taking a decade or more. It is profoundly distressing, therefore, to learn that the vast majority of promising compounds that have emerged from animal research have failed in clinical trials.

For example, Kola and Landis (2004) examined the fate of all the drugs developed by the 10 largest pharmaceutical companies between 1991 and 2000. The average rate of success for all these drugs going from first-in-human testing to regulatory approval in the United States and Europe was just 11%. Cardiovascular therapies were approved at a rate of 20%, but drugs for cancer and neurological disorders succeeded only 5% and 8% of the time, respectively. Most of the attrition occurred at relatively late stages of clinical development, when major costs had already been incurred; the main reasons for the failures were a lack of efficacy, toxic side effects, or other safety concerns. A larger analysis of 1,738 drugs developed by the top 50 pharmaceutical companies between 1993 and 2004 revealed a slightly higher mean success rate of 16% (DiMasi et al., 2010). Promising as this increase appeared, an even larger study of 4,451 drugs developed by 738 biotech and pharmaceutical companies between 2003 and 2011 once again yielded an overall success rate of merely 10%, with cancer, cardiovascular, and neurological drugs exhibiting the lowest approval rates (Hay et al., 2014).

The picture is even bleaker when you focus on specific disorders. Most depressing is the failure rate for drugs targeting Alzheimer's disease, which was 99.6% for the period between 2002 and 2012 (Cummings et al., 2014). Many additional Alzheimer's drug trials have been conducted since then, but all of them have failed. (The drug aducanumab received accelerated approval by the US Food and Drug Administration [FDA] while this book was in press, but the FDA's own panel of experts had previously recommended against its approval [Hoffman, 2020; FDA Commissioner, 2021].) In light of this high failure rate, several large pharmaceutical companies have closed their neurological divisions. Indeed, across all therapeutic areas, the number of new drugs approved per dollar spent on research and development has declined substantially, and rather steadily, since 1950 (figure 1.1A). This does not bode well for the future of drug companies or, more importantly, patients. There are some indications that approval rates have recently increased in a few therapeutic areas (e.g., rheumatoid arthritis and cancer immunotherapy), but the overall success rate for novel therapies remains distressingly anemic.

Multiple reasons likely account for this persistent crisis of biomedical translation. For one thing, novel drugs must generally be more effective (or have fewer side effects) than older drugs that are already available, which makes it harder for the newer drugs to get approved. In addition, the approval process has generally become stricter, although the FDA has occasionally loosened a few rules. Other likely explanations are that some of the clinical trials may have been flawed in some way (e.g., in the composition of their subject pool), or that some of the preclinical models were not, in fact, good

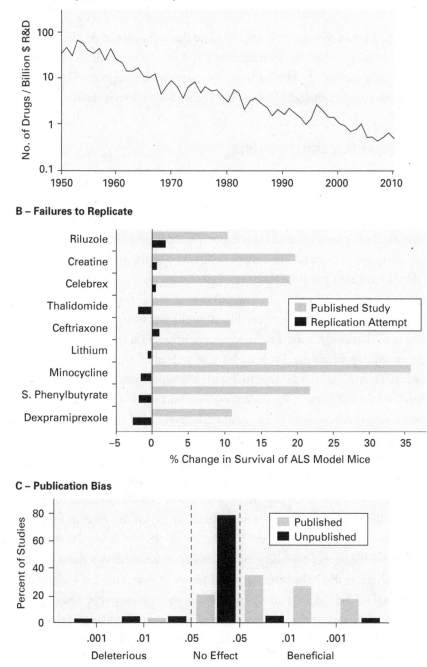

A – Declining Research Efficiency

B – Failures to Replicate

Published Study
Replication Attempt

% Change in Survival of ALS Model Mice

C – Publication Bias

Published
Unpublished

Deleterious   No Effect   Beneficial

Figure 1.1

Crises in biomedical research. (*A*) Declining research efficiency. The number of new FDA-approved drugs has decreased substantially over the last 60 years, relative to the inflation-adjusted investment in research and development (R&D). (*B*) Failures to replicate. The ALS Therapy Development Institute tested nine compounds in amyotrophic lateral sclerosis (ALS) model mice that had previously been reported to be beneficial in mice. The graph shows how they could not replicate the previously published findings. Eight of the illustrated compounds also failed in human clinical trials; only riluzole is approved for ALS. (*C*) Publication bias. A comparison between 42 published studies on motor (rotarod) performance in R6/2 mouse models of Huntington's disease and analogous unpublished studies, performed by PsychoGenics, Inc., shows that the published studies are biased toward more positive results (*x*-axis shows *p* values). Adapted from (A) Scannell et al. (2012); (B) Perrin (2014); (C) Brunner et al. (2012).

models for the human condition. We will explore this topic thoroughly in later chapters. Finally, it is quite possible that in many cases the preclinical research was simply not strong enough to support the clinical trials; this hypothesis is supported by the dawning realization that many findings in biology (and other fields) do not replicate reliably.

1.2 THE REPLICABILITY CRISIS

Since pharmaceutical companies stand to lose large sums of money if their clinical trials do not succeed, it is not surprising that they usually attempt to validate the preclinical research before proceeding to trials. When scientists from Bayer Health Care examined their own efforts in this regard for a period of four years and including 67 separate projects (focused on cancer, women's health, and cardiovascular therapies), they found that the published preclinical data were "completely in line" with their in-house preclinical results in only 20% to 25% of their projects (Prinz et al., 2011). More than half of the cases produced significant inconsistencies. According to the authors of that study, these findings are consistent with the conventional wisdom among venture capitalists that "at least 50% of published studies, even those in top-tier academic journals, can't be repeated with the same conclusions by an industrial lab" (Prinz et al., 2011, para. 7). An analogous effort at another large biotechnology company revealed that only 6 of 53 "landmark" preclinical studies on blood diseases and cancer could be replicated in-house (Begley & Ellis, 2012).

Even more frustrating results were reported by a nonprofit biotechnology company that focuses exclusively on the development of therapies for amyotrophic lateral sclerosis (ALS). They screened more than 70 drugs in 18,000 mice and found "no statistically significant positive (or negative) effects for any of the 70 compounds tested, including several previously reported as efficacious" (Scott et al., 2008, p. 5). The authors blamed the irreproducibility mainly on experimental design flaws in the earlier studies, which generally lacked the statistical power to overcome the variability of the studied animal models. A later study from the same company reported more specifically on eight compounds that had emerged as especially promising in the preclinical research but ultimately failed in clinical trials (figure 1.1B). When the company tried to replicate those original studies, the failure rate was 100% (Perrin, 2014). Similarly, a large-scale effort funded by the National Institutes of Health (NIH) to replicate preclinical studies on spinal cord injury revealed that the majority could not be replicated by other investigators (Steward et al., 2012).

All these failures to replicate previous studies are part of a general reproducibility crisis in science (Kafkafi et al., 2018). In psychology, for example, several large-scale efforts to replicate major findings yielded success rates of only 39% to 54% (Open

Science Collaboration, 2015; Klein et al., 2018). In the social sciences, the replication rate for 21 studies published in *Nature* and *Science* between 2010 and 2015 was 63% (Camerer et al. 2018). Even highly cited clinical studies are not immune to this problem, although it does appear that the replicability of clinical studies is greater than that of the preclinical research (Ioannidis, 2005b; Collins & Tabak, 2014). In aggregate, these findings prompted the leadership of the NIH to declare that "the irreproducibility of significant numbers of biomedical-research publications demands immediate and substantive action" (Collins & Tabak, 2014, p. 613). Fighting this problem is not an easy matter, however, as it stems from multiple causes.

Some of the blame falls on poor experimental design, such as experimenters assigning subjects to treatment groups nonrandomly, not being "blinded" to those assignments when they perform their data analyses, or performing interim statistical analyses to determine whether the sample size should be increased (Landis et al., 2012; Steward, 2016). Moreover, the majority of preclinical studies, as well as many clinical trials, are underpowered statistically, meaning that they do not include enough experimental subjects to reach robust conclusions on the hypothesized effects (Ioannidis, 2005a; O'Collins et al., 2017). These insufficient sample sizes do not just make it difficult to demonstrate true positive effects but also increase the likelihood of reporting false-positive results (Button et al., 2013a, 2013b).

A related problem is that experimenters usually perform many different experiments, as well as multiple analyses, but they tend to publish only their positive results (figure 1.1C) (Turner et al., 2008; Brunner et al., 2012; Drucker, 2016). Those results are typically reported as having a less than 5% chance of being false, but the statistical calculations tend not to include the failed efforts (Simmons et al., 2011; Tsilidis et al., 2013). A similar problem plagues high-throughput studies that test thousands of hypotheses at the same time (e.g., transcriptome analyses). Most of these studies do titrate the rate of likely false discoveries to some generally acceptable level (Benjamini & Hochberg, 1995), but the statistical procedures for combining the results of such studies across multiple iterations (i.e., replications) are nontrivial (Amar et al., 2017).

One way to fight this crisis of replication is to train young scientists more thoroughly in proper experimental design and statistics (Howells et al., 2014), which should help them realize that some time-honored methods for obtaining good results are in the long run misleading. In addition, scientists should be encouraged to preregister their experiments, describing what will be done and how, and then report even negative results. This approach extends the methodology of clinical trials to preclinical research and is becoming more popular in a few disciplines (Nosek et al., 2018).

However, even with preregistration, a non-negligible number of the statistically positive results would likely be due to chance (Colquhoun, 2014). Moreover, a strict requirement for preregistration may stifle exploratory inquiry by reducing the frequency of unexpected but ultimately important discoveries. To mitigate the latter dilemma, some authors have proposed that extreme experimental rigor should be required only for preclinical research that is about to become the basis for a clinical trial (Mogil & Macleod, 2017). If such preclinical trials fail, then canceling the clinical trials would save a great deal of effort and money and, importantly, minimize false hope.

1.3 RECKONING WITH BIOLOGICAL VARIATION

Although it may be tempting to attribute the current translatability crisis solely to replicability issues, this is probably wishful thinking. A pernicious additional problem is that many scientists are insufficiently cautious about extrapolating findings from their particular model systems to a much broader target population. To combat this bad habit in psychology, some authors have suggested that each published research paper should be accompanied by an explicit "constraints on generality" statement (Simons et al., 2017).

In the biomedical context, the analogous problem is overly optimistic generalization from the preclinical models to human patients. For example, researchers (and press releases!) often fail to highlight that the manipulations used to generate a disease-like state in animal or cell culture models tend to be poor or merely partial imitations of what causes the human disease. More specifically, a transgenic mouse carrying a human "disease gene" may not accurately model the human disease, if that disease has other or additional causes (Garner, 2014).

Unfortunately, the idea that species differences may be relevant to the preclinical modeling of human diseases is often neglected. As a former director of the NIH once observed, it is not unusual for biological psychiatrists to "assume that a mouse is a small rat, a rat is a small monkey, a monkey is a small human, and that all of these are 'models' for studying abnormal behavior or abnormal brain function in humans" (Insel, 2007, p. 1337). More generally, many biologists seem to think that "the fish is a frog . . . is a chicken . . . is a mouse" (Kimmel, 1989) and that, looking back on the last few decades of biological research, "it did not matter which animal you chose—fundamental processes were fundamentally conserved" (Grunwald & Eisen, 2002, p. 722).

This strong belief in our ability to generalize from specific findings in a few species to biology in general was boosted enormously by the success of genetics and molecular biology in revealing a number of broadly conserved principles of life. This optimism began to weaken, however, when biologists started to realize that, even within the human species, unrestrained generalization can cause real harm.

1.3.1 Taking Sex Differences Seriously

For many years, women were rarely included in clinical trials, largely because scientists wanted to protect them and their potential fetuses from the possibly detrimental effects of novel drugs. It eventually became apparent, however, that this approach shortchanged women by neglecting the possibility of biological sex differences and simply assuming that the results obtained in men would generalize (Clayton & Collins, 2014). As we now know, this assumption is quite often false: women occasionally exhibit different drug side effects than men, require different drug dosages, experience different symptoms for the same health issues (e.g., a heart attack), and benefit maximally from different therapies (Neigh & Mitzelfelt, 2016; Westergaard et al., 2019). Furthermore, some diseases that affect predominantly women were relatively neglected in the male-dominated clinical research. In recognition of these inequities, the NIH Revitalization Act of 1993 mandated the inclusion of women in clinical research (Mazure & Jones, 2015). Pursuant to this legislation, just over half of NIH-funded clinical research participants were women by 2014 (Clayton & Collins, 2014).

The inclusion of female subjects in preclinical research has proceeded at a much slower pace, especially in pharmacology and neuroscience. It was only in 2016 that the NIH officially mandated that all preclinical research must include sex as a biological variable, unless strongly justified otherwise. This edict received some pushback at the time, in part because scientists believed that cyclical variations in hormone levels made the results obtained from females more variable than those obtained from males. This assumption has now been falsified for several research domains (Beery, 2018).

However, researchers also complained that analyzing their results for males and females separately would force them to increase their animal numbers, which would increase experimental costs and duration (as well as animal suffering). Indeed, testing for sex differences can require more animals when differences exist, but being unaware of those differences can cause important, sex-specific findings to be missed (Cahill & Hall, 2017).

Fortunately, it has now become increasingly common for preclinical studies to include both males and females, and to report sex differences when they exist (Arnegard et al., 2020). In fact, it seems to me that the existence of sex differences is currently more broadly accepted for nonhuman animals than for humans, especially in domains other than reproduction (Cahill, 2014).

1.3.2 Personalized Medicine

The NIH Revitalization Act of 1993 also specified that NIH-sponsored clinical trials should include "members of minority groups and their subpopulations." This clause was included, at least in part, because minority groups were, like women, underrepresented in many earlier clinical trials (Nazha et al., 2019). This lack of proportional representation probably caused minorities to miss out on some therapies that might otherwise have been developed for them. Moreover, the underrepresentation surely led to a relative lack of information about drug dosages and side effects for some of those minorities.

Indeed, significant health disparities for some racial or ethnic groups have been reported frequently. For example, self-described Blacks/African Americans in the United States die of cancer far more frequently than non-Hispanic Whites, and their infant mortality is more than twice as high (Office of Minority Health, 2020). These health disparities arise in large measure from socioeconomic and environmental inequalities, but more inclusive clinical trials would surely provide more data relevant to those inequities.

A case in point is the discovery that a heart failure treatment, called BiDil, improves survival rates and time-to-hospitalization in self-identified Black patients (Temple & Stockbridge, 2007). This drug was first tested in a large clinical trial that revealed no statistically significant benefits for the patient population as a whole. However, a post hoc analysis indicated a significant reduction in heart attack mortality for the Black/African American subpopulation. Based largely on this finding, the FDA suggested that the company should do a follow-up trial specifically on Black patients (because a more inclusive trial would have been prohibitively expensive). This subsequent trial showed that BiDil reduced mortality by 43%, which then prompted the FDA to approve the drug specifically for African Americans. Critics complained that this selective approval risked stereotyping African Americans and was premature in any case, given that it remains unclear why the treatment has different effects in different subpopulations. The FDA responded that the mechanisms of action remain unknown for many drugs and that this ignorance is not enough to withhold approval, especially when such a withholding might well be deemed unethical.

A more complicated issue is that self-reported race or ethnicity is, in the words of NIH director Francis Collins, an "imperfect proxy" for biological or environmental differences that might be medically relevant (Collins, 2004). Indeed, contemporary biologists have pointed out that genetic variation between racial or ethnic groups is significantly smaller than the variation within these groups (Christensen, 2004), suggesting that the biological concept of race in humans is blurry at best. In fact, some "race-specific" diseases, such as sickle cell anemia, are less specific than commonly assumed and, therefore, often fail to be diagnosed in people of other races (Yudell et al., 2016). Because of those blurred boundaries between human races, and genetic heterogeneity among humans in general, biologists have increasingly focused not on race but on specific molecular features that correlate with disease risks and therapies.

This search for biomarkers that can guide medical research and treatments has been called precision medicine (National Research Council, 2011a). However, its original name—personalized medicine—seems more appropriate because it emphasizes that humans differ from one another in a wide variety of ways and may, therefore, benefit from personalized medical treatment.

Personalized medicine has been most successful in cancer research and therapy, mainly because the DNA of tumor biopsy samples can nowadays be sequenced rapidly, revealing mutations that are likely to have transformed the cells (i.e., made them cancerous). Scientists can then examine which of the many available cancer cell lines have similar mutations and how they had responded to diverse treatments in cell culture (Keshava et al., 2019). Such in vitro information can then be translated back to the human patient, giving them the treatment that worked best in the cultured cells and, hopefully, will be ideal for the patient as well. This approach is not without problems, but it works reasonably well for some types of cancer (see chapter 5).

Whether personalized medicine will be equally successful for other diseases remains unclear because the associations between most diseases and specific genes are relatively weak and do not readily suggest ideal treatments. In addition, extensive subdivision of the subject pools in clinical trials reduces the statistical power of the individual analyses, which weakens the evidence underlying highly personalized therapies (Djulbegovic & Ioannidis, 2018; Kimmelman & Tannock, 2018). Still, the number of drugs approved for personalized therapies has increased over the last decade, suggesting that the general approach is promising (Personalized Medicine Coalition, 2019).

1.3.3 The Differences That Broke a Clinical Trial

If sex and population differences deserve consideration in clinical trials, then the possibility of differences between humans and the animals used in preclinical research should also be considered carefully. Much of this book is about such differences, but a powerful introduction to this topic is provided by a famous clinical trial that ended disastrously, namely the trial of drug TGN1412. This trial is sometimes cited as evidence for the general claim that animal research cannot predict how humans will respond to test compounds, but the case is more nuanced and interesting than that.

TGN1412 was a genetically engineered monoclonal antibody against a signaling molecule (called CD28) that is located on the surface of specialized white blood cells, namely T cells. Initial experiments in laboratory rats showed that injections of TGN1412 caused a subset of these T cells (regulatory T cells) to proliferate and release molecules that reduce inflammation. This finding suggested that TGN1412 might be an effective treatment against autoinflammatory diseases such as rheumatoid arthritis. This hypothesis was supported by experiments on cultured T cells from humans. The investigators then injected various doses of TGN1412 into macaque monkeys and found that the animals tolerated at least 50 mg/kg of the drug. On the basis of these preclinical data, the company developing TGN1412 received approval to conduct a small clinical trial. On March 13, 2006, six men were injected with 0.1 mg/kg of the drug.

Sadly, in the words of a participating researcher, "on this day, scientific excitement turned into a nightmare" (Hünig, 2016, p. 3325). Within an hour of receiving the drug, all six patients developed headaches, nausea, and strong back pain. In the next few hours they developed high fevers and multiple organ failure. All were transferred to intensive care units; two of them required hospitalization for more than a week, and some lost fingertips and toes. Further analysis revealed that the drug had triggered a cytokine storm, defined as a massive release of inflammatory molecules (cytokines). Thus, the drug's effect was functionally the opposite of what the researchers had expected on the basis of their animal and in vitro studies. The company soon went bankrupt, but the question remained: what went wrong?

Follow-up studies revealed that the response of cultured human white blood cells to TGN1412 depends on how the drug is presented to the cells. In the original experiments, both the drug and the cells had floated freely in the culture medium, and under those conditions the cells do not release inflammatory cytokines. However, when the drug was attached to the inner surface of the culture dish before adding the cells, presumably mimicking the natural condition more closely, the cells released large amounts of the inflammatory cytokines (Stebbings et al., 2007). A separate set of experiments revealed that TGN1412 can also activate white blood cells when they are precultured at high density (Hünig, 2012). Either way, we can conclude that, had the original in vitro experiments been designed just slightly differently, they would have raised safety concerns much earlier, preventing the calamity.

Even more interesting for present purposes is that white blood cells from macaques do not, in contrast to their human counterparts, release inflammatory cytokines in response to TGN1412, even when the drug is bonded to the culture dish (Stebbings et al., 2007). The explanation for this species difference lies in a specific subset of the white blood cells, namely the memory effector T cells. In humans these cells express CD28 and release inflammatory cytokines upon activation by TGN1412. Macaques also have memory effector T cells, but they do not express CD28, so they cannot be activated by TGN1412 (Eastwood et al., 2010). Putting it all together, we can conclude that the safety testing of TGN1412 in macaques failed to trigger the cytokine storm observed in the human volunteers because the drug activates the cytokine-releasing T cells only in humans, not macaques. Because the in vivo testing of TGN1412 in rats had generated the desired anti-inflammatory response without triggering a cytokine storm, it appears that rats and monkeys have similar T cells, making humans the odd-balls here. This hypothesis has not been tested directly, but later studies showed that human immune cells tend to be hyperreactive even in comparison with those of chimpanzees (Soto et al., 2010).

One lesson scientists took away from the TGN1412 trial was that experimental drugs should not be given to multiple subjects at the same time. The drugs should also be injected more incrementally so that the trial can be aborted before unanticipated negative effects become severe. In addition, regulators began to recommend that dosages for first-in-humans trials should be based on minimum effective doses, rather than tolerability (Stebbings et al. 2009).

Less widely recognized is what the TGN1412 trial taught us about the dangers of extrapolating from animal and in vitro models to humans. Clearly, subtle differences in culture conditions can have profound effects on experimental results and so can species differences, even when the species are as closely related as humans and macaques. One may ask, therefore, how we can improve our ability to predict which preclinical findings will generalize to human patients, and which will not. More generally, how can we get better at predicting which species or in vitro preparations will be good models for any given human condition?

1.4 WHICH MODEL IS "BEST"?

One response to failed clinical trials is to declare preclinical research worthless and recommend that novel therapies be developed solely in humans (Horrobin, 2003). Although good arguments for more extensive clinical research can certainly be made, abandoning safety and efficacy testing in animals is unlikely to increase the rate of drug discovery, except perhaps in cases of an ongoing pandemic (these words were written in the midst of the COVID-19 scourge). The suffering of laboratory animals would certainly decrease if animal research were substantially curtailed, but veterinary medicine would be held back and human suffering would surely escalate. Nor is working with cultured human cells the panacea that some claim it to be. As TGN1412 exemplified, in vitro experiments often fail to mimic the in vivo conditions. So, if we accept that animal experiments are indispensable, then one must ask, as Claude Bernard did in the chapter's opening quotation, which species and laboratory strains should scientists select for their research?

The most common answer to this question is that it depends on the research question, but what exactly does that mean? Often it simply means that a researcher wants to do a particular type of experiment that can only be done, or is most conveniently performed, in a particular species, strain, or in vitro system. This is a reasonable rationale, but if the results from that experiment cannot be generalized as broadly as the researcher had hoped, is the rationale still good? If the larger aim of the research is to improve human health but the experimental findings do not translate to humans, is the research still justified? Many scientists would argue that the answer is yes, because such findings would still contribute to a larger store of biological knowledge that may yield unexpected dividends at some point down the road. This, too, seems fair enough, because the benefits of pure, basic research are widely recognized (Comroe & Dripps, 1974; Fricker, 2016; Flexner, 2017; Spector et al., 2018). Still, all experimental biologists, especially those just starting out in their careers (Yartsev, 2017), would likely benefit from thinking more comprehensively about how and why they selected their preferred model systems. This book will hopefully assist them in this task.

1.5 THIS BOOK'S APPROACH AND ORGANIZATION

My aim in this book is to review the question of model selection in biomedical research from a variety of perspectives, ranging from philosophy and history to the perspective of practicing biologists. Balancing the sometimes competing perspectives on biological model systems is difficult, and most readers are likely to object to some of the stated viewpoints. However, my overarching goal in this book is not to convince you of any specific position, but to provide you with a broad array of information, putative diagnoses, and food for thought that you can use to reach conclusions of your own.

As a first step in this effort, chapter 2 reviews some philosophical concepts relating to model systems. Many biologists have little patience for philosophy, and the famous physicist Richard Feynman supposedly once said that "philosophy of science is about as useful to scientists as ornithology is to birds" (quoted in Trubody, 2016). However, dismissing philosophy is not the same as having no philosophy, and Feynman actually knew the subject well. Personally, I have often found philosophical discussions of biology to be enlightening (e.g., LaFollette & Shanks, 1993; Bolker, 1995, 2017; Ankeny & Leonelli, 2011; Noble, 2011; Parkkinen 2017), and some insights from this work are distributed throughout this book. At a minimum, being aware of one's own assumptions about the scientific process can certainly not hurt (Burian, 1993).

To clarify what working with a model means, I distinguish between abstract models and material models. Both tend to be simpler than their target (i.e., the system being modeled), but the former are explicitly stripped down to some core elements, whereas the latter (which include all cell culture and animal models) come with their own complexities, which are partly unknown and may be undesirable. The chapter also explores some of the main assumptions biologists tend to make about how animals resemble one another and how those similarities vary with phylogenetic distance and biological level. These factors sometimes come into conflict with ethical concerns about animal suffering, which leads to complex compromises and cognitive dissonances that are often highly personal and rarely expressed.

Chapters 3 and 4 trace the history of the most widely used animal and in vitro models, respectively. Instead of focusing on the role of individual scientists in developing these models, the chapters foreground the models themselves: how their popularity has waxed or waned, and how the animals and cultured cells were sometimes modified to suit the research purposes. Special emphasis is placed on the kinds of research questions that the model species helped to answer, how those questions changed over the years, and why the species were selected initially. These chapters also address historical changes in societal attitudes toward animal welfare and the legislation that was passed to regulate some types of animal research. Overall, the various animal and in vitro models compete with one another in a sort of ecosystem of models. In such a system, the rise and dominance of specific models is usually adaptive, but it can, at times, hinder medical progress.

Chapters 5 and 6 review how animal and cell culture models have been used to develop therapies for bacterial and viral infections, cancer, cardiovascular diseases, and (in chapter 6) a variety of neurological disorders. None of these disease groups is discussed in detail; instead, emphasis is placed on how research on these types of diseases has taken advantage of, or been hampered by, model system differences. In the fight against infectious diseases, for example, the development of novel in vitro models greatly reduced the need for research animals, though treatments are still tested for safety and efficacy in animals, including nonhuman primates. By contrast, the development of surgical treatments for heart disease relied more heavily on animal research, especially large animal models (e.g., dogs). Cancer research has employed a variety of in vitro and animal models, including hybrid models that combine the two. Its special challenge is that cancer consists of many different cancer types; nonetheless, some good treatments for select cancer types are now available. The neurological disorders have been very difficult to study in model systems, and therapy development has been frustratingly slow. It is often said that the various models are partial or incomplete, but calling them "imperfect" may be more accurate.

The book's final chapter attempts to diagnose the basic problems that make biomedical research on model systems so challenging. This diagnosis is presented in the form of four different perspectives on the challenges and their potential solutions. Although these four perspectives are distinct, they are largely compatible with one another. In fact, the field as a whole can accommodate them all. I finish the book with some specific recommendations: (1) know your animals and cells, (2) standardize, but not too much, (3) learn from clinical trial failures, (4) embrace diversity, and (5) reckon with complexity.

Overall, the book offers no facile solution to the translatability crisis in biology, but it explores the problem and its underlying causes in considerable depth and from a variety of perspectives. Hopefully some of the presented information and analyses will stimulate you, the reader, to engage in further thought. Perhaps some of those thoughts will make a difference.

2 PHILOSOPHY OF MODELS IN BIOLOGY

All models are wrong, but some models are useful. So the question you need to ask is not "Is the model true?" (it never is) but "Is the model good enough for this particular application?"

—BOX ET AL. (2009), P. 61

When scientists say they are using a model, they mean that they are examining one thing, the model, in hopes of better understanding not just the model itself but the modeled system—the target. As the prominent statistician George Box articulated in the chapter's opening quotation, models of this kind are never identical to the target. However, they should be analogous to the target in some respects and more accessible to examination (Oreskes, 2007). Moreover, the analogies should be more than superficial. In particular, good models should resemble the target at a causal level so that the responses of the model to experimental manipulations can generate hypotheses about how the target should respond to similar interventions (Hesse, 1963; LaFollette & Shanks, 1995; Shapiro, 2004). If these predictions are confirmed, scientists will have learned something about the target; if they fail, then modifications of the model are called for.

Most philosophers and scientists probably agree on the preceding statements (e.g., see Creager et al., 2007), but scientists use many different kinds of models, and philosophers have offered a variety of classification schemes for them. For example, they have distinguished "theoretical models" from "model organisms" (Levy & Currie, 2015; Parkkinen, 2017), "exemplary models" from "surrogate models" (Bolker, 2009), "models of" from "models for" (Keller, 2000), and "model organisms" from other organisms that experimental biologists may study (Ankeny & Leonelli, 2011). The lack of perfect correspondence between these distinctions likely stems from the fact that different philosophers have emphasized different kinds of biological research, which differ in their aims. Particularly important is that some biologists try to extrapolate from their experimental model to a very limited set of targets, while others seek broad generalizations or even universals (Ankeny & Leonelli, 2011; Parkkinen, 2017). This variation in model usage and philosophical distinctions is important to recognize, but difficult to review without going into more details than this book can accommodate (or its intended audience would likely tolerate). Interested readers may consult the cited references to gain entry into this stimulating literature.

For present purposes, it is sufficient to distinguish between *abstract models*, which are purely conceptual, and *material models*, which in biology comprise mainly cultured cells, plants, and animals (Rosenblueth & Wiener, 1945; Pease & Bull, 1992). Although biologists increasingly use both types of models, the bulk of their research involves material models that are studied as proxies for one or more target systems (Bolker, 2009). Some biologists use such material models to improve livestock or agricultural products, but this book focuses mainly on the application of material models to human health and disease, which is why model plants (e.g., maize and *Arabidopsis thaliana*) are given short shrift. The book also neglects the use of material models to better understand nonhumans, which is central to veterinary medicine (Cunningham et al., 2010).

2.1 ABSTRACT MODELS

Abstract models mimic key elements and relationships of the target system and, importantly, do not include extraneous detail. Paradigm examples of abstract models are mathematical models, which typically contain multiple differential equations. Such models are not meant to be complete representations of the target system; rather, they include only those factors and relationships that are thought to be necessary for explaining the target system's behavior. If subsequent testing reveals that the model's behavior fails to mimic that of the target in some important respects, then the model should be modified.

This is, in fact, the principal purpose of abstract models: they reveal deficiencies in our thinking about the target and, therefore, require us to modify our assumptions (Gunawardena, 2014). In addition, abstract models can drive discovery by generating novel predictions that are then confirmed in the target system, but were previously unknown (Hesse, 1963). Although abstract models are ideally framed in the language of mathematics, they may also take the form of less precise conceptual models (Thompson, 1917). However, all abstract models are constructed to be as simple and elegant as possible, but no simpler (as Einstein supposedly once said).

Mathematical models have long been central to physics, but they were slow to take hold in biology. For example, the mathematician and early computer scientist Alan

Turing in 1952 published a mathematical model to explain the formation of structural patterns in organismal development, but his model was ignored by biologists for many years, partly because of questionable assumptions (Keller, 2002). More successful was Hodgkin and Huxley's (1952) mathematical model of the neuronal action potential using experimental data obtained from squid giant axons.

Mathematical models are still relatively rare in biomedicine, but pharmacology has long emphasized mathematical models (van der Graaf et al., 2016), and quantitative models have recently been developed for several aspects of cancer biology (Altrock et al., 2015). Abstract models that do not employ mathematics have, of course, been common in all realms of biology for many years, even if those conceptual models were sometimes rather fanciful (Arikha, 2007).

2.2 MATERIAL MODELS

As defined here, material models are concrete, physical entities that are analogous to the target system in important respects but have numerous properties that are not shared with the target and cannot easily be stripped away. The history of geology, for example, includes various material models of the earth's crust that were built from layered materials that could be compressed laterally to see if they would produce folds analogous to mountain ranges. These models "made the inaccessible accessible" (Oreskes, 2007).

Staying within geology, mathematical models of how the earth's magnetic field is generated from interactions between the earth's rotation and subterranean currents of molten metal are too complex to solve, and the entire earth is obviously too large for controlled experiments. Therefore, geophysicists have created a 1-meter diameter, rapidly spinning metallic sphere filled with liquid sodium to study its induced magnetic field (Lathrop & Forest, 2011). Among other things, they hope to discover whether the magnetic field of this material model will occasionally flip polarity, as the earth's field regularly does. In this example, the physical properties and dynamics of the material model were selected to match those of the earth closely, but differences clearly persist; the extent to which they are negligible remains to be seen.

Biologists likewise use material models to make the relatively inaccessible accessible. Most obviously, they perform experiments on animals that would be unethical to carry out on humans. In addition, biologists seek principles that are conserved across a large number of species. Of course, they cannot study all the earth's species and must, therefore, focus their studies on a select few, easily accessible species, which then function as representatives for a larger taxonomic group (Ankeny & Leonelli, 2011). The selected species are generally easy to breed and maintain in laboratory environments, and experimentally tractable. As the Nobel Prize–winning physiologist August Krogh put it in 1929, some species are more "convenient" than others for experimental investigation. In fact, Krogh exhorted experimentalists to consult zoologists to help them find the most convenient species for whatever problem they are studying. Hans Krebs (1975) later referred to this observation and advice as "the August Krogh principle" (for more on this, see section 2.6 and chapter 3; for a list of all winners of the Nobel Prize in Physiology or Medicine, as well as their principal model systems, please consult the appendix).

A very general problem with the use of model species in biology is that evolution can modify any aspect of a species, which means that no one species in any particular study is a priori guaranteed to be representative of any (or all) other species, even at the molecular and cellular levels. Biologists may work with "simple animals" in the hope that those models exhibit the target's key elements and interactions without additional complexities, but evolution is not always conservative. Indeed, supposedly simple animals are often more complex than people imagine, and their complexity may be quite different from our own (Sterling & Laughlin, 2015). An analogous problem complicates in vitro research: cell and tissue culture systems inevitably differ from their in vivo counterparts in multiple respects. As critics of cell culture work like to proclaim, *in vivo veritas!* (Matarese et al., 2012).

Scientists can modify their animal and cellular models to make them more similar to the target, but it remains essentially impossible to strip away from such models all their extraneous, idiosyncratic features. The closest approximation to a stripped-down material model in biology is the recently created "synthetic bacterium" whose genome contains only essential genes that are conserved across bacteria (Hutchison et al., 2016). However, such a minimal organism is not a good model for research on complex human diseases.

2.2.1 Model Scope and Validation

An important aspect of material models in biology is that they vary considerably in scope: some serve as a model for a very small subset of organisms (e.g., only humans), whereas others are meant to represent taxonomically broad swaths of species, including potentially all life. Unfortunately, as noted in the previous section, it is impossible to know a model's scope ahead of time. Only after examining the target do we know whether our model selection served our purpose. This dilemma, which has been called the "extrapolator's circle" (Steel, 2007), is sometimes cited as an argument against all animal experiments, but this conclusion would be warranted only if findings obtained in animals predict human results at levels near or below chance.

As we discussed in chapter 1, the high failure rate of clinical trials is certainly concerning, but animal research has, over the long course of history, made an undeniably positive contribution to medical progress (see chapters 5 and 6). Therefore, biological models have clearly sometimes been useful (or, as George Box would say, good enough for their particular applications). The real problem is that a model's utility is an empirical question and cannot, therefore, be assumed a priori (Levy & Currie, 2015). Moreover, even models that have proven to be useful for some research questions may fail when they are used to address other, different questions. In short, model systems can never be validated once and for all.

Within the field of biological psychiatry, researchers have proposed a multidimensional set of model validation criteria (McKinney & Bunney, 1969; Willner, 1984; Belzung & Lemoine, 2011). *Face validity* specifies that the model and its target should be similar "on the face of it," which in practice means that they ought to exhibit similar symptoms. By contrast, *construct validity* refers to a similarity of underlying causal mechanisms. The distinction between symptoms and mechanisms can sometimes be difficult to draw, especially for so-called biomarkers, but the distinction is valuable because models with strong construct validity may sometimes be useful even if they exhibit low face validity (van der Staay, 2006). The third main criterion is *predictive validity*, which refers to how well a model predicts the target's response to established or, ideally, new therapies (e.g., Fossat et al., 2014). Some models with high predictive validity may be little more than assays to find new drugs that are similar to those already known to be effective in the target system—a phenomenon dubbed "receptor tautology" (Geyer et al., 2012)—but the generation of successful predictions is clearly desirable for any model.

Although these three validity criteria have thus far failed to resonate beyond their field of origin, analogous criteria are likely used implicitly by most biologists as they evaluate material models in their research area. In general, construct and predictive validity seem more useful than face validity, mainly because assessments of the latter are often superficial or downright fanciful (Garner, 2014).

As mentioned previously, a major constraint on all material models is that they may harbor unsuspected, uncontrolled features that break their analogy to the target and thus limit their utility. This is true even for models that satisfy multiple validity criteria. Moreover, model selection is heavily influenced by factors that are rarely considered publicly, such as ethical concerns, costs, regulations, and societal pressures. Indeed, Dietrich et al. (2020) have identified a set of 20 interacting criteria that biologists typically consider, to varying extents, in selecting their models. Therefore, determining the "best" model for any particular question is usually a highly personal and subjective affair, heavily influenced by scientific tradition. This may be unavoidable, but the process would likely benefit from more extensive discussions (see chapter 7).

2.2.2 Model Modification

Biologists often try to mitigate the limitations of their material models by modifying their research organisms. For example, they frequently "standardize" their animals by creating highly inbred lines (see chapter 3). Although inbreeding tends to yield reduced fertility and other abnormalities in the short term, strong artificial selection over many generations of inbreeding can create viable lines that are genetically uniform. The reduced variability of these inbred strains makes it much easier to obtain statistically significant effects when the organisms are manipulated experimentally and compared with controls.

In addition, biologists often create modified strains in which one or more genes have been mutated, knocked out, or inserted into the genome. Once these genetically modified strains have been created, they can be maintained in central repositories and shipped to individual investigators as needed. The creation of such mutant libraries is a major reason why some species—notably mice, rats, yeast, and zebrafish—have become so widely used that they are sometimes called "model organisms" (Ankeny & Leonelli, 2011, 2020; Preuss & Robert, 2014; Katz, 2016) (figure 2.1). Researchers also tend to modify the genes of cultured cells, which is why cell lines in which such modifications are relatively easy to perform have become widely used (see chapter 4).

Aside from manipulating the genomes of their models, researchers often manipulate the environment in which those models live. For instance, the solutions in which cultured cells are grown (i.e., the media) can have profound effects on how the cells proliferate and develop. Similarly, raising animals in small cages with monotonous food is quite different from raising them in physically, socially, and nutritionally enriched environments, which in most cases more closely resemble the animals' natural state. For many studies it also matters whether the animals are raised in an artificially sterile environment or exposed to immune system challenges and allowed to grow their own intestinal microbiome. Even the temperature at which animals (or cells) are kept can have wide-ranging effects. For example, mice kept at the standard housing temperature of ~22°C, which is human room temperature, are chronically cold-stressed (Kokolus et al., 2013; Fischer et al., 2019). Although such environmental factors often receive less attention than genetic manipulations, they can have profound effects (figure 2.2) and should be considered integral to any animal or in vitro model.

Biologists sometimes study diseases that occur naturally in their research animals, but more frequently they modify the organisms to induce the disease. They may do so via genome manipulation or by means of other modifications that trigger processes thought to be involved in generating the disease (see chapter 5). The advantage of these artificial disease inductions is that the manipulated animals tend to develop the disease reliably and with relatively uniform symptoms. As in the case of inbreeding,

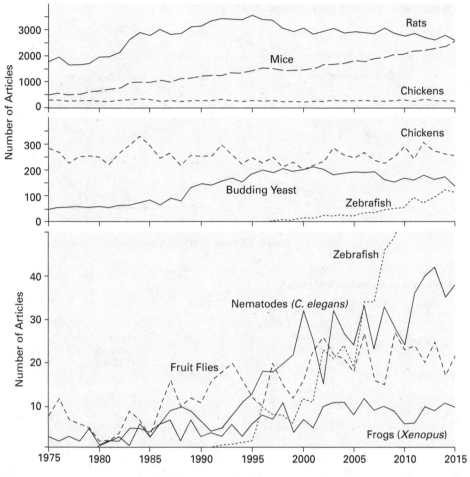

Figure 2.1

Variation in the popularity of select model organisms. Peirson et al. (2017) analyzed more than 3 million abstracts in the PubMed database for mentions of the principal National Institutes of Health–designated model organisms (spread evenly across 26 disease categories). The three graphs are from the same analysis, but show the data at different scales. Shown here are only data for the top eight of the 13 models in the analysis. Adapted from Peirson et al. (2017).

this reduction in interanimal variability makes it much easier to obtain statistically significant results in subsequent experiments, including tests of potential therapies. However, this research strategy assumes that experimenters have correctly identified what causes the human disease and that their manipulation triggers a cascade of disease processes that mimics human disease progression. If this assumption is false, then there is no guarantee that reversing the effects of an experimenter's manipulation will lead to therapies that are effective in humans. As we shall see in later chapters, this possibility should not be ignored.

A – Tumors Grow Faster in Cold-stressed Mice

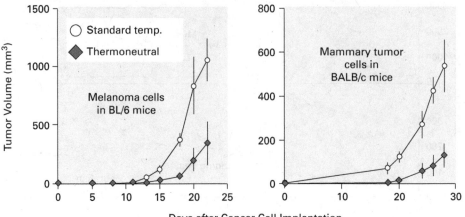

Days after Cancer Cell Implantation

B – But Not in Immunodeficient Mice

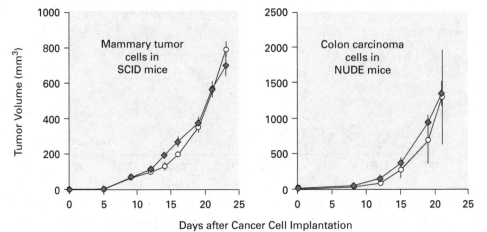

Days after Cancer Cell Implantation

Figure 2.2

Mice are cold-stressed at temperatures that humans find comfortable. (*A*) Tumors grow faster in cold-stressed mice. Various types of cancer cells injected under the skin of mice (of the same strains used to derive the cancer cells) form tumors that grow faster when the mice are kept at a standard human room temperature (22–23°C) than when they are maintained at the higher temperatures (30–31°C) that are thermoneutral (i.e., comfortable) for mice. (*B*) This difference in tumor growth rates is not observed in severe combined immunodeficient (SCID) or nude mice, which lack a functional immune system, suggesting that the difference seen in (A) results from an effect of cold stress on the immune system's ability to fight cancer (see chapter 5). Adapted from Kokolus et al. (2013).

Because biologists routinely modify their research organism, it is not surprising that those organisms are sometimes viewed as tools, fashioned to facilitate experiments. Philosophers, too, sometimes refer to material models as "instruments for investigation" (Morrison & Morgan, 1999). It is important to remember, however, that the cells and animals used in biomedical research cannot be built entirely to experimenter specifications. They come with natural baggage that may interfere with the experiments. As noted in the next section and throughout the book, models vary in the amount of this natural baggage, and this is (and should be) a major factor in model choice.

2.3 ASSUMPTIONS ABOUT MODEL FIDELITY

Many biologists may not have an explicit philosophy of science, but they all make assumptions about their research enterprise. Particularly relevant to the use of material models in biology are assumptions about how the similarities between a model and its target vary with phylogeny and across levels of biological organization.

2.3.1 Similarity versus Phylogenetic Distance

It is widely assumed that the difference between any two species increases steadily with phylogenetic distance, such that close relatives tend to be more similar than distant relatives. Indeed, randomly evolving traits tend to diverge as lineages branch, and in aggregate the degree of similarity between any two species decreases with the square root of the phylogenetic distance between those species, defined as the time since their evolutionary divergence (Letten & Cornwell, 2014). However, this square root rule holds only when the traits evolve randomly and when the degree of similarity is averaged over many traits. Once natural selection is added to the mix, individual traits may evolve faster or slower than expected under random evolution, which then disrupts the correlation between similarity and phylogenetic distance (Striedter, 2019).

In addition, species with short generation times tend to diverge more rapidly than species that reproduce more slowly. This is important because most of the model organisms used by biologists reproduce quite rapidly and, therefore, diverge from other species (including humans) more quickly than one would otherwise expect (Bolker, 1995; J. A. Thomas et al., 2010). The relatively short generation time of rats and mice also explains why the primate genomes, overall, are more similar to those of cats, dogs, pigs, and cows than they are to rodent genomes, even though rodents are more closely related to primates than those other species (i.e., carnivores and artiodactyls) (figure 2.3). Given that evolutionary change tends to accumulate across generations, it makes sense that the rapidly reproducing rodents "evolved away" (i.e., diverged) from their nonrodent relatives so fast and so much that they are now less similar to us than our more distant, less prolific relatives.

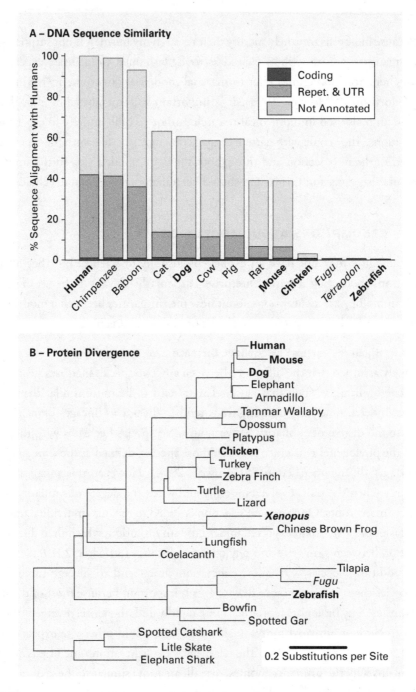

Figure 2.3

Genetic similarity versus phylogeny. (*A*) DNA sequence similarity. This histogram depicts the overall similarity of the human genome to that of various other vertebrates, using percent sequence alignment as a measure. By this measure the large animal models are more similar to humans than are rodents. UTR, untranslated region; Repet, repetitive DNA sequences. (*B*) Protein divergence. The topology of this dendrogram represents the phylogenetic relationships of various vertebrate species. The degree of protein sequence divergence (based on 243 conserved proteins) is represented by the lengths of the horizontal line segments. The dog proteins are more similar to their human counterparts than the mouse proteins, even though mice are more closely related to humans. Adapted from (A) Thomas et al. (2003); (B) Braasch et al. (2016).

Despite these complications, it still seems reasonable to believe that human diseases should, in the absence of ethical considerations (which we will discuss), be modeled in animals closely related to us, notably nonhuman primates or other mammals. Russell and Burch (of 3R fame; see chapter 1) referred to this view as the "high-fidelity fallacy" (1959) and argued, instead, that mammals do *not* always make better models for human diseases than non-mammals. Unfortunately, the examples they used to illustrate their argument were poor. A better way to address this question is to ask whether rodents are always the best nonprimate models for human diseases.

One might think that rodents are ideal nonprimate models for humans because rodents, together with rabbits and hares, comprise the mammalian lineage most closely related to primates. However, as I just noted, rodents have genetically diverged from primates more than some of our more distant relatives (figure 2.3). Furthermore, rodents tend to be much smaller than humans, and body size correlates with many therapy-related traits. It is well established, for example, that drug doses scale more reliably with body weight than phylogenetic relationship, at least among mammals (see chapter 7). Similarly, dogs and pigs are frequently used as models for cardiovascular research not because they are very closely related to humans, but because their hearts and blood vessels are very similar to ours in size (Lee et al., 2017); performing analogous experiments in mice would be extremely difficult. The existence of such size-dependent similarities between humans and dogs or pigs probably explains why the latter are often considered "higher vertebrates," even though the supposedly lower rodents are more closely related to humans.

Thus, we cannot offer a blanket statement on how similarities vary with phylogenetic relationship. It is generally true that similarities decrease with phylogenetic distance, but there are exceptions to the rule. Some of these exceptions produce greater than expected similarities (i.e., convergent similarities); others yield greater differences (e.g., in lineages with short generation times or in the form of special adaptations). Most biologists have learned to live with such exception-riddled rules.

2.3.2 Biological Hierarchies and "Disease Genes"

Despite the many obvious differences between species at the level of behavior and morphology, molecules and their interactions are often said to be highly conserved across broad swaths of phylogeny. One of the first scientists to articulate this conviction was the biochemist Alvert Jan Kluyver, who proclaimed that "from the elephant to butyric acid bacterium—it is all the same" (Friedmann, 2004). Or, as rephrased by the Nobel Prize–winning biochemist and molecular biologist Jacques Monod, "Anything that is true of *E. coli* must be true of elephants—only more so" (quoted in Friedmann, 2004). These statements were made during the early days of molecular biology but were reinforced by many subsequent discoveries of homologous genes in widely divergent species. To quote a more recent molecular biologist, "At the outset, no one could assume that vertebrates shared so much of their molecular biology with yeast, worms and fruit flies. However, the universality of the genes and circuits that govern developmental and regulatory phenomena has since become an article of faith" (Davis, 2004, p. 73). In contrast, higher level features, especially those relating to morphology and behavior, are thought to be less broadly conserved. As Hans and John Krebs wrote in an influential paper, "it appears that at the molecular level generalizations are often valid across a very wide range of species. But at higher and more complex levels such as those of ecology and behaviour (where specialised functional adaptations have evolved) generalisations must necessarily be more restricted" (1980, p. 380).

This widespread belief in molecular conservation is evident in the notion of "deep homology" (Shubin et al., 1997). The term was coined to indicate instances in which higher level characters (e.g., behavioral or morphological features) that were not considered to be evolutionarily conserved—that is, not homologous to one another—are discovered to involve the action of homologous genes or proteins. A widely cited example is the discovery that the eyes of insects and vertebrates, which are not homologous as eyes, involve the action of *pax-6*, a broadly conserved gene (Wagner, 2007). Such findings are certainly interesting, but deep homology is sometimes interpreted as meaning that the higher level characters *must* be homologous (as higher level characters) simply *because* they involve homologous genes (Strausfeld & Hirth, 2013). Such extensions of the deep homology concept are dubious, because the deep homology might simply represent independent co-option of homologous genes into the causal networks underlying nonhomologous higher level traits (Shubin et al., 2009; Scotland, 2010). After all, it is well established that genes and proteins may change their functions during phylogeny (e.g., Liao & Zhang, 2008). In short, evolution may proceed independently at different levels of biological organization, and the homology of genes need not imply that all their functions are conserved (Striedter, 1998, 2019; Striedter & Northcutt, 1991).

A more specific and medically relevant example of scientists assuming that evolutionary conservation is more extensive at the molecular level is the notion that homologs of human "disease genes" can be studied in distantly related species and still provide important insights into the corresponding human diseases, even if the examined animals do not themselves fall ill with the disease. Thus, Rubin et al. (2000) reported that fruit flies possess homologs to 177 of 289 human disease genes, which they defined as genes that are "mutated, altered, amplified, or deleted in a diverse set of human diseases." More recent studies have reported that 80% of the human disease genes have homologs in sea anemones, and that 52% of them originated prior to the origin of animals (Sullivan & Finnerty, 2007; Maxwell et al., 2014). The broad conservation of these disease genes is often taken to imply that important discoveries about the mechanisms underlying human diseases can be made in fruit flies, sea anemones, and even more distant relatives of *Homo sapiens*, but the extent to which this assumption is warranted remains very unclear.

By contrast, it is clear that homologous disease genes and proteins have often diverged considerably in terms of their molecular structure. Most of the proteins associated with Alzheimer's disease, for example, have homologous counterparts in mice, but roughly half of them are less than 80% identical to the human proteins in terms of their amino acid sequences (Hasselmann & Blurton-Jones, 2020). In addition, molecular nonhomology is more common than one might think. For example, only 20% of all the genes in 61 different strains of the *Escherichia coli* bacterium are conserved across all of the strains (Lukjancenko et al., 2010). Nor is it simple to determine whether homologous genes and proteins are more similar than homologous morphological traits and behaviors; homologous molecules are easy to compare in terms of their percent sequence identity, but analogous comparisons for higher level traits are much more difficult and controversial. Yet another problem is that molecular functions clearly can change, not just across stages of development (e.g., Wang et al., 2014) but also across phylogeny.

The latter problem is nicely illustrated by Huntington's disease, which is a neurodegenerative disease caused by a dominant genetic mutation in a gene called *huntingtin* (*htt*). The disease appears to occur naturally only in humans, but homologs of *htt* have been identified in other mammals and several invertebrates, including fruit flies and sea urchins (Rubinsztein et al., 1994; Djian et al., 1996; Tartari et al., 2008). Research on mouse models revealed that mutant htt protein (mHtt) tends to form intracellular aggregates, which were subsequently discovered in human brains as well (Davies et al., 1997; DiFiglia et al., 1997). Experiments in fruit flies revealed additional information about how mHtt is processed inside fly cells, and how these post-translational modifications affect the protein's toxicity (Barbaro et al., 2015).

Thus, scientists have certainly learned useful information about mHtt and its downstream effects from animal models (Marsh et al., 2012). However, those effects vary considerably across species (Yu-Taeger et al., 2017) (figure 2.4). Indeed, such differences are to be expected because the *htt* gene sequences have diverged considerably over the course of evolution. Human and mouse *htt* are only about 90% identical (Barnes et al., 1994), and the fly gene has fewer than half of the exons found in human *htt* (Li et al., 1999). Moreover, both the normal and the mutant forms of Htt interact with many other proteins and genes (Langfelder et al., 2016) that themselves diverged across phylogeny. These differences help to explain why most of the potential therapies that have emerged from animal research on Huntington's disease have not borne fruit

A – Gene Expression in HD Patients vs. Model Rats

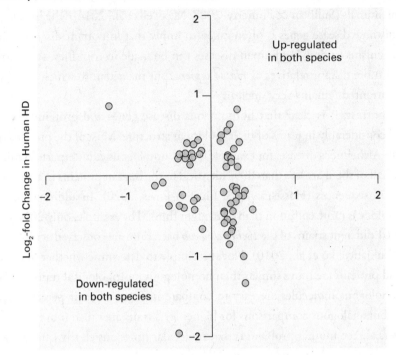

B – HD Patients vs. HD Model Rats vs. HD Model Mice

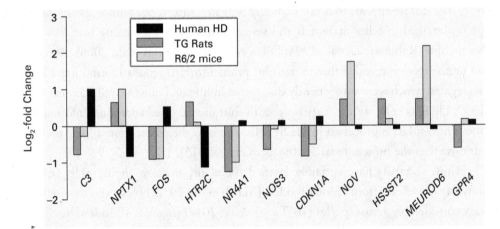

Figure 2.4

Differential gene expression in Huntington's disease (HD) patients and animal HD models. (*A*) Gene expression data from 61 genes that are up- or down-regulated in the striatum of human patients (versus human controls) and in transgenic rats expressing mutant human *huntingtin* (versus control rats). The data show that 61% of these genes were differentially expressed in opposite directions (i.e., discordantly) in the two species. (*B*) HD patients versus HD model rats versus HD model mice. This histogram compares the direction and magnitude of differential expression for 11 genes that are discordant between humans and rats (as shown in A) and also differentially expressed in HD model mice (of the R6/2 line). The mouse data consistently align with the rat data rather than the human data, as one would expect given the close relationship between the two rodent species. Adapted from Yu-Taeger et al. (2017).

in clinical trials (Wild, 2016), and why the currently most promising trials are aimed at suppressing mutant huntingtin, rather than modulating its downstream effects (for more on Huntington's disease, see chapter 6).

The important open question is, to what extent does the study of homologous disease-linked genes and proteins in diverse model systems allow for reliable extrapolations to humans? To answer this question, we must consider that disease-linked genes and proteins may change across phylogeny in terms of how they interact with other molecules, which themselves may have diverged. Such changes in gene and protein networks are likely to accumulate across phylogeny, just like evolutionary changes at other levels of biological organization. Therefore, the utility of disease models is likely to increase, at least on average, with their phylogenetic proximity to humans. However, as noted earlier, the genome as a whole has diverged at a significantly faster rate in mice than in some of the larger carnivores and artiodactyls (see figure 2.3). Thus, if the goal is human translation, it would—on average—be better to study human disease genes in nonhuman primates or, at least, other large mammals than in mice or more distant, non-mammalian relatives. Of course, as discussed later, this conclusion must be balanced against ethical and practical concerns.

2.4   DEALING WITH BIOLOGICAL COMPLEXITY

As research has progressed, the number of genes linked to human diseases has expanded to over 4,000 (Online Mendelian Inheritance in Man, 2020). Some diseases result from mutations in specific genes and run in families, but these Mendelian disorders tend to be quite rare, collectively affecting less than 5% of the population (Brunham & Hayden, 2013). The more common diseases, by contrast, tend to be associated with a large number of genes that, individually, contribute relatively little to disease risk and may be swamped in their effect by nongenetic factors such as smoking and nutrition (Joyner, 2011). Therefore, even diseases that have a substantial genetic component often have relatively common "sporadic" forms that cannot be predicted from family histories. Indeed, most minor risk genes were identified only through very large genome-wide association studies that examined thousands of patients.

In general, it remains quite difficult to specify how individual genes contribute to their associated diseases. Even in Huntington's disease, where mutation of a single gene causes the disease, it remains a serious challenge to explain how this mutation leads to neurodegeneration and, ultimately, death. The reductionism of molecular biology excels at identifying disease-related genes and proteins, as well as their potential molecular interactions, but explaining how you get from the molecular level to the level of organismal physiology and disease progression is more difficult. In the words of Salvador Luria, a founding father of molecular biology, "Everyone keeps going down, down, down, trying to be more reductionist, trying to see finer and finer, to find the basis of structure and function. . . . I think it's time start going up again—going in the opposite direction" (quoted in Weiner, 1999, p. 213). This is easier said than done, of course. As Seymour Benzer, who became famous for his work on the genetic mechanisms underlying learning and memory in flies, reportedly once said, "It takes a long time of going down before you start looking to go up again. Down is a much easier way to go" (quoted in Weiner, 1999, p. 213).

2.4.1 Endophenotypes

In an effort to manage this problematic aspect of traditional genetic reductionism, some biologists have proposed a middle-out approach that starts at some middle level of the biological hierarchy and then extends downward to the level of genes as well as upward to higher biological levels. Features at this middle level are called endophenotypes because they "occupy the terrain between disease symptoms and risk genotypes" (Insel & Cuthbert, 2009). An explicit focus on these endophenotypes was recently endorsed by the National Institute of Mental Health for the study of mental disorders (Kozak & Cuthbert, 2016). Part of the motivation for this move was the recognition that the diagnostic criteria for most psychiatric disorders (aka mental or behavioral disorders) do not align in any simple way with the findings emerging from genetics and neuroscience, thereby impeding the classic reductionist approach. Another motivator was that "animals will never have guilty ruminations, suicidal thoughts, or rapid speech. Thus, animal models based on endophenotypes that represent evolutionarily selected and quantifiable traits may better lend themselves to investigation of psychiatric phenomena than models based on face-valid diagnostic phenotypes" (Gottesman & Gould, 2003, p. 641).

Schizophrenia, for example, is essentially impossible to model fully in nonhumans because so many of its symptoms involve higher mental processes (see chapter 6). Therefore, researchers have focused on creating animal models of simpler processes that are thought to be components of human schizophrenia, such as deficits in sensorimotor gating, learning and memory, as well as anxiety and sleep disturbances (Tarantino, 2000). The hope is that the animal studies will elucidate the mechanisms underlying these endophenotypes and that the aggregate of all this information will explain human schizophrenia and lead to more effective treatments. Moreover, researchers hope that some of the same endophenotypes will also contribute to other complex mental disorders, such as autism spectrum disorder. Whether these aspirations will be fulfilled remains unclear. One may doubt, for example, that complex mental disorders are caused by linear combinations of endophenotypes; it seems more likely that some endophenotypes, like many of their underlying genes, can interact in a nonadditive manner.

An interesting twist on the endophenotype approach is provided by research on narcolepsy, a disease that is defined by chronic drowsiness and sudden attacks of sleep. Several breeds of dogs suffer from a very similar, likely homologous disease, and those dogs all have mutations in the gene for a receptor of a signaling peptide called hypocretin (Lin et al., 1999). Subsequent research showed that deletion of the hypocretin (aka orexin) gene in mice correlates with increased sleepiness (Chemelli et al., 1999). Importantly, later studies showed that humans with narcolepsy tend to lack hypocretin-expressing neurons (Thannickal et al., 2000). Thus, the animal research in this case made an accurate prediction about an endophenotype underlying a human disease.

The thought-provoking twist is that human narcolepsy patients generally do not exhibit mutations in the hypocretin or hypocretin receptor genes (Bućan & Abel, 2002). Indeed, the human disease is generally not what geneticists call familial (i.e., inherited), and considerable evidence suggests that the major type of narcolepsy in humans results from immune cells attacking the body's own hypocretin neurons (Kornum & Jennum, 2020). In short, the animal models and human narcolepsy patients share the disease and a disease-causing endophenotype, but they differ in the endophenotype's causal basis; apparently there are multiple ways to kill the hypocretin-expressing neurons or render them dysfunctional.

2.4.2 Systems Biology

Biological research has also been transformed by the rise of systems biology, which may be defined as "studying the interrelationships of all of the elements in a system rather than studying them one at a time" (Hood, 2003, p. 9). A common aim of systems biology is to integrate vast amounts of data at various levels of biological organization, especially at the molecular levels, by using statistical methods to identify pathways and modules of coregulated or interacting molecules (e.g., Langfelder et al., 2016; Ament et al., 2017, 2018). These molecular network features are then linked to higher level functions through correlational analyses, targeted manipulations, and gene ontology databases that collate information about the functions of individual genes. Some aspects of this systems biology paradigm have been criticized as "low-input, high-throughput, no-output science" (see Friedberg, 2008). However, systems biology is changing rapidly and promises to yield useful output by identifying critical modules, hubs, or choke points at which a system can be manipulated most effectively for therapeutic purposes (Langfelder et al., 2013; Swarup et al., 2019).

The basic tools of systems biology have been employed for more than a decade in the field of environmental toxicology (National Research Council, 2007). During this time, the field largely moved away from testing potentially toxic compounds in animals toward examining their effects on cultured cells, preferably of human origin. In particular, researchers examined whether those chemicals activated molecular toxicity pathways, also known as "adverse outcome pathways" (Jeong & Choi, 2018). A major benefit of this widely heralded paradigm shift was that it greatly reduced animal suffering and increased the rate at which compounds could be tested. The major drawback thus far has been a paucity of studies examining how well the various in vitro toxicity assays predict toxicity in animal models or humans. As a result, confidence in the new approaches appears to remain relatively low (Ginsberg et al., 2019). Indeed, regulatory decisions still tend to be based on both in vitro and animal tests as well as human epidemiological data. Despite these challenges, there have been calls to extend the adverse outcome pathways approach from toxicology to disease-oriented research (Langley et al., 2015). At that point, this approach merges with general systems biology, although the latter is less committed to the replacement of animal research with in vitro and in silico studies.

Another promising direction for systems biology is to develop mathematical models of complex molecular systems. The mathematical models allow investigators to make predictions about how the target system should respond to perturbations (Hood, 2003). The mathematical models themselves can also be examined for unexpected properties. A particularly interesting finding to emerge from such studies is that the models of many biological systems tend to be insensitive to variations in most of their parameters, a phenomenon referred to as "sloppy control" (Gutenkunst et al., 2007). This discovery implies that most systems can be controlled by varying a relatively small set of key parameters. Moreover, a system's sensitivity to changes in any one parameter depends to a surprisingly large degree on the values of the system's other parameters (Wagner, 2015), which is to say that the importance of any given parameter is highly context dependent. Finally, simulated evolutionary changes in a model's parameters can quickly change the distribution of parameter sensitivities, such that previously very important parameters become unimportant. This observation might explain why genes that are essential for survival or reproduction in one species or cell line are often nonessential in others (Liao & Zhang, 2008). More generally, these findings reinforce the notion that extrapolations of gene functions across species should be made cautiously (see chapter 7).

2.5 ANIMAL WELFARE AND COGNITIVE DISSONANCE

Experimental biologists are "often portrayed by animal rights groups as having little or no regard for animal welfare" (Franco et al., 2018, p. 14), but most biologists do care about the welfare of their research animals. Not all of them are as eager as the environmental toxicologists we just discussed to replace animal models with nonanimal alternatives, but most accept that doing so is a good long-term goal (Franco et al., 2018). In several countries (especially Japan) researchers even hold memorial ceremonies for their research animals (Narver et al., 2017). In that sense, biology has come a long way.

2.5.1 From "Can They Think?" to "Can They Suffer?"

The Romans killed thousands of animals in their circuses, including bears, tigers, and elephants, forcing them to fight with one another or with gladiators (Singer, 1990). The early Christians showed occasional flickers of concern for animals, but St. Thomas Aquinas represented the traditional Christian position when he proclaimed that "it matters not how man behaves to animals, because God has subjected all things to man's power . . . and it is in this sense that the Apostle says that God has no care for oxen" (1947, I-II, Q102, A6, ad.8). Sustained concern for animal welfare and rights did not emerge in Europe until the 18th century (French, 1975), though it arose earlier in other parts of the world with other religious traditions (Caruana, 2020). Particularly important was the contribution of the philosopher Jeremy Bentham. Before him, most Western philosophers had thought that animals deserved much less consideration than humans because, as Plato had argued, only humans have a "rational soul" (Rosenfield, 1940; Cottingham, 1978; Smith, 2010). Bentham changed this calculus by arguing that "the question is not, Can they *reason?* nor, Can they *talk?* but, Can they *suffer?*" (1789, p. 311; emphasis in original). He concluded that it would be acceptable to slaughter animals rapidly and relatively painlessly, at least for food, but that there was no justification for, as he put it, tormenting them. But which animals are capable of suffering? More generally, which ones have feelings? To use more technical language, which ones are sentient?

Answers to these questions cannot be obtained by direct observation of animal behavior because expressions presumed to be of pain or joy are not necessarily associated with conscious feelings (Hatfield, 2007). Language can give us some insight into the feelings of other humans, but without linguistic communication, the "other minds problem" becomes a Gordian knot (Avramides, 2019). Many people are willing to give other mammals and birds the "benefit of the doubt" (Harnad, 2016) with regard to sentience, perhaps because those animals are warm-blooded and cute, especially when they are young. But what about cold-blooded reptiles, fishes, or invertebrates? Does sentience come in degrees that correlate with brain size or behavioral complexity, as some have suggested (de Waal, 2019)? I tend to think so, but this idea is difficult (if not impossible) to demonstrate rigorously. In general, our willingness to attribute sentience to animals seems to fall off with their phylogenetic distance from us, but this is not exactly true—birds are more closely related to cold-blooded reptiles than to mammals. It is more accurate to state that most people attribute sentience in accordance with where they place animals along a vaguely defined continuum commonly known as the "phylogenetic scale." The obvious problem with this idea is that the notion of a phylogenetic scale has long been debunked (Hodos & Campbell, 1969).

Evolution produces bushes rather than ladders, and different ranking criteria lead to different arrangements of species along putative scales. To illustrate, consider cephalopods (octopuses and squids): as invertebrate mollusks, they are only distantly related to vertebrates, yet they have remarkably complex nervous systems and behavior. So where do they fit on the supposed phylogenetic scale? Recognizing this nonlinearity problem, the European Union recently labeled cephalopods "an exceptional invertebrate class" and gave them special research protections (Berry et al., 2015).

Another problem with the phylogenetic scale is that most humans rank pests much lower than pets, irrespective of their phylogenetic position. Most glaringly, most humans are far more inclined to protect cats and dogs than mice or rats, even though cats and dogs are more distantly related to humans (Foley et al., 2016). Perhaps we should replace the phylogenetic scale with a genetic divergence scale in which species are ordered by their overall genetic similarity to humans; according to such a scheme, dogs and pigs would rank higher than mice and rats (see figure 2.3). However, such a genetic divergence scale has problems of its own. Moreover, many genes and other traits clearly do follow phylogeny, for those "phylogenetic signals" are what scientists use to reconstruct phylogenies.

2.5.2 Animal Welfare Laws and Regulations

Given the complicated and evolving nature of our collective views on animal sentience, it is not surprising that the laws and regulations concerning animal experiments have changed over the years and are not completely rational. For example, the first significant animal welfare legislation, the United Kingdom's Cruelty to Animal's Act of 1876, extended protections to all vertebrates, but its initial draft had excluded all cold-blooded animals, including reptiles, fishes, and frogs. By the time the act was passed, it applied to all vertebrates and excluded all invertebrates. This changed in 1993, when the United Kingdom decided to regulate experiments on the common octopus; those regulations were extended to all cephalopods in 2013 and, as mentioned in the preceding section, now apply throughout most of Europe.

The situation in the United States is more complex. The current version of the Animal Welfare Act (US Department of Agriculture, 2020) protects all warm-blooded animals but includes the following proviso:

> The term "*animal*" means any live or dead dog, cat, monkey (nonhuman primate mammal), guinea pig, hamster, rabbit, or such other warm-blooded animal, as the Secretary [of the USDA] may determine is being used, or is intended for use, for research, testing,

experimentation, or exhibition purposes, or as a pet; but such term excludes *(1)* birds, rats of the genus Rattus, and mice of the genus Mus, bred for use in research, *(2)* horses not used for research purposes, and *(3)* other farm animals, such as, but not limited to live-stock or poultry, used or intended for use as food or fiber . . . (7 U.S.C. §2132(g))

In response to public pressure, the US Congress in 1970 gave the US Department of Agriculture (USDA) the authority to include mice and rats in its regulations, but the USDA declined. Because the USDA collects animal numbers only for the species under its purview, it is difficult to obtain precise estimates of the total numbers of animals used for research in the United States. The best one can currently do is to request records from individual research institutions about how many animals of the various species they are housing on any given day, on average (figure 2.5A). By comparison, the United Kingdom publishes annual reports on how many procedures were approved for the various vertebrate species, plus cephalopods (figure 2.5B).

Despite the species gaps in record keeping by the USDA, the use of all vertebrates in biological research in the United States is regulated by the Public Health Service, which administers the National Institutes of Health (NIH) and thus provides most of the support for US animal research. As a result, all research institutions that receive money from the NIH must follow its regulations regarding animal research. Still, mice and rats continue to be exempted from federal inspections and reporting requirements.

2.5.3 Conflicting Attitudes

Ethical concerns frequently conflict with practical concerns, especially for experimental biologists, farmers, and people who eat meat or need medicines derived from animal research. Even Mary Beth Sweetland, a former vice president of People for the Ethical Treatment of Animals (PETA), regularly took insulin for her diabetes, even though the discovery of insulin was clearly based on animal experiments. She may have justified this choice by arguing that she needs her life to fight for the rights of animals, but the case still illustrates the internal conflicts that animal experiments may spawn. Importantly, many people harbor such conflicts without being fully aware of them, creating cognitive dissonance.

In general, animal rights advocates tend to believe that animals and humans have roughly the same capacity for suffering and thought, but many also claim that animals are so different from humans that experiments on animals can tell us nothing about human biology. These beliefs are inconsistent with one another, unless we are prepared also to believe that feelings and thought are based on something other than biological processes.

Biologists, in contrast, tend to have the opposite problem (Shapiro, 2004). They often think that animals and humans are so similar to one another that the former can

A – Animal Inventory in US Research Institutions

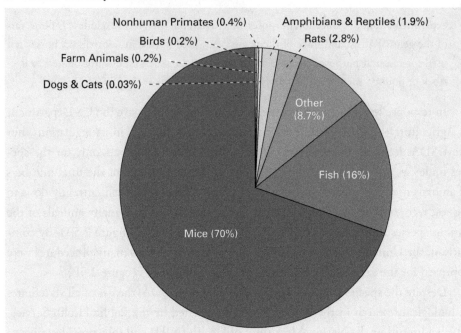

B – UK Experimental Procedures by Species, 2018

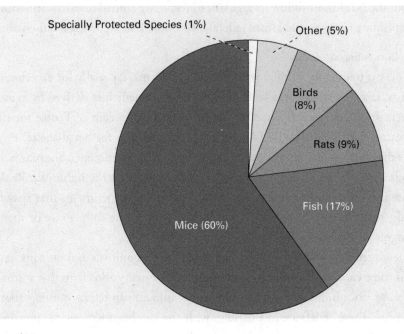

Figure 2.5

Frequency distribution of vertebrates used in research. (*A*) Animal inventory in US research institutions. Because the US government does not track the number of rodents and birds in research, Goodman et al. (2015) obtained average daily animal inventory numbers from 21 major US research institutions from 2008–2012. Mice clearly dominate this data set, which is limited to vertebrates. (*B*) Distribution of vertebrate species used for experimental procedures in the United Kingdom in 2018, as reported by the Home Office. The specially protected species include nonhuman primates, dogs, cats, and horses. Based on data in (A) Goodman et al. (2015); (B) Home Office (2019).

serve as models for the latter in most instances. Yet biologists also tend to believe that the suffering of a mouse, for example, is not as ethically troubling as that of a human. The resulting dissonance was clearly recognized by Anna Kingsford, one of the first women ever to obtain a medical degree, as well as an influential vegetarian and anti-vivisectionist. As she wrote in 1883,

> it is precisely the subtle but enormous differences existing between the manifestations and character of the nervous system as we see them in man and as we see them in other animals, which distinguishes the former from the latter, and which endows vivisectors with the legal right they now possess to inflict on anthropoid apes injuries and mutilations which, if they inflicted the same on men, would be held to render the perpetrators guilty of crime. When, therefore, it is understood that this occult nervous differentiation is capable of constituting a distinction so vast, how is it possible to suppose that the study of biological function in the beast is capable of explaining satisfactorily the mysteries of human life? (cited in French, 1975, p. 316)

Compared with the concerns expressed over animal experiments, sympathy for animals we use for food is rather limited, even though farm animals outnumber laboratory animals by at least two orders of magnitude. Despite major gaps in the reporting of animal numbers, the number of vertebrate animals used in research worldwide is probably below 100 million per year (Statista Research Department, 2016). By comparison, humans consume approximately 50 billion chickens per year, excluding all the male chicks that are killed right after hatching because they cannot lay eggs and are not needed for breeding (Thornton, 2019).

Importantly, virtually all these chickens are killed without anesthesia, which is not the practice in animal research. Particularly brutal is the mass killing of animals when herds or flocks get infected with serious diseases. For example, during the 2003 outbreak of Newcastle disease, the State of California ordered 3.1 million chickens to be killed within a few weeks. One farmer killed 30,000 of his chickens in a wood chipper, an act that was later judged to be "standard industry practice" (Gillick, 2003). Similarly, farmers in just one US state (Minnesota) are estimated to have killed 90,000 excess hogs during six weeks in early 2020 because local meatpacking plants were closed on account of the COVID-19 pandemic (Bailey, 2020); by comparison, US biologists kill roughly 50,000 pigs for research every year (APHIS Annual Report Animal Usage by Year, 2019).

A cognitive dissonance that is of more direct and practical concern to experimental biologists is the push from federally mandated institutional animal care and use committees (IACUCs) to *reduce* the number of animals used in research, while statisticians are telling the biologists that their studies are underpowered and should include *more* animals, including both sexes (see chapter 1). Similarly, IACUCs are instructing biologists not to duplicate existing research, while the ongoing replicability crisis has highlighted the need for more replication. In short, biologists are regularly confronted with a variety of conflicting exhortations.

2.6 COMPARATIVE BIOLOGY AND AUGUST KROGH REVISITED

Although popular views and regulations concerning animal welfare are imbued with the idea of a phylogenetic scale, which is centered on species differences, many biologists are more engrossed with species similarities. Especially molecular biologists have traditionally been interested only in "universals" and considered variation to be uninteresting noise (Orzack & McLoone, 2019). As Rowland Davis (2003) put it in his book on the use of model microbes in biology,

> The rise of model organisms had an ironic effect. The reductionist approach, in favoring simple systems, brought scientists to model microbes, and genetic rationales trapped its professionals in their use. The result was a concentration on a few organisms explicitly used to discover generalizable principles. Indeed, geneticists, in standardizing stocks and media, provided artificial representatives of various species living in an artificial environment. . . . the appreciation of diversity and complexity was lost for some time among molecular biologists. A curious construction of biology arose in which many biologists spoke of "the cell" or "the organism" when in fact they were studying *N. crassa* or *E. coli* or *S. cerevisiae*. A Platonic view of life emerged in which model organisms were the reality and the rest of the living world was a chaos of variants and exceptions. The irony is that even as neo-Darwinian views of evolution freed us from such thinking, molecular biology acquired an even more rigorous typological stance than that of Linnaean taxonomists of the early 19th century. (p. 254)

Modern molecular biology has broadened its focus, but a strong emphasis on similarities continues to pervade much of biology. It is even inherent in Krogh's principle (see section 2.2) because John Krebs, who named this principle, expected the discoveries made in the "most convenient" experimental species to generalize across species, at least at the molecular and cellular levels (Krebs & Krebs, 1980). This is why he considered the study studied species to be "examples" rather than "models" (Krebs, 1975). It is true that "Krogh organisms" (Green et al., 2018) are typically selected for having extreme adaptations, which is what makes them so accessible to inquiry, and that the study of these creatures "may reveal general principles not readily observable in less extreme species" (Pollak, 2014, p. 442). However, the discovery of general principles does not preclude the existence of fundamental species differences. To underscore this point, consider Krogh's own research.

August Krogh was a comparative physiologist who studied a wide variety of organisms, including many invertebrates, and focused much of his research on the topic of gas exchange (Krogh, 1941). How do animals absorb oxygen and shed carbon dioxide? How do the mechanisms they adopt vary with the environment in which they live (e.g., aquatic versus terrestrial habitats), with body size, and with specific aspects of their anatomy? To answer such questions, Krogh used general principles, such as those governing gas diffusion and absorption, as well as scaling laws that govern surface-to-volume ratios. His principal goal was not to discover new universals but to reveal how such principles were implemented differently in different species. He wanted to know what solutions evolution brought forth, given the constraints of physicochemical laws and biological structure-function principles. Given this background, it becomes clear why Krogh wrote that "the route by which we can strive toward the ideal [state of general physiology] is by a study of the vital functions in all their aspects throughout the myriads of organisms" (1929, p. 202). He sought to explain variation in terms of general principles, which is quite different from studying the most convenient species to understand them all.

In short, Krogh advocated for a truly comparative approach that focuses on differences as much as similarities. Physiology has generally retained this orientation, but other fields have not. For example, Daniel Lehrman lamented the loss of a truly comparative approach in behavioral research during the heyday of behaviorism: "The value of comparison comes not from the merging of different levels into a misleadingly unified conception of behavior but from the development of an evolutionary perspective which enables us to appreciate the emergence of new qualities without neglecting the underlying continuities and their transformations" (Lehrman, 1971, p. 468). Of course, this tension between those who seek universals and those who emphasize variation is hardly new—nor confined to biology. As Francis Bacon wrote in 1620,

> The greatest and, perhaps, radical distinction between different men's dispositions for philosophy and the sciences is this, that some are more vigorous and active in observing the differences of things, others in observing their resemblances . . . each of them readily falls into excess. (cited in Friedmann 2004)

As detailed in chapter 7, the kind of truly comparative perspective advocated by Lehrman and Krogh helps explain the current translatability crisis and suggests a more productive way forward. But first, let us explore how biologists have used diverse material models to address human diseases and other frailties.

3 A HISTORY OF ANIMAL MODELS

> For a large number of problems there will be some animal of choice or a few such animals on which it can be most conveniently studied. . . . I have no doubt that there is quite a number of animals which are . . . "created" for special physiological purposes, but I am afraid that most of them are unknown to the men for whom they are "created" and we must apply to the zoologists to find them and to lay our hands on them.
>
> —KROGH (1929), PP. 202–203

The menagerie of animals that biologists have used in their research is a highly nonrandom set. More than a million animal species inhabit the earth, but the vast majority of animal research is concentrated on just a dozen or so species. This taxonomic bias is present in most areas of biological research (Troudet et al., 2017; Rosenthal et al., 2017), but it is especially severe in biomedical research, where the terms "model species" and "model organism" are encountered frequently (Fields & Johnston, 2005; Katz 2016). It is difficult to quantify how frequently researchers employ the various species in their studies, but one can attempt to do so by examining either their publications or regulatory reports (figure 3.1). Both approaches have serious limitations. For one thing, many publications do not mention the names of the examined species in their title (nor, quite frequently, their abstract). For another, the United States does not report animal numbers for the most commonly studied species: mice, rats, birds, and all cold-blooded animals (see chapter 2). Nonetheless, the available data clearly show that a relatively small number of species account for most of the research (see also figure 2.1). They also show that the popularity of the various species in biological research, and the financial support for such research (Farris, 2020), has varied over time. Mice, for example, have become vastly more popular among biologists since the 1960s, while rabbits, hamsters, and guinea pigs have become less frequently studied.

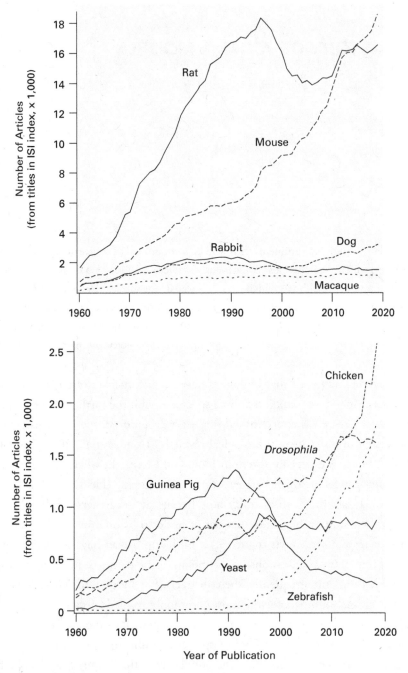

Figure 3.1

Relative distribution of model species featured in journal publications. These graphs indicate the frequency with which several frequently studied species were mentioned in the titles of journal articles (as indexed in the ISI Citation Index), broken down by year of publication. Note the difference in *y*-axis scales. The data were gathered by the author using procedures analogous to those described in Dietrich et al. (2014) but include additional years and species.

All of which raises interesting questions about how and why biologists select their research species. Why did some species become extremely popular among biologists, and why did some of them eventually fade? I here explore these questions in the hope that this exercise may bring to light some general principles of model selection and help inform some model choices going forward. I focus here exclusively on multicellular animal models, but unicellular organisms (notably bacteria and yeast) are covered in chapter 4, together with other in vitro models. The chapter begins with the early history of animal models in biology and then covers the major model species used today, emphasizing their principal contributions to biomedicine as well as their limitations. Due to space constraints, relatively little attention is paid to the biologists driving this research, but this history is well covered in other books (Kohler, 1994; Rader, 2004; Finger, 2004; Endersby, 2007).

3.1 ECLECTIC BEGINNINGS OF BIOLOGY

Early biologists studied a wide variety of species. This was true especially in ancient Greece and Rome, but it remained the case through the rise of experimental physiology in 19th century Europe. Many of those early studies used domesticated animals, including dogs, horses, rabbits, and guinea pigs, but biologists back then usually examined many different species to identify those that were optimally suited to the problem at hand (Logan & Brauckmann, 2015). This approach was championed most effectively by August Krogh, whom we discussed in chapter 2 (section 2.6) and who gave us the opening quotation of the present chapter; but let us start farther back in time.

3.1.1 Ancient Greece and Rome

Aristotle, who lived in Greece during the 4th century B.C., was one of the world's greatest philosophers and the founding father of comparative biology. He examined more than 500 species of animals, including numerous invertebrates (Blits, 1999) and then looked for patterns of similarity and difference. Aristotle was particularly interested in the distinctions between different types of animals and in the various categories of causes that could explain those differences. Thus, Aristotle constructed the first systematic framework for comparative anatomy and a philosophy for explaining nature's diversity.

Roughly 500 years later, Galen made comparative anatomy relevant to human medicine and became Rome's greatest physician. Because of cultural prohibitions against touching human corpses, Galen did not dissect human beings. He did, however, dissect a great number of different animals, including macaques, cats, dogs, weasels, camels, lions, wolves, deer, bears, mice, and even an elephant (Finger, 2004). He also experimented on diverse living animals, cutting various nerves or parts of the spinal cord to examine their functions, even though surgical anesthetics were not yet available. Collectively, Galen's dissections and vivisections allowed him to make numerous discoveries, which he then extrapolated to humans.

Galen's teachings formed the foundation of medicine for many centuries, but his accounts of the human body were challenged in the first half of the 16th century by Andreas Vesalius, who dissected human cadavers as well as diverse nonhumans. At some point in 1540, Vesalius compared the skeleton of a human and a macaque monkey side by side and suddenly realized that Galen had described the monkey skeleton quite precisely, but had mistakenly assumed that the human skeleton would be the same. Vesalius concluded that Galen's lack of direct experience with human bodies caused him to make numerous unwarranted extrapolations from monkeys to humans. Overall, Vesalius found more than 200 such mistakes in Galen's body of work (Finger, 2004). One famous example was Galen's description in humans of a *rete mirabile*, which is a network of small blood vessels at the base of the brain. Vesalius pointed out that such a *rete mirabile* exists in several mammals that Galen had dissected, such as oxen, but is absent from humans.

3.1.2 The Rise of Experimental Physiology

Galen and Vesalius had done some experiments on live animals, but most of their research was anatomical. Eventually, however, experimental physiology became more prominent. Especially influential was William Harvey's work on the "motion of the heart and blood," which was originally published in 1628 (Harvey, 1628/1928). Harvey carefully studied the anatomy of the circulatory system in more than 100 species (Cole, 1957) but also performed a large number of experiments on living animals (i.e., vivisections), asking, for example, how contractions of specific chambers of the heart correlate with arterial pulsations. By studying a large number of different species, Harvey was able to infer general structure-function principles about how blood is pumped through the body. He fully recognized that these principles are instantiated differently in different species. This is an important point, to which we shall return in chapter 7 (section 7.1.4).

Experimental physiology expanded rapidly toward the end of the 19th century, shortly after the development of ether as a surgical anesthetic (Wood Library Museum of Anesthesiology, 2021) and in tandem with the invention of more sophisticated research equipment. Many of these physiological studies used cold-blooded vertebrates, especially frogs. A major advantage of these species was that they stayed alive longer than warm-blooded animals during invasive procedures. Even isolated nerves and muscles continued to function for hours after the frogs had died. In addition to cold-blooded vertebrates, experimental physiologists in the late 1800s worked with a wide variety of warm-blooded animals, including pigeons, rabbits, and dogs. As Cheryl Logan pointed out in her thorough review of published studies from this period, the early physiologists "attempted to include as many animals as possible" in their research, and for all of them "generality was the goal" (2002, p. 347).

One of the hot topics during that time was the question of functional localization within the cerebral cortex. Some physiologists had stimulated or destroyed specific brain regions in various animals and found that the behavioral effects varied across cortical areas (Young, 1990). Others argued, instead, that functions were more globally distributed across the cerebral cortex. Ultimately, the proponents of functional localization prevailed, but resolution of the controversy required a large number of animal experiments that surely induced a fair amount of animal suffering. Many of the experiments were performed on rabbits, dogs, or monkeys, and some of them were demonstrated during public lectures to convince a skeptical audience. In retrospect, it is not surprising that these experiments spawned anti-vivisectionist sentiment (especially toward experiments on unanesthetized animals) and the emergence of animal welfare regulation.

The principal motivation underlying this research had been to discover general structure-function principles at work in vertebrate brains, rather than to solve a specific human medical problem. However, the principle of functional localization is applicable also to human brains, as first shown by Paul Broca and Carl Wernicke with regard to language (see Finger, 2004) and later extended to other brain functions. Ultimately, this knowledge became clinically relevant because clinicians often use it, for example, to assess brain damage after stroke and guide brain cancer surgery.

3.2 EARLY EXPERIMENTAL MEDICINE: RESEARCH ON PET SPECIES

Because experimental biology in the late 1800s and early 1900s led to several major medical advances, especially with regard to the treatment of infectious diseases (de Kruif, 1926), the number of experiments on living animals increased exponentially during this period. In the United Kingdom, 250 vivisections were performed in 1880, 17,000 in 1900, 90,000 in 1920, and 700,000 in 1940 (French, 1975). It is unclear what fraction of these studies were performed on anesthetized animals, but after the turn of the century the use of anesthetics for invasive procedures on animals became quite widely accepted (Franco, 2013).

Also of interest is that the dramatic rise in animal experiments during this period involved mainly research on mammals, especially guinea pigs, rabbits, and dogs. Not coincidentally, these species were widely kept as pets.

3.2.1 Guinea Pigs

Guinea pigs are neither pigs nor from Guinea. Instead, they are small rodents native to South America. Guinea pigs were domesticated more than 3,000 years ago by the Incas and became popular as pets in Europe during the 16th century. During the 19th century, guinea pigs became popular experimental animals because they were docile and readily available through the pet trade. Moreover, guinea pigs reproduce readily and rapidly in laboratory conditions, becoming sexually mature at three months and able to bear litters every two to three months after that.

Louis Pasteur used guinea pigs (as well as rabbits and dogs) to develop some of the first vaccines in the 1870s (see section 3.2.2 and chapter 5), and Robert Koch a few years later used guinea pigs to isolate the bacteria that cause tuberculosis. Curiously, Koch had initially injected dogs and mice with tuberculosis-causing extracts, but these animals did not get sick. By contrast, guinea pigs were similar to humans in their sensitivity to the tuberculosis bacterium. Also interesting is that Koch claimed to have developed an antitoxin therapy against tuberculosis, which ultimately proved ineffective. His general approach did, however, bear fruit in 1890, when van Behring and Shibasaburo discovered that infected guinea pigs produced antitoxins (later called antibodies; see chapter 5) that were effective at treating and preventing diphtheria when injected into other guinea pigs.

Another line of research that depended heavily on guinea pigs focused on nutritional deficiencies, notably scurvy. It had long been known that sailors on long voyages developed scurvy unless they supplemented their bland diet with citrus fruit, but the specific cause of the disease remained unclear. Researchers had tried to induce scurvy in pigeons and rats by feeding them pure grain diets, but these animals did not get sick. In contrast, impoverished diets rapidly induced scurvy in guinea pigs (Holst & Frölich, 1907), which is why these animals were used in subsequent experiments that led to the discovery of vitamin C (Endersby, 2007). We now know that this species difference in disease susceptibility arises because pigeons and rats can synthesize their own vitamin C, whereas humans and guinea pigs must obtain it from food.

Guinea pigs also played a role in the 1949 discovery of lithium for the treatment of bipolar disorder (see chapter 6) and in a few additional research areas. The use of guinea pigs as "guinea pigs" in biomedical research continued to increase after 1960 but then decreased over the last 30 years (figure 3.1). The most likely explanation for this decline is the increased reliance on mice over this same period. In addition, guinea pigs are difficult to anesthetize (e.g., compared with dogs, rabbits, and rats) (Brodbelt et al., 2008), and efforts to create transgenic guinea pigs are still nascent. Aside from such technical limitations, many biologists may have turned away from guinea pigs simply because they are so endearing and popular as pets; in contrast, mice and rats elicit weaker ethical concerns.

3.2.2 Rabbits

Rabbits were domesticated thousands of years ago, largely for food, but the details of that process remain unclear. By the 16th century humans had created several distinct breeds of rabbit, and by the 19th century rabbits had become popular also as pets and research animals. Their popularity among experimental biologists probably derived mainly from their notorious proclivity for rapid reproduction (females reach sexual maturity in about six months and can then bear around 12 offspring every month). In addition, rabbits are larger than guinea pigs, which makes it easier to perform surgeries on them and draw substantial amounts of blood.

One important early use of rabbits in biomedical research was the development of a rabies vaccine. Pierre-Victor Galtier in 1879 infected rabbits with ground-up neural tissue from rabid dogs and observed that the rabbits fell ill quite rapidly, roughly twice as fast as similarly injected dogs ("Pierre-Victor Galtier," 2020). Louis Pasteur and his collaborators took this idea further by taking spinal cords from the infected rabbits and exposing them to dry air for several days, which made the virus progressively less virulent (Rappuoli, 2014). When extracts of these dried spinal cords were then injected into dogs, it protected them from later rabies infections and could even stop ongoing infections. Seeing the method work in dogs, Pasteur in 1885 tested his vaccine on a boy who had been bitten by a rabid dog, and the boy lived. Although Pasteur had previously developed vaccines for several other diseases, notably swine flu, cattle anthrax, and chicken cholera (Geison, 1995), the antirabies vaccine was his first vaccine directed at a human disease.

Syphilis is another disease that rabbits helped to treat. In the early 1900s, this devastating sexually transmitted disease affected more than 60 out of every 100,000 persons and accounted for more than 10% of the patients in some insane asylums. In 1905 biologists discovered that syphilis is caused by a bacterium called *Treponema pallidum*. It also became clear around that time that rabbits are more susceptible to this microbe than other nonprimates (Esteves et al., 2018). Paul Ehrlich and his Japanese student Sahachiro Hata then infected countless rabbits with syphilis and tested hundreds of arsenic-related compounds in the hope of finding a treatment (Frith, 2012). They eventually succeeded in 1909 with compound #606 and in 1910 with Salvarsan, which became known as the first "magic bullet" drug for its effectiveness and enormous impact. These drugs required multiple injections over more than a year, but they worked quite well. They became obsolete only in 1943, when it was discovered that penicillin, the world's first antibiotic (see chapter 5), effectively cures syphilis.

Rabbits have continued to be used in diverse research areas, especially in immunology. Many studies have described basic aspects of the immune system in rabbits, revealing a complex pattern of similarities and differences among rabbits, humans, and other mammals (Flajnik, 2002; Haley, 2003; Pinheiro, et al., 2016). Rabbits have also been used extensively in the production of polyclonal antibodies, mainly because their size allows investigators to collect ample amounts of blood. These antibodies were then used in many different research applications, such as immunohistochemistry and Western blot analyses.

In addition, rabbits were heavily used in toxicology, especially in studies using the Draize test, which was developed to predict whether novel cosmetic compounds might cause skin or eye irritation in humans (Wilhelmus, 2001). After both scientists and animal rights advocates pointed out that the Draize test could be replaced with in vitro studies (see chapter 4), the US Food and Drug Administration in 1981 began to accept validated alternatives to the Draize test. These newer models cannot mimic human eyes or skin perfectly, but neither can rabbits (Verstraelen & Van Rompay, 2018). As a consequence of these developments, eye irritation testing in rabbits decreased by 87% between 1982 and 1991. By 2011 rabbits accounted for less than 4% of all animal research procedures in the United Kingdom (van der Staay et al., 2017). In the United States as well, the use of rabbits in research has declined significantly over the last 20 to 30 years (figure 3.1).

3.2.3 Dogs

As noted in the section 3.1, dogs were used quite frequently in the early days of experimental physiology and medicine. Not yet mentioned was that dogs were often used to develop and learn new surgical techniques, which are more difficult to execute in smaller animals. They were also frequently employed in studies of the circulatory system because their heart and major arteries are relatively large and anatomically similar to those of humans. An early example of this work was the use of dogs to develop a hemodialysis machine that filtered nitrogenous waste out of the blood, which later became routine therapy for kidney failure (Abel et al., 1914). Another good example is Vivien Thomas's development of a treatment for blue baby syndrome, which is chronicled in the movie *Something the Lord Made* (Sargent, 2004). Thomas himself reported having used around 200 dogs over several months, first to induce the syndrome in his animals and then to correct the problem (Smith, 2013).

Given that dogs are widely considered "man's best friend," it is not surprising that their use as research animals engendered significant anti-vivisectionist sentiment. Research on dogs was at least partly responsible for passage of the 1876 Cruelty to Animals Act in the United Kingdom. In the United States, passage of the Animal

Welfare Act in 1966 was motivated by a photographic essay in *Life* magazine titled "Concentration Camps for Dogs" (Wayman, 1966). This essay documented deplorable conditions at dubious dog dealers who sold animals to research institutions. The public was most concerned that some of these animals might have been lost or stolen pets. In response, the Animal Welfare Act gave special protections to dogs (as well as cats). Nowadays, the vast majority of dogs used in research are bred specifically for that purpose, either by specialized (class A) dog dealers or in colonies at research institutions. Most of these dogs are beagles, which were selected as the preferred breed primarily because they are relatively small, docile, and social.

Research publications on dogs have become more frequent over the last 20 years (figure 3.1A), but 36% of these publications are veterinary reports that examined dogs for their own sake, rather than as models for other species. By contrast, the number of dogs used at research institutions has remained relatively steady over the last decade or two, both in the United States and the United Kingdom (figure 3.2). Most of these animals were used in mandatory safety tests for "new chemical entities," such as putative new therapeutic drugs. These safety tests generally require that the novel compound be tested in two different types of animals; the first is usually a rodent species, and the second one is often a dog. That said, the ability of studies in dogs to predict human toxicity is imperfect. For example, chocolate is far more toxic to dogs (and cats) than to humans. Overall, the concordance of toxicity results between humans and dogs has been estimated at 63% (versus 43% for rodents), but even this finding remains controversial (Olson et al., 2000; Matthews 2008). Similarly controversial is whether any nonanimal replacements can yield better predictions of human toxicity for novel compounds (see chapter 4, section 4.3).

Due to space constraints, I do not here discuss research on cats except to note that, overall, they have served as research animals less frequently than dogs (Institute of Medicine & National Research Council, 2012), yet contributed significantly to Nobel Prize–winning research on the brain and spinal cord during the 1950s and early 1960s (Eccles et al., 1962; Lienhard, 2017, 2018).

3.3   FRUIT FLIES: THE FIRST SUPERMODEL

Guinea pigs, rabbits, and dogs were widely used as animal models, but the fruit fly *Drosophila melanogaster* became the world's first "supermodel," which Rowland Davis defined as "an organism that reveals and integrates many and diverse biological findings applying to most living things or to most members of a kingdom" (2003, p. 199). This definition is flexible, but the term is certainly catchy (it also clearly applies to *Escherichia coli* and yeast, which we discuss in the next chapter). *Drosophila* are called

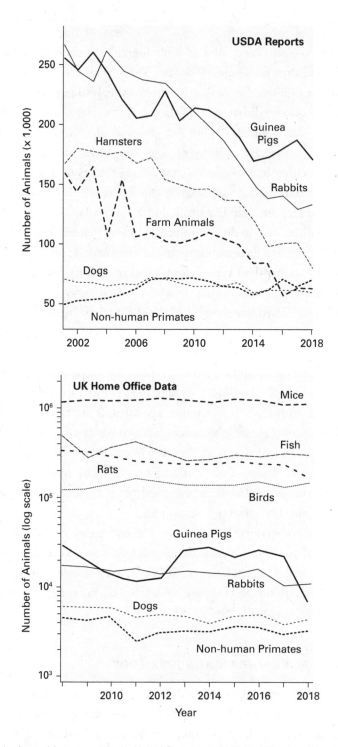

Figure 3.2

Numbers of animals used for research in the United States and the United Kingdom. Shown on the top are the numbers of animals, by species or kind, reported by the US Department of Agriculture as having been used for research between 2001 and 2018. Note that mice, rats, fish, and birds are not counted by the USDA and that pigs and sheep are here lumped with other farm animals. Shown on the bottom are data on equivalent animal numbers reported by the UK Home Office. These data include mice, rats, fish, and birds (which are considered animals in the UK; see chapter 2).

fruit flies because they like to feed on the yeast that grows on fruits. Their ascension to supermodel status began with the work of Thomas H. Morgan and his collaborators in the first few decades of the 20th century (Kohler, 1994; Endersby, 2007).

Morgan had been interested in determining whether new species form by the accrual of many minor mutations, as Darwin had proposed, or by major mutations that cause large changes in the phenotype. To answer this question, Morgan studied a diverse array of species, including pigeons, chickens, mice, and rats, but he eventually settled on *Drosophila* for a number of reasons. Particularly important was that fruit flies breed year-round and rapidly, with a generation time of 10 to 20 days; each female is able to produce around 300 offspring during her lifetime. With this high reproduction rate, Morgan was able to breed tens of millions of flies (Endersby, 2007), which allowed him to identify mutants even when mutation rates were relatively low. Fruit flies were also easy to feed (with bananas in the early days) and maintain as separate breeding colonies in small spaces. In addition, fruit flies tolerate extensive inbreeding better than many other species, which allowed Morgan's team to create very uniform strains in which mutants are more apparent than in heterogeneous populations.

Tracking the inheritance of mutations across multiple generations allowed Morgan and his colleagues to discover that some mutations are sex-linked. Subsequently, they discovered that multiple mutations tend to be inherited together, forming linkage groups. Finally, they discovered that the occasional crossing over (i.e., recombination) of homologous chromosomes causes systematic variations in the strength of those linkages. This last observation ultimately led Morgan and his team to conclude that genes are concrete entities that are arranged linearly along the chromosomes and could be "mapped" by careful analyses of their linkage frequencies (Morgan, 1915). These discoveries were facilitated by the fact that fruit flies have only four pairs of chromosomes and that crossing-over does not occur in the males of this species. The latter trait is relatively rare (John et al., 2016) but made genetic mapping much easier; it is a good example of the kind of experimental "convenience" that August Krogh emphasized in the chapter's opening quotation.

Once Morgan and his fellow Drosophilists had established that genes on chromosomes are the basis of Mendelian inheritance, interest in fruit flies as research animals faded, and geneticists turned their attention to yeast, bacteria, and viruses (see chapter 4). However, fruit flies attracted attention again in the 1960s when Seymour Benzer began to generate fly mutants that were impaired in specific behaviors, such as phototaxis, circadian rhythms, and memory tasks (Weiner, 1999; Greenspan, 2008). In addition, Christiane Nüsslein-Vollhardt and Eric Wieschaus in the 1970s generated more than 130 fruit fly mutants with defects in larval development. Research on these developmental mutants revealed a complex network of genes and proteins that together control critical aspects of *Drosophila* development. Many of these genes and proteins have vertebrate homologs that are important for development. It is important to note, however, that some of the mechanisms responsible for fruit fly development do not generalize to vertebrates. For example, the mechanisms underlying body segmentation are quite different in fruit flies and vertebrates and involve many non-homologous genes (Blair, 2008). Still, many of the principles derived from the work on fruit flies, both with regard to behavior and morphological development, have been broadly applicable.

After researchers developed techniques for creating transgenic fruit flies in the 1980s (Rubin & Spradling, 1982), fruit flies came to be used as models of human disease. For example, researchers created a fruit fly model of Huntington's disease by inserting into flies a mutated fragment of the human *huntingtin* gene. These transgenic flies develop abnormal intracellular accumulations of the Huntingtin protein, neuronal degeneration, and behavioral symptoms that are somewhat reminiscent of human Huntington's disease (Warrick et al., 1998). Similarly, researchers in 2000 inserted into flies a mutated version of the human *alpha-synuclein* gene, which has been linked to familial Parkinson's disease (Feany & Bender, 2000). Although the symptoms of these disease-model flies are often only superficially similar to those of human patients (see chapter 6), the transgenic flies allow researchers to test thousands of potential therapies in high-throughput studies. Another rarely discussed but important advantage of working with flies is the lack of animal welfare regulations for insects (Jennings, 2011).

Although the translation of potential therapies from fruit flies to humans remains more of a promise than an obvious success, work on *Drosophila* has provided vast amounts of information on molecular interactions and cellular processes in flies, and much of that basic science has made significant contributions to research on other species. For instance, the molecular mechanisms underlying circadian rhythms were first discovered in *Drosophila*, but this work then facilitated analogous studies in other species, revealing both similarities and differences (Glossop & Hardin, 2002; Rubin et al., 2006; Tomioka & Matsumoto, 2010). Thus, not everything one finds in *Drosophila* extrapolates to other species, but some of the discovered principles are general; in any case, the fruit fly work has often guided subsequent research. It is for those reasons that calling it a supermodel seems appropriate.

3.4 LABORATORY RATS: THE FIRST STANDARDIZED MAMMALS

Rats have long been regarded as thieves of our food and as dirty creatures that live in sewers, spreading disease. However, experimental biologists have frequently used Norway rats (*Rattus norvegicus*) in their research, starting in the 19th century and then accelerating in the first half of the 20th century. Why did biologists like to experiment with rats? One reason is that rats, as mammals, are more closely related to humans than non-mammalian vertebrates, or invertebrates such as fruit flies. This phylogenetic proximity is generally presumed to correlate with overall similarity and, therefore, to favor translation (see chapters 2 and 7). A second major reason is that rats breed even more rapidly than guinea pigs and are easy to maintain in large numbers. Rats are also a convenient size for many surgical and histological procedures—not too big and not too small. Another factor in favor of rats is that they, in contrast to guinea pigs, are born in a very immature (altricial) state, which makes it easier for biologists to study their early development. For example, biologists can study the effects of early castration without having to castrate embryos in utero. Studies on aging are also relatively easy to perform in rats, because they only live for two to three years.

Another major impetus for the increasing popularity of laboratory rats in the early 20th century was the desire of biologists to minimize the variability in their experiments. During the early days of experimental physiology, novel pieces of equipment had been invented at a rapid clip, but the variability between different machines made it difficult to compare the results obtained in different laboratories. To combat the resulting crisis of replicability, experimentalists began to standardize their equipment and techniques (Logan, 2002). It made sense, therefore, to standardize the research animals as well. Accordingly, researchers created several strains of rats expressly for research. Especially popular were albino rat strains because any cross-contamination from wild brown rats was relatively obvious. Albino rats also lessen the inherent aversion most people have toward wild rats; letting the imagination run, one might even think of these albino rats as wearing white lab coats. The most famous of these albino rat strains was created at the Wistar Institute at the beginning of the 20th century (King, 1918; Clause, 1993). Because these Wistar rats were highly inbred, they exhibited relatively little variation, which meant that their phenotypic traits could be summarized in numerous standard tables and descriptions (Donaldson, 1924). By 1912, the Wistar Institute was shipping around 6,000 Wistar rats per year to scientists around the world (Endersby, 2007).

Because of their strong sex drive, rats were often used to study reproductive behaviors and hormonal mechanisms. In addition, rats were used in many nutritional studies because their dietary requirements are, in some respects, quite similar to those of humans (Richter, 1968). The popularity of laboratory rats extended also to psychology and neuroscience, mainly because rats are relatively fast learners, at least compared to mice. In the words of Ian Whishaw, "At a fork in the evolutionary road . . . the brown rat chose complexity as a survival trait whereas the mouse chose simplicity. The rat became social, intelligent, complex, and skilled, all of which are attributes it shares with humans" (1999, p. 411). Particularly advantageous is that rats construct and navigate complex tunnel systems in nature (Calhoun, 1963). The laboratory equivalent of these burrows is the various mazes that experimentalists have devised to study learning and memory in rats (Small, 1900; Watson, 1903). These maze experiments were a key building block in the development of behaviorism and other general theories of animal psychology (Watson, 1913; Skinner, 1938; Tolman, 1948). They also anchored many studies on the neural mechanisms of learning and memory (O'Keefe & Nadel, 1978).

As the scope of rat research expanded during the 20th century, biologists increasingly considered the rat to be the ideal animal model for asking many different types of questions and obtaining answers that could be generalized to many other species—a biological Rosetta stone (Gest, 1995). In words of one influential biologist who had research experience with many different mammals, "If someone were to give me the power to create an animal most useful for all types of studies on problems concerned directly or indirectly with human welfare, I could not possibly improve on the Norway rat" (Richter, 1968, p. 403). In parallel with the emergence of the Norway rat as the dominant laboratory animal, the interest in diversity that had characterized earlier generations of biologists gradually waned. Instead of testing whether findings from one species would generalize to others, their generality became widely assumed (Logan, 2001).

Some scientists became alarmed at this heavy reliance on laboratory rats. Frank Beach (1950), for example, became concerned that comparative psychologists were learning a great deal about the behavior and mental processes of laboratory rats but losing sight of the larger goal, which was to learn about animal minds in general. Citing a poem by Lewis Carroll, Beach argued that comparative psychologists had set out to "hunt snarks" but instead found a very dangerous "boojum," which makes the hunters disappear. At the end of his essay, Beach switched stories and compared the laboratory rat to the Pied Piper of Hamelin, who used his flute to lure the city's rats into a river where they drowned. "Now," Beach wrote, "the tables are turned. The rat plays the tune and a large number of human beings follow. . . . Unless they escape the spell that *Rattus norvegicus* is casting over them, experimentalists are in danger of extinction" (1950, p. 117). Other psychologists echoed Beach's concern. Daniel Lehrman, for example, recommended that psychologists should focus on "questions arising from the natural life of a particular species (rather than questions applied to an arbitrarily selected species from a generalized theoretical framework)" (1971, p. 464). To him, generality was something that should be inferred from comparative data, rather than assumed.

That said, many of the findings obtained from rats, both in psychology and other disciplines, have proven to be quite general. The larger question, then, is not whether some general principles exist, but whether species differences are important to recognize. For example, is it important to recognize that "isolation housing may produce symptoms of psychopathology in rats, but in mice, it reveals normal species-typical behavior" (Miczek & de Wit, 2008, p. 296)? Similarly, is it important that rats, in contrast to humans, mice, and most other mammals, lack gall bladders (McMaster, 1922)? Does it matter that sleep deprivation is more lethal in rats than in humans or mice (Siegel 2008)? The answer to all of these questions is "sometimes yes" or "it depends" (see chapter 7). In short, rats and mice can serve as good animal models for many purposes, yet differ from humans and one another in some important respects (Ellenbroek & Youn 2016).

3.5 LABORATORY MICE: FROM CANCER FIGHTER TO MODEL ORGANISM

Mice played a relatively minor role in the early days of experimental biology, but they attracted interest around the same time that fruit flies became interesting to biologists and for the same reason: the study of genetics. In particular, several biologists in the early 1900s were asking whether inbreeding with strong artificial selection would lead to stable, viable lines. They explored this question in diverse species, including both fruit flies and mice. Most influential was C. C. Little, who founded and directed the Jackson Laboratory, which eventually became one of the world's most prolific producers of laboratory mice. Starting with just a few strains in 1929, the Jackson Laboratory now maintains more than 8,000 genetically defined mouse strains and ships millions of mice to laboratories around the world. The black-6 mouse (C57BL/6) eventually emerged as the world's most common mouse strain, not for any obvious reason but simply because it was already widely used; as one author put it, "use begat ubiquity" (Engber, 2011).

Why did Little and his colleagues focus so heavily on mice as research animals? They liked that mice take up less space per animal than rats and multiply more rapidly (up to 10 litters per year versus around six per year). Little also appreciated that rodents do not attract as much anti-vivisectionist sentiment as other mammals that might be used for research (Little, 1935). However, Little's main selling point for mice as research animals was that he considered them to be ideal for research on cancer, which by the 1930s had emerged as major threat to human health. Little and his colleagues had observed that mice get cancer just as humans do and that some strains of mice develop cancer more frequently than others. Energetically promoting his mice as heroic miniature troops in the war against cancer (Little, 1935), Little thought that breeding such mice could help identify the genes that cause cancer (Rader, 2004).

As we will see in chapter 5, mice have indeed played a major role in understanding and treating cancer, but Little had underestimated the many nongenetic factors (e.g., viruses, tobacco smoke [Brandt, 2012], and other carcinogens) that play a role in tumor formation.

Whereas *Drosophila* research was driven primarily by the discovery of mutant flies, mouse mutants were relatively rare during the early days of mouse model development, presumably because mice breed much more slowly than fruit flies. By 1931, for example, only 25 mutant mouse stocks were recognized (Rader, 2004). This changed after World War II, when scientists exposed enormous numbers of mice to ionizing radiation (Russell, 2013), which greatly increased the animals' mutation rate. The resulting mouse mutants were studied by many different investigators and led to useful insights. However, the project's original mission had not been to find interesting mutants but to inform public policy about what levels and duration of radiation exposure would be harmful to humans. Indeed, the "mouse-house" work did lead to policy revisions. Most interesting for present purposes is that scientists had extensive debates about whether one could extrapolate exposure effects from mice to humans. For example, Alfred Sturtevant (a major figure in early fruit fly research) argued that, despite quantitative species differences, the data from nonhumans were perfectly sufficient to predict that high-energy radiation causes genetic damage in humans (Sturtevant, 1954).

Mouse research again advanced substantially after the invention of genetic engineering in the 1980s. Especially transformative was the ability to create transgenic mice by injecting recombinant DNA into the nuclei of fertilized eggs and then implanting those eggs into female mice (Gordon et al., 1980); for reasons that are still somewhat unclear, this technique had a much higher success rate in mice than rats (Pradhan & Majumdar, 2016). Many of the early transgenic mice contained insertions of human disease genes or parts thereof, but those transgenes varied in copy number and tended to insert at unpredictable locations in the genome. In most of the published cases the transgenes were significantly overexpressed, raising questions about whether the phenotypes were due to mutations in the transgenes or their unusually high expression level (Fisher & Bannerman, 2019). These limitations were overcome by the invention of techniques for modifying the sequence of specific genes by homologous recombination (Thomas et al., 1986). This same technique could also be used disrupt or "knock out" specific genes (Capecchi, 2005), which allowed biologists to investigate the functions of specific genes without having to wait for mutants to arise by chance. In addition, researchers developed mice that express a transgene only under certain conditions. More recently, they started to create "designer mice" using gene editing techniques (such as CRISPR-Cas9; Fischman, 2020).

Because of the ready availability of so many mutant mice and tools to make additional varieties, genetically modified mice came to dominate vast swaths of biological research. Researchers working on bacteria, yeast, and flies also had access to many mutant strains, but mice are far more closely related to humans and, therefore, more similar to humans, at least when averaged across all traits (see section 2.3.1; see also chapter 7). Already in 1952, C. C. Little had referred to mice as "miniature human beings" (see Rader, 2004), and the yeast geneticist Ira Herskowitz reportedly once said, "I don't consider the mouse a model organism. The mouse is just a cuter version of a human, a pocket-size human" (quoted in Rader, 2004, p. 267). These statements may have been tongue in cheek, but by the beginning of the 21st century mice clearly reigned supreme in biomedical research. In the minds of some observers, this dominance created a sort of "group think about the ready translatability" of data from mice to humans, spawning "murine 'model' monotheism" (Libby, 2015).

Of course, biologists readily admit that mice differ from humans in numerous respects (Perlman 2016). For example, the genome's "regulatory landscape" is quite different between humans and mice (Fisher & Bannerman, 2019), as are many aspects of their immune system (Mestas & Hughes, 2004) and nervous system development (Finlay, 2019). Mice also live accelerated lives compared with us, and this affects many different aspects of their biology, ranging from metabolic rate and normal aging to diverse disease processes (Agoston, 2017). Another good example came to light in 2005, when "hydrogen sulfide was reported to be revolutionary for emergency trauma care and treating soldiers on the battlefield based on its extraordinary ability to place a mouse in suspended animation for many hours with apparent complete recovery of function. . . . However, the hydrogen sulfide concept failed to translate into larger animal models because the mouse was the wrong animal model for the translation of that particular question" (Dobson, 2014, p. 480). The reason for this particular translational failure was that mice have a naturally evolved ability to enter torpor, which is a sort of suspended animation similar to hibernation; humans, of course, do not share this ability.

Because mice clearly are not "miniature human beings," researchers sometimes "humanize" their mice: they often insert human genes into mice or edit the endogenous mouse genes so that they possess the human DNA sequence. These modifications are usually limited to single genes or parts of genes, but it is becoming increasingly feasible to humanize multiple mouse genes at the same time. In addition, biologists can humanize mice by injecting them with human cells. For example, they can partially humanize the immune system of mice by injecting human immune cells into mice that are otherwise immunodeficient (Shultz et al., 2007). Similarly, human gut microbiomes can be transplanted into mice that lack a microbiome of their own (Arrieta et al., 2016).

3.6 NONHUMAN PRIMATES

Nonhuman primates have long been used as models for humans in the belief that their close phylogenetic relationship to us would make them "high fidelity" models. As we discussed in section 2.3.1, we cannot assume such high model fidelity for any given trait; however, on average, traits are more similar between humans and their closest relatives than between, say, humans and fruit flies (see also chapter 7). Unfortunately, this realization simultaneously increases the ethical concerns over performing experiments on nonhuman primates because their neural and mental traits are also likely to be more similar to our own, at least on average. Thus, any discussion of how much benefit has been derived from research on nonhuman primates, or can be gained in the future (Phillips et al., 2014), must be tempered by questions about the costs in terms of animal suffering. These are not simple discussions to have.

The number of nonhuman primates used in biological research is relatively small, compared with the total number of experimental animals, and has been relatively constant over the last 10 years (figures 3.1 and 3.2). Many of these studies have examined the immune system of nonhuman primates (Messaoudi et al., 2011) because it differs substantially from that of rodents and other model animals. Nonhuman primates are also used in neurobiological research, because nonhuman primate brains share many features with our own brains that are not found—or are very different—in other species (Striedter, 2005).

However, the vast majority of nonhuman primates are used for legally mandated safety and efficacy tests of novel drugs and therapies, including vaccines (Weatherall, 2006). For example, macaques and African green monkeys have been used extensively to understand and develop vaccines against the Zika virus, which can cause serious neurological birth defects in humans (Osuna & Whitney, 2017; Haddow et al., 2020). An even better example of how nonhuman primates were used to advance medical research is provided by the development of vaccines against poliomyelitis.

3.6.1 Polio Vaccine Development

Poliomyelitis, or polio, is a viral disease that paralyzes many of its victims (often children) and is frequently fatal. Polio reached epidemic proportions in 1916, when it killed 6,000 people in the United States. Recurring annually during the summer months, polio epidemics peaked in the 1940s and early 1950s, with the 1952 outbreak yielding more than 57,000 cases in the United States, including 3,000 deaths and 21,000 cases of permanent paralysis. Treatments for the disease were largely ineffective, and thousands of victims with respiratory paralysis could be kept alive only by means of specialized ventilators (the so-called iron lungs). Fortunately, the first vaccines against polio became available in 1955, and extensive vaccination programs then produced an exponential decline in the incidence of polio (Nathanson & Kew, 2010). Although polio has now been eradicated from many countries, outbreaks occasionally recur in some parts of the world.

Nonhuman primates were critical to the discovery of the virus that causes polio and to the early stages of vaccine development. Specifically, researchers in 1908 showed that injecting neural tissue from human polio victims into monkeys caused the monkeys to develop the paralysis and motor neuron degeneration typical of polio. On the heels of this discovery, others used tissue from those monkeys to infect additional monkeys, thus propagating the virus and allowing the scientists to study its behavior. Eventually, some researchers managed to breed a strain of the polio virus that causes the disease in mice (which the human polio virus does not), and this substantially reduced the need for monkeys in polio research (Armstrong, 1939).

Even more important was the 1949 discovery that polio viruses could be grown in cultured human or monkey cells, which made it possible to harvest large quantities of virus with relative ease (Enders et al., 1949). Initially, the cultured cells were derived from human embryos, but immortalized monkey kidney cells later eliminated the need to sacrifice living organisms entirely (Furesz, 2006). Plus, the virulence of the harvested virus could be determined by measuring the extent to which it damaged the cultured cells (Enders et al., 1980), rather than having to inject the virus into whole animals. Thus, the case of polio nicely illustrates both the crucial role of nonhuman primates in biomedical research and the potential impact of replacing animals generally thought to be highly sentient (see chapter 2, section 2.5.1) with other animals or cell culture.

Animal rights advocates sometimes point out that some of the early research on polio in monkeys had been misleading. Indeed, a series of influential early studies on monkeys suggested that the polio virus enters the human brain via the olfactory system, whereas later studies revealed that in humans the virus typically enters through the gut. The discrepancy arose because the researchers working with monkeys had unwittingly evolved a strain of the polio virus that is highly neurotropic, meaning that it preferentially infects neural tissue, which is most directly accessible through the nose. This early misstep caused some delays in vaccine development, but, in the words of Albert Sabin, who ultimately developed one of the successful polio vaccines, "without the use of animals and of human beings, it would have been impossible to acquire the important knowledge needed to prevent much suffering and premature death not only among humans but also among animals" (quoted by Speaking of Research editor, 2011). Indeed, Sabin reported that, during the four years leading up to his successful vaccine, his laboratory had used "approximately 9,000 monkeys, 150 chimpanzees, and 133 human volunteers" (Sabin, 1956, p. 1589). Importantly, the human volunteers only received vaccines that had been previously tested on monkeys and chimpanzees (Sabin, 1965).

A frequently unappreciated aspect of polio vaccine development is that some early attempts to produce these vaccines ended up giving humans polio, rather than making them immune to it (Horstmann, 1985). These early failures underscored the need for extensive testing of vaccines before they are given to humans. Such tests do not always have to be performed in nonhuman primates, but they should be performed in animals that are susceptible to the disease, which not all species are (Rivera-Hernandez et al., 2014).

Although animal tests are usually mandated by regulatory agencies, those mandates can be waived in the midst of pandemics. During the COVID-19 outbreak of 2020, for example, some vaccines went into clinical trials before animal test results had been obtained (Boodman, 2020). Thus, societies are sometimes willing to risk human health in the hope of reducing overall suffering. In normal times, however, humans are more risk-averse.

In fact, many people do not trust vaccines even when they are widely regarded as safe and effective. It will be interesting to see whether these antivaccine advocates will change their mind now that COVID-19 vaccines have been demonstrated to save numerous lives, or whether they will find their suspicions confirmed if a few individuals get sick after receiving a COVID-19 vaccine. Most likely, both views will persist, as they have for other vaccines (DeStefano et al., 2019).

3.6.2 The Silver Spring and U Penn Monkey Affairs (1980s)

Neurobiological studies on nonhuman primates are less common than toxicological or vaccine-related studies, but they tend to attract more attention from animal rights advocates, in part because their relevance to human health is not as apparent. A prime example is the case of the "Silver Spring monkeys," which in 1981 facilitated a significant expansion of the animal rights movement in the United States and provided the impetus for major changes in US laws and regulations concerning animal welfare.

The case involved the laboratory of Edward Taub at the Institute for Behavioral Research in Silver Spring, Maryland. Taub had been cutting the sensory nerves in one arm of macaques, which led to long-term motor impairments. Taub was doing these experiments to test whether his animals could learn to use their damaged arm if they were prevented from using their "good" arm (Taub, 1980; Taub et al., 1994). In the midst of this research, Taub's laboratory was infiltrated by an animal activist who was supposed to help care for the animals. This activist secretly documented that some of the animals gnawed on the limbs lacking sensory nerves, had poorly bandaged wounds, and lived in small, filthy cages. While Taub was on vacation, the activist alerted the police and press. As a result, the monkeys were confiscated, and Taub was convicted on several counts of cruelty to animals. The convictions were later overturned, but the case attracted enormous media attention, in part because the confiscated monkeys were briefly kidnapped by individuals who did not want them returned to Taub.

Although this case revealed real problems with monkey research at the time, Taub's original hypothesis has turned out to be largely correct and led to the successful development of "constraint-induced movement therapy" for humans with nerve damage in their arms (Wolf et al., 2008; Fritz et al., 2012). In addition, later studies on Taub's monkeys revealed a dramatic reorganization of their cerebral cortex (Pons et al., 1991). This finding then inspired many later studies on cortical plasticity, which is now recognized as being crucial to rehabilitation after nervous system damage (Kleim & Jones, 2008).

A second good example of the controversies engendered by neurological research on nonhuman primates involves baboons that were used at the University of Pennsylvania to study traumatic brain injury. An activist group broke into this laboratory in 1984 and stole 60 hours' worth of videotapes documenting the experiments. These tapes were then edited down to a 30-minute movie that shocked most viewers (Dusheck, 1985), mainly because many of the monkeys were not fully anesthetized when the trauma was applied, and because their subsequent behaviors were pitiful.

Since the activists' tactics had been blatantly illegal and the need for animal research on traumatic brain injury was widely accepted among scientists, the authorities responded more slowly to this episode than the Silver Spring monkey case. Still, the researchers were ultimately cited for violations of the Animal Welfare Act, the National Institutes of Health (NIH) temporarily halted the relevant grant, and the university stopped supporting nonhuman primate research on head injuries. In conjunction with the Silver Spring monkey case, the episode contributed substantially to the passage of the 1985 amendment of the Animal Welfare Act and to changes in NIH policy.

In the wake of the fracas at U Penn, research on traumatic brain injury became more focused on rats and mice (Xiong et al., 2013; Shultz et al., 2017). These rodent studies have provided extensive information about the cellular and molecular consequences of various types of brain damage, and they produced a number of promising therapies that minimize brain damage in rodents (see chapter 5). Unfortunately, when these therapies were tested in humans, all of them failed (Narayan et al., 2002;

Stein, 2015). One reason for these clinical trial failures is that human brain injuries are multifaceted and variable, whereas the animal models all focus on a specific type of injury and standardize all variables as much as possible. It has also been suggested that rodents, with their small, smooth brains, are poor models for humans when it comes to traumatic brain injury.

To circumvent the latter problem, minipigs (see section 3.7.1) have become an increasingly popular model for this type of research (Kinder et al., 2019). The brains of pigs are 50 to 90 times larger than those of rats and have a highly folded cerebral cortex. Moreover, the public tends not to be as concerned about the welfare of pigs as of monkeys, though the rationale for this differential empathy is rarely spelled out. Whether research on pigs will ultimately lead to treatments for brain trauma in humans remains to be seen.

3.6.3 The Chimpanzee Research Debate

The vast majority of biological research on nonhuman primates has been conducted with macaques, specifically rhesus macaques (*Macaca mulatta*) or crab-eating macaques (*M. fascicularis*, aka cynomolgus monkeys). In addition, biologists have studied some New World monkeys (notably marmosets; see section 3.6.4) and, sometimes, the common chimpanzee *Pan troglodytes*. The latter work has been extremely controversial, because chimpanzees are our closest living relatives. Their DNA sequence is often said to be 99% identical to ours, but that number drops to 96% to 97% when one takes deletions and insertions into account (Chimpanzee Sequencing and Analysis Consortium 2005; Varki & Altheide 2005). Still, chimpanzees and humans share more than 13,400 proteins, and about 29% of them have identical amino acid sequences in the two species; the rest differ by just one or two amino acids.

Consistent with this high level of genetic similarity, chimpanzees and humans are similar in many morphological and physiological respects, as well as mental and emotional capacity (de Waal, 2016, 2019). Some of the cognitive similarities may be "overzealous efforts to dismantle arguments of human uniqueness" (Povinelli & Bering, 2002, p. 115), but it seems fair to say that no other animals are as broadly similar to humans as chimpanzees.

Most important for present purposes is that chimpanzees are very similar to humans in many medically important features (Institute of Medicine & National Research Council, 2011; Phillips et al., 2014). For example, the rates at which several drugs are cleared from the body after oral administration correlate quite well between humans and chimpanzees, but not between humans and monkeys, dogs, or rats (Wong et al., 2004) (figure 3.3). Similarly, several viruses, including human immunodeficiency virus (HIV) and the hepatitis C virus, infect humans and chimpanzees but not most

A – Drug Clearance Rates Compared to Humans

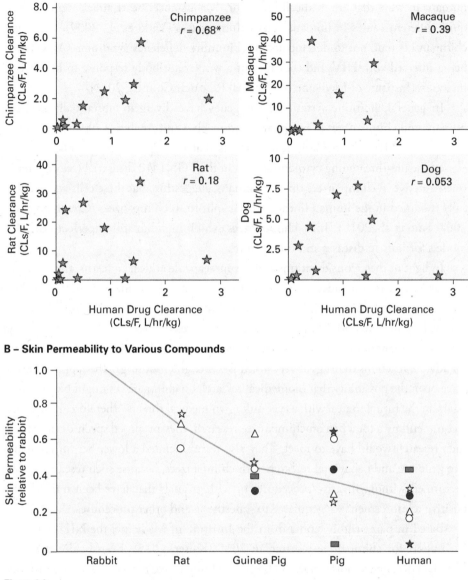

Figure 3.3

Variation across species in drug clearance and absorption. (*A*) Comparison across five species shows the rates at which various (orally administered) drugs are cleared from the animal's system. Clearance rates are significantly correlated between humans and chimpanzees (the asterisk indicates a statistically significant Pearson's *r* coefficient at $p = 0.015$) but not between humans and macaques, beagles, or rats. (*B*) The skin of humans and pigs is much less permeable to various chemicals than the skin of rabbits, rats, or guinea pigs (the data are normalized to each compound's permeability in rabbits; not all compounds were tested in all species). Data such as these are used to justify the use of pig skin as a model for human skin. Adapted from (*A*) Wong et al. (2004); (*B*) Calabrese (1991, figure 2-17).

other species (Haigwood & Walker, 2011). Of course, chimpanzees also differ from humans in ways that are medically relevant. For instance, heart attacks tend to have quite different causes in humans versus chimpanzees (Varki et al., 2009). Moreover, chimpanzees tend not to develop acquired immunodeficiency syndrome (AIDS) after being infected with HIV, and they mount a weaker antibody response to hepatitis C infection (Institute of Medicine & National Research Council, 2011).

In general, immune system–related genes have diverged substantially between humans and chimpanzees (Chimpanzee Sequencing and Analysis Consortium, 2005; Bitar et al., 2019). Particularly interesting is that the type of white blood cell that caused the life-threatening cytokine storms in the TGN1412 drug trial (see chapter 1) is less reactive in chimpanzees than in humans, suggesting that these cells were genetically modified in the human lineage after it split from chimpanzees (Chapman et al., 2007; Soto et al., 2010). Thus, chimpanzees exhibit high but still imperfect fidelity as models for human diseases and therapeutics.

In light of these considerations and the inescapable ethical concerns, the US Institute of Medicine concluded in 2011 that research on chimpanzees is no longer necessary for invasive biomedical research (Institute of Medicine & National Research Council, 2011). Instead, it highlighted as alternatives the use of humanized mice and the development of extremely sensitive techniques that make it possible to test humans for adverse reactions to drugs at very low doses (i.e., microdosing). The rapporteurs did leave open the possibility that biomedical research on chimpanzees might become necessary in the future to deal with as yet unknown medical threats. Therefore, rather than recommending a total ban on chimpanzee research, they proposed specific criteria that such research would have to meet. They also recommended a lower bar for comparative genomic and behavioral research with chimpanzees, because such research can be performed without invasive procedures or with animals that have been trained with positive reinforcement to "acquiesce to" anesthesia and other procedures.

Based in part on this report from the Institute of Medicine, the NIH withdrew its support for chimpanzee research in 2015 (Kaiser, 2015). Even before then, the NIH had declared a moratorium on breeding chimpanzees for research and transferred many of its captive chimpanzees to animal sanctuaries. An outright ban on research with great apes, called the Great Ape Protection Act, was introduced in the US Senate repeatedly between 2008 and 2012, but it never passed. Nonetheless, access to chimpanzees for research purposes has become extremely difficult in the United States (VandeBerg & Zola, 2005). Similar barriers to chimpanzee research have been erected in many other countries that carry out significant amounts of biological research. In the United Kingdom, for example, no great apes have been used in research since 1986 (Weatherall, 2006). Some important comparative research is being done with archival tissue from chimpanzees—especially their brains (see chimpanzeebrain.org)—but even this research has suffered from inconsistent grant support.

3.6.4 Marmosets

Relatively new to the scene of nonhuman primate research is the common marmoset (*Callithrix jacchus*). In contrast to humans, chimpanzees, and macaques, which all belong to the catarrhine lineage of primates, marmosets are platyrrhine primates (aka New World monkeys). Given this phylogenetic position, it is not surprising that marmosets lack some of the features that are unique to catarrhine primates and that their protein coding sequences are, on average, less similar to our own than those of macaques (Burkart & Finkenwirth, 2015; Preuss, 2019). Marmosets also differ from many other primates in that they tend to give birth to twins and are unusually small, weighing an order of magnitude less than a typical macaque. Because of this small body size, marmosets require less vivarium space than other primates, which in turn reduces housing costs. Marmosets also reach sexual maturity after just 1.5 years (versus three to five years for macaques), which facilitates breeding and the creation of transgenic marmosets (Sasaki et al., 2009; Park et al., 2016; Sato & Sasaki, 2017). Yet another important consideration is that marmosets do not carry the macacine herpes virus, which can be transmitted from macaques to humans and cause serious illness or death.

Because of these advantages, marmosets have become increasingly popular as research animals in several countries. Japanese researchers, in particular, have developed a large research program focused on marmosets (Okano et al., 2016). Much of this research is centered on the brain, which is far more similar between humans and marmosets than between humans and nonprimates (Preuss, 2019). Although Japan has clearly taken the lead, the People's Republic of China is also investing in marmoset research. However, China has opted for an even greater focus on macaques, including genetically modified macaques (Hao, 2007; Poo et al., 2016; Park & Silva, 2019). In fact, China is luring prominent primate biologists away from other countries, where nonhuman primate research is more difficult, expensive, and controversial (Abbott, 2014; Cyranoski, 2016; Vogel, 2020).

3.7 THE REST OF THE MENAGERIE

In addition to the previously mentioned species, biologists study a wide array of other species, most of which were selected according to Krogh's principle, which is to say that they are either especially convenient for experimental investigations or highly specialized in other, interesting ways. Many of them are studied to answer basic science questions that have no immediate biomedical relevance. I will return to these nontranslational research efforts at the end of this section. First, however, let us review four species that have played major roles in translational research, namely pigs, chickens, zebrafish, and the nematode worm *Caenorhabditis elegans*.

3.7.1 Pigs

Roughly 50,000 pigs are used for research purposes in the United States per year, and their prevalence seems to be increasing globally (Brown et al., 2013). Especially popular are several strains of minipigs; weighing 20 to 90 kg, they require less space and food than the large pigs bred in agriculture (Gutierrez et al., 2015). Although pigs reproduce more slowly than mice or rats, they mature relatively fast (four to five months), can have multiple litters per year, and produce five to eight offspring per litter (van der Staay et al., 2017; Luo et al., 2018). Thus, they reproduce significantly faster than nonhuman primates.

The pig genome was sequenced in 2012 and found to be more similar to that of humans than the mouse and rat genomes are, which is surprising given that rodents are more closely related to primates (see figure 2.2). A closer look reveals that it is the mouse genome that has diverged unusually fast, mainly by undergoing large chromosomal deletions (Thomas et al., 2003). Genetic engineering is not as easy in pigs as in mice, but biologists have now developed a variety of ways to modify porcine genomes, and the use of genetically modified pigs has risen substantially (Vidinská et al., 2018; Wolf et al., 2019).

Aside from being similar to humans in some genetic respects, pigs resemble humans in diverse other ways. In the words of one pig model enthusiast, "pigs are not just pigs but almost human" (Douglas, 1972, p. 226). Indeed, several organs are quite similar between humans and pigs, including the skin, the eyes, the immune system, and the gastrointestinal tract. These similarities have encouraged the use of pigs as models for a variety of human maladies, ranging from sunburn to brain injury and depression (Gieling et al., 2011; Gutierrez et al., 2015; Kinder et al., 2019). The large size of pigs also makes them frequent subjects for surgery technique development and medical device testing (Swindle et al., 2012; Hennessy & Goldstein, 2019).

The most common use of pigs in biological research, however, is as a "second non-rodent species" in toxicology and therapy testing (Bode et al., 2010; Brown et al., 2013). Regulatory agencies usually require such a second species, and experiments on dogs and primates tend to raise more ethical concerns than those on pigs. One should note, however, that this disparity arises mainly because pigs are so widely consumed for food. Numerous studies have suggested that pigs are actually quite intelligent and emotionally complex (Marino & Colvin, 2015).

3.7.2 Chickens

Domestic chickens descended from Red Jungle Fowl more than 8,000 years ago. The number of chickens used for research in the United States is difficult to estimate because those numbers are not reported to the US Department of Agriculture (USDA). In the United Kingdom, however, researchers typically perform experiments on just over 100,000 birds per year (mostly chickens; see figure 3.2). Much of this research deals with agricultural issues, notably selective breeding, nutrition, and animal health. However, chickens also serve as models for the study of more general biological questions.

At least since Aristotle, biologists have realized that chicken embryos are easy to study because they are easy to maintain and breed, develop outside of the mother, can be incubated artificially, and are relatively large. They also float on top of the yolk, which makes them easy to observe (through a window in the egg shell) and manipulate experimentally. Researchers have also learned how to create genetically modified chickens (van de Lavoir et al., 2006; Collarini et al., 2015; Woodcock et al., 2017). Collectively, these studies have revealed many features of vertebrate development that seem to be broadly conserved (Stern 2005).

In addition, chicken embryos and chicken cell lines have been used extensively for vaccine development; in fact, approximately 82% of all flu vaccines in the United States are grown in chicken eggs (this technique does not work for coronavirus vaccines [Yeung, 2020]). One may further note that chickens have contributed significantly to cancer research, notably through the discovery of cancer-causing viruses (see chapter 5) and the study of ovarian cancer, which develops spontaneously in chickens and humans, most likely because both of these species ovulate more frequently than other species (Bahr, 2008). Finally, chickens and their eggs are widely used as models in environmental toxicology, where the aim is to determine the effects of diverse chemicals on humans or wildlife (Giesy et al., 2003; E. G. Xu et al., 2019).

3.7.3 Zebrafish

The zebrafish (*Danio rerio*) is a small teleost fish native to southern Asia. It is popular in the pet trade, hardy, and inexpensive to maintain. Zebrafish become sexually mature by 10 to 12 weeks, lay on the order of 100 eggs at a time, and can mate every couple of weeks (Meyers, 2018). Thus, they reproduce extremely rapidly (for vertebrates) and can be housed cheaply in very large numbers. These features make zebrafish very convenient for genetic studies, especially when mutagens are used to increase the rate of mutant production. Indeed, tens of thousands of mutant zebrafish lines have now been created and can be ordered from centralized facilities (e.g., zebrafish .org). Although inbreeding significantly suppresses fertility in zebrafish (Monson & Sadler, 2010), mutant lines can be maintained by experimentally inducing the females to reproduce parthenogenetically (i.e., without DNA from the male) or, more recently, by combining frozen sperm, in vitro fertilization, and strategically designed breeding regimes (Geisler et al., 2016).

The zebrafish genome was sequenced in 2013, and 70% of all zebrafish genes were found to have human homologs (Howe et al., 2013). However, more than half of these genes do not have simple one-to-one homology (i.e., orthology) relationships to their mammalian counterparts, mainly because teleosts apparently duplicated their entire genome early in their history and then lost a subset of the duplicated genes (Inoue et al., 2015). Because of these complex genetic differences, the effects of single gene mutations in zebrafish can be difficult to discern if the duplicate gene remains intact and compensates for the mutated gene. However, duplicated genes sometimes diverge in such a way that each of the duplicates assumes a subset of the ancestral gene's functions (Force et al., 1999). In such instances, identifying those functions may actually be easier in zebrafish than in species where the genes are not duplicated (Kleinjan et al., 2008).

Another difference between zebrafish and the more traditional model organisms—notably fruit flies and mice—is that the wild-type lines of zebrafish are not well standardized, both because of inbreeding suppression and because of divergence between strains maintained in different facilities, which are periodically outcrossed to wild zebrafish to increase their vigor. In addition, the various wild-type zebrafish lines currently in use were derived from different subsets of the wild population, which is very heterogenous genetically (Suurväli et al., 2020).

Starting in the 1960s, zebrafish were used extensively for the study of embryonic development. Zebrafish were especially convenient for such research because their embryos develop outside the mother's body and do so very rapidly (most of their organs are formed within four days). Furthermore, zebrafish embryos are transparent, which makes it possible to trace the lineage and movements of individual cells in living embryos, be they mutant or wild type. In addition, researchers have developed a variety of methods for surgical or genetic manipulation of zebrafish embryos. Collectively, these studies have revealed a variety of developmental mechanisms that apply not only to zebrafish but also, at least in principle, to other vertebrates (Grunwald & Eisen, 2002).

A second area where zebrafish are used extensively is toxicology, mainly because it is relatively easy to expose multiple zebrafish larvae or embryos to multiple test compounds "in parallel" (in multiwell cell culture plates) (Horzmann & Freeman, 2018; Cassar et al., 2020). As explained more thoroughly in the next chapter, such high-throughput screens are desirable because regulatory agencies have a large backlog of compounds that have never been tested for toxicity. Indeed, several zebrafish toxicology screens have shown at least moderate concordance with data from mammalian species. For example, a set of drugs to known to affect mammalian heart function has somewhat similar effects in zebrafish (Milan et al., 2003; Dyballa et al., 2019).

Collectively, these data indicate that the predictive validity of the zebrafish model in toxicology is far from perfect, especially when it comes to toxin sensitivity, but no worse than that for other species—and sometimes better than that of human cell culture systems (Dyballa et al., 2019). At the very least, the results from zebrafish screens can be used to prioritize some chemicals for additional scrutiny. Although this tiered approach is widely endorsed, the risk of false negatives (i.e., missing adverse effects) appears to be relatively high even in mammalian screens (Olson et al., 2000; Monticello et al., 2017). One should also note that terrestrial mammals are typically exposed to toxins via inhalation, ingestion, or across the skin, whereas the zebrafish in typical screens are simply (and very conveniently!) immersed in water-soluble chemicals; such different routes of exposure may well produce divergent results.

Over the last two decades, improved techniques for targeted mutagenesis and genome editing have allowed researchers to model a variety of human diseases in zebrafish (Bradford et al., 2017; Davis & Katsanis, 2017). To create these models, researchers typically insert the human disease genes into the zebrafish genome or, using gene editing, mimic the disease-linked mutations in the endogenous zebrafish homologs. When the resulting fish exhibit symptoms similar to those of the human disease, the model is usually considered successful, and the models are then used for high-throughput screening of potential therapeutics (Wiley et al., 2017).

Some of this work has led to novel therapies for diseases in zebrafish models, which were then tested in humans (Cully, 2019). For example, a wide variety of cancers have been genetically induced in zebrafish and found to be quite similar to cancers in other species (Kirchberger et al., 2017). Some researchers have even implanted zebrafish with cells derived from human tumors and then tested how well the tumors in the fish respond to various potential therapies (Fior et al., 2017); the drugs found to be most effective in the "avatar fish" can then, at least in theory, be administered to the patient (Fazio et al., 2020). These uses of zebrafish to fight cancer are very promising, but they are not yet in clinical use. One anticancer drug that had emerged from zebrafish research (i.e., ProHema) passed phase II clinical trials but then "fell by the wayside" (Cully, 2019) as other forms of cancer therapy emerged.

Similarly, zebrafish are being used to model Dravet syndrome, a rare but severe form of epilepsy that involves mutations in a specific sodium channel gene. Zebrafish with analogous mutations exhibit abnormal patterns of brain activity as well as swimming movements that are "reminiscent" of epileptic seizures in humans (Griffin et al., 2016). These abnormalities abate when the fish are treated with some known antiepileptic drugs (Baraban et al., 2013). Thus, this model scores relatively high in terms of construct, face, and predictive validity (see chapter 2, section 2.2.1). Moreover, a high-throughput drug screen performed on these fish revealed several effective treatments, the most promising of which is clemizole. A phase II trial testing whether this compound (aka EPX-100) is safe for children with Dravet syndrome is ongoing (no results have been published as of July 2021).

In short, zebrafish show a great deal of promise as models for the study of human disease, but the extent to which such work will deliver new medicines remains an open question. This is not surprising given that the use of zebrafish to model human diseases only began in earnest 20 years ago and clinical trials often take a decade or more to come to fruition.

3.7.4 "The Worm" *Caenorhabditis elegans*

The roundworm *C. elegans* is just over 1 mm long as an adult and can live at high density in petri dishes containing agar and bacteria. Most individuals are females that reproduce as hermaphrodites via self-fertilization, and each female can produce approximately 300 offspring, which reach adulthood in three to five days. *C. elegans* anatomy is quite simple, with each adult female containing 959 cells, including 302 neurons.

Because the animals are transparent, researchers have been able to trace the developmental lineages of all these cells. Remarkably, they found the entire cell lineage map and most of the neuronal connections to be highly stereotyped across genetically identical animals (Sulston, 2002; Ankeny, 2007). Laser ablation of specific cells subsequently revealed that the fate of many cells depends on interactions with other cells and is not, therefore, rigidly preprogrammed (Kenyon, 1988). The cell lineage studies also showed that a significant number of cells die during normal development, which is to say that they undergo programmed cell death. More recent studies have combined the natural transparency of *C. elegans* with the use of calcium indicator dyes to monitor neuronal activity during behavior (Larsch et al., 2013), something that is much more difficult to do in larger animals.

A major aim of *C. elegans* research has long been to study the functions of genes (Brenner, 2002). It was convenient, therefore, that *C. elegans* breeds so rapidly and can be maintained in enormous numbers. The use of chemical mutagens and the ability to rapidly screen thousands of animals further facilitated the identification of mutants, and the worms' asexual mode of reproduction made it easy to create homozygous mutant lines. Having identified interesting mutants, researchers were able to identify the altered genes and compare their sequences to those of genes in other species. Using this general approach, researchers have identified numerous *C. elegans* genes with important developmental functions, including programmed cell death (Horvitz, 2002).

A number of these genes have homologs in humans, and some of their functions appear to be conserved as well, at least at the cellular and molecular levels. Because this degree of conservation is surprising given the enormous phylogenetic distance between humans and worms, it helped to spawn the widespread belief that all the truly "fundamental" features of animal life are broadly conserved (Horvitz, 2002). However, sequencing of the *C. elegans* genome revealed that only about 35% of its genes have human orthologs (i.e., one-to-one homologs) (Shaye & Greenwald, 2011).

Despite the significant amount of genetic divergence between roundworms and humans (Zdobnov et al., 2005), *C. elegans* make an attractive model for toxicology research because high-throughput screens are so easily performed with this species. Indeed, like zebrafish, roundworms are often viewed as "a bridge between *in vitro assays* and mammalian toxicity testing" (Hunt 2017, p. 56). Unfortunately, the data obtained from the *C. elegans* screens do not closely track the results obtained in zebrafish assays (Boyd et al., 2016). Indeed, the concordance rates between different model systems in toxicology are generally not very high (see chapter 4), which is why regulatory agencies usually request tests to be conducted in at least two different models and why positive test results in non-mammalian systems are mainly used to prioritize *potential* toxins for further testing. In general, toxicological decision making must balance the need to avoid false negatives, which would endanger public health, against the problem of false positives, which can create needless economic distress for chemical or pharmaceutical companies. Achieving this balance is very difficult when concordance rates are low.

Analogous issues arise when *C. elegans*—or, for that matter, other invertebrates—are used to model human diseases. In the words of Titus Kaletta and Michael Hengartner,

> Given that even mammalian models are often not reliably predictive of drug action in humans, it is—from a preclinical model perspective—unrealistic to expect an invertebrate system to give enough confidence to predict drug action and safety in humans. Non-mammalian model organisms will be typically used in early research and should deliver fast answers to a discovery problem, such as the function of a gene, or pioneer medical research to define novel therapeutic entry points. Of the animal models, *C. elegans* is certainly the fastest and most amenable to cost-effective medium/high-throughput technologies. *C. elegans* is a valuable disease model if the disease can be defined on a molecular basis. (2006, p. 387)

Indeed, biologists generally accept that invertebrate or in vitro models of human diseases never fully replicate the human condition (see chapter 6, section 6.4.3). Press releases aside, the principal goal of such studies is to learn how the modified genes function in the model system and then to use this information to guide experiments in humans or, more frequently, in other models that more closely resemble humans.

This is a reasonable, time-honored approach. However, the success of the cross-species extrapolations depends on the degree of sequence similarity between the human disease genes and those of the model, as well as the molecular, cellular, and organismal contexts in which those genes operate.

Given the large genetic differences between *C. elegans* and humans, it is not surprising that attempts to extrapolate between these two species do not always succeed. This moderation of expectations explains why invertebrate disease models are generally considered successful when genetic manipulations produce cellular or behavioral symptoms that bear at least a superficial similarity to the modeled human disease, not when they lead to novel therapies that succeed in clinical trials.

To the best of my knowledge, few if any therapies of human diseases have emerged directly from *C. elegans* research. Perhaps this is why Sydney Brenner, the founding father of *C. elegans* research, reportedly suggested in 2008,

> Throw out the animal models for the moment and focus on man as the primary organism of study. I started on *C. elegans* as a model system because humans were not experimentally accessible. But now we have the human genome. And if man's genes are accessible, it's man's genes that we should be focusing on. (Friedberg 2008, p. 9)

This view is probably not widely shared among biologists, but it is worth considering. We will come back to it in chapters 6 and 7.

3.7.5 Animal Models in Nontranslational Research

Many biologists study species other than those I review in this book, and they frequently do so for reasons that are independent of any translational ambitions. For example, they study how songbirds learn their songs, how some fish use weak electrical signals to communicate and navigate in muddy waters, how hummingbirds or bumblebees fly, how locusts swarm, or bees and ants communicate (Camhi, 1984; Catania, 2020). Such work tends to focus on species that exhibit the traits of interest either uniquely or most robustly than other species, and are well suited for experimental inquiry. Small research communities usually form around the study of these specialist species—sometimes called "Krogh organisms" (Green et al., 2018)—and community members often share research-facilitating resources, such as husbandry techniques and genomic data. Thus, these species qualify as "model species" and may even aim for "model organism" status insofar as the research on them spans diverse levels of analysis (National Research Council, 1985; Ankeny & Leonelli, 2020).

However, most of these nontraditional model species were not initially selected because researchers thought that findings obtained in them would generalize to humans or other distant relatives. On the contrary, they were generally selected for some fascinating trait that is *not* present in all species. The ensuing research is sometimes said to be curiosity driven, but it can certainly yield unexpected societal benefits (e.g., via bio-inspired engineering or unexpected medical applications). Recalling Pasteur's metaphor of science as a fruit tree (see chapter 1), we can say that this nontranslational, curiosity-driven research is aimed at growing the tree, rather than reaping a harvest.

Unfortunately, a full discussion of nontranslational research and the species it relies on would burst the limits of this book. However, I believe that curiosity-driven research should be valued more highly than is currently the case (Zoghbi, 2013; Lindsley, 2016). It should be viewed in concert with applied/translational science as being part of an overarching search for biological principles that are quite general, even if the details of their implementation are somewhat species-specific (see chapter 7, section 7.1.4).

3.8 MODEL SYSTEM ECOLOGY

As shown in figure 2.1 and in several figures of the present chapter, the popularity of the various species used for research has waxed and waned over the years. Since the various model systems are competing with one another for research funds and attention from researchers, one can think of them as being players in an ecosystem of models. We will discuss this notion of competing model systems more thoroughly in chapter 4, but already some trends and principles are clear.

As the use of guinea pigs, rabbits, dogs, and other mammals in research declined, laboratory rats and mice became the dominant research animals. This shift was largely driven by the fact that rats and mice breed more rapidly than other mammals and are cheaper to house in large numbers. It was also propelled by a shift toward "big science" biology (Logan, 2019) and its attendant view of biomedical research as an industrial enterprise that would proceed much more efficiently with highly standardized and "pure" research materials (Rader, 2004; Kirk, 2012).

Particularly important in this transformation was the push by C. C. Little and others in the late 1930s to view inbred mice as "the biological equivalents of standardized, interchangeable parts in a well-oiled machine of biomedical research production" (Rader, 2004, 155). As funding for the centralized production of laboratory mice (mainly at the Jackson Laboratory) increased, the abundant supply created (or at least supported) more demand by researchers. A postwar effort at one of the US National Laboratories to study the effects of radiation on enormous numbers of mice (Russell, 2013) further entrenched inbred mice as the go-to animals for large-scale biological research. With the rise of experimental techniques for manipulating mouse genomes in the 1980s, the popularity of mice in research laboratories became unparalleled.

Not to be neglected is that rats, mice, and flies are pests, so they evoke relatively little compassion from either scientists or the general public. In general, biologists have moved away from species widely held as pets to species that are pests or bred for food. Worms do not fit neatly into this general rule, but killing them is widely considered acceptable. Zebrafish also occupy an interesting middle ground because, as vertebrates, they have intricate nervous systems, complex behaviors, and, according to some, the capacity for feeling and remembering pain (Sneddon, 2009; Braithwaite, 2010; Key, 2015; Striedter, 2016). However, the small, transparent, and abundant zebrafish larvae used in many studies are unlikely to trigger much empathy in most observers; even adult fish are usually considered low on the (supposed) phylogenetic scale. In the words of two young zebrafish researchers, "if you can study it in fish, all things equal, you'd rather do that than in a mouse or a dog . . . I'm happy working on fish. . . . I'm not sure I'd be happy working on mice. I definitely would not be happy working on primates" (quoted in Endersby, 2007, pp. 407–408). Similar sentiments are widely shared among biologists, though rarely verbalized.

The research questions and aims pursued by biologists have also changed over the years. In the early days, biologists were keenly aware of species differences and sought to discover principles that could explain and organize those differences. This was true even in the early days of genetics, where Mendelian inheritance and chromosome maps emerged as some of the key principles. The pattern began to change with the rise of molecular biology, which unearthed principles so general that variation became largely uninteresting (e.g., the genetic code, which actually does exhibit some interesting variation) (see chapter 4, figure 4.2). The guiding assumption became that all truly important findings would be "fundamentally conserved" (Grunwald & Eisen, 2002). As a result, model selection became almost exclusively a matter of experimental convenience and ethical concerns about animal welfare; species differences became an afterthought, at least for most medically oriented biologists.

Now, however, biologists are reluctantly discovering that many of their findings do not generalize across models as well as they had hoped (Yartsev, 2017). This disappointment is evident both in toxicology, where cross-species concordance is relatively low (see chapter 4), and in disease modeling (see chapters 5 and 6), where therapies that work in animals fail at alarming rates in clinical trials. As I will argue in chapter 7, these observations suggest that biologists should pay more attention to model differences and, once again, take up the search for principles that can accommodate variation.

4 A HISTORY OF IN VITRO MODELS

Anything that is true of *Escherichia coli* must be true of elephants, only more so.
—JACQUES MONOD, 1967 (SEE FRIEDMANN, 2004, P. 49)

Instead of working with intact animals or plants, many biologists perform their research on single-celled microbes or on cells and tissues that are grown in petri dishes or culture flasks. That is, they work with *in vitro* systems (meaning "in glass" in Latin). In this chapter, we first discuss the major microbial models, which comprise select bacteria, viruses, and fungi. Then we move on to tissue explants, primary cell cultures, immortal cell lines, stem cells, and the creation of complex cell cultures, including organoids. These discussions are followed by a brief review of how microbial and cell culture models have been used in toxicology. The chapter ends with a discussion of how and why the popularity of the various model systems, be they in vitro or in vivo, waxes and wanes. As we shall see, this dynamic involves both biological and sociological factors.

4.1 MICROBIAL MODELS

Microbes are often defined as any organisms that are too small to see with the unaided eye, but for present purposes they comprise viruses and unicellular organisms. Because multicellularity evolved independently from unicellular life at least 25 times, today's microbes are scattered across the phylogenetic tree (figure 4.1). Most microbes are prokaryotes, which means that they lack a distinct nucleus, intracellular organelles, and several other features typical of cells in the eukaryotes, which include the fungi (e.g., yeasts and mushrooms) and all animals. Viruses are often considered inanimate because they cannot replicate without the help of their host cells, and their phylogenetic relationships remain a major puzzle because viruses can evolve rapidly and engage in extensive horizontal gene transfer (Nasir & Caetano-Anollés, 2015).

A – Multicellularity Evolved Repeatedly

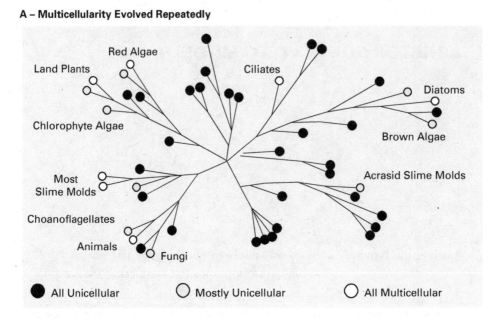

B – Global Distribution of Biomass

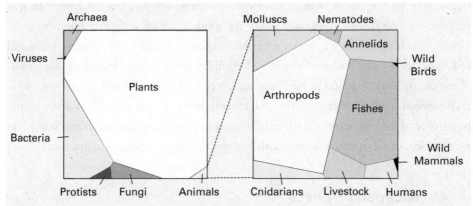

Figure 4.1

Unicellular life forms are ancestral to multicellular life and account for more biomass than animals. (*A*) Multicellularity evolved repeatedly. A phylogeny of eukaryotes indicates the distribution of multicellularity; the most parsimonious conclusion is that it evolved repeatedly from unicellular life. Multicellularity also evolved a few times among prokaryotes, which are not shown here. (*B*) Relative distribution of global biomass. For reference, bacteria account for about 15% of global biomass, whereas humans account for 0.01%. Adapted from (*A*) Grosberg & Strathmann (2007); (*B*) Bar-On et al. (2018).

4.1.1 Bacteria

Bacteria are typically less than 10 micrometers long (for elongate bacteria) and 0.5–2.0 micrometers in diameter. They are estimated to comprise about 1 million species (aka operational taxonomic units) and constitute 15% of the globe's biomass (Bar-On et al., 2018; Louca et al., 2019) (figure 4.1B). They have been studied for many years, especially after Louis Pasteur, Robert Koch, and others in the second half of the 19th century discovered their causal role in numerous infectious diseases (see chapter 5). Biologists have studied many different bacteria, but a few species have attracted disproportionate attention. One of these heavily studied bacteria is *Salmonella typhimurium*, which causes food poisoning. It was in this bacterium that researchers first described the process of transduction, defined as the virus-mediated exchange of genetic material between bacteria (Zinder & Lederberg, 1952). However, the bacterium that has had the most profound effect on biological research is *Escherichia coli*. As one enthusiast put it, "It is not hyperbole to say that *E. coli* is now the most important model organism in biology" (Blount, 2015, p. 1).

E. coli is named for Theodor Escherich, who first cultured it from human feces, and for the lowest part of the intestine, the colon, where *E. coli* lives roughly half the time (the remainder being spent in water, sediment, or soil, waiting to infect another warm-blooded vertebrate host) (Savageau, 1983). Most strains of *E. coli* are benign and actually help humans obtain vitamins K and B12; they also make it easier for other bacteria to thrive in the human gastrointestinal tract by consuming any oxygen there that might harm anaerobic bacteria. As you probably know, however, some strains of *E. coli* are toxic. The O157:H7 strain, for example, normally lives in cattle but can cause severe diarrhea and other complications in humans. Fortunately, the most widely used laboratory strain of *E. coli* (K-12) has lost the ability to thrive in the human colon and is, therefore, quite safe.

Aside from being relatively safe, *E. coli* is well suited for laboratory research because it can synthesize all amino acids and most other organic essentials, which means that it can grow in very simple culture media. Moreover, *E. coli* cells divide every 20 to 60 minutes under ideal conditions. This generation time is not the shortest among bacteria—that distinction goes to *Vibrio natriegens*, which can divide every 10 minutes—but it still allows researchers to grow cultures containing billions of *E. coli* overnight (Sezonov et al., 2007). Such a high rate of population growth facilitates the search for mutations and many other kinds of experiments. It even allows for studies in experimental evolution, where changes in specific traits are followed over thousands of generations (Tenaillon et al., 2012). Another factor that simplifies research with *E. coli* is that its genome contains only 4.6 million base pairs, including approximately 4,300 protein-coding genes (Blattner et al., 1997). By comparison, the human genome contains 3 billion base pairs and at least

20,000 genes. Therefore, *E. coli*'s genome should be easier to unravel. An interesting complication is that, in a comparative study of 61 *E. coli* strains, only 20% of genes were found to be conserved across all strains (Lukjancenko et al., 2010). Given this observation, it is not surprising that humans and *E. coli* share relatively few homologous genes (O'Brien et al., 2004).

An early obstacle to genetic studies with *E. coli* was the belief among many biologists that bacteria exhibit Lamarckian inheritance, with mutations appearing in response to environmental needs rather than by chance. In the words of a leading *E. coli* biologist, "bacteria could not be adopted as models for genetic research until there was some substantiation of the view that they had a genetic system like other organisms" (Lederberg, 1987, p. 23). This problem was overcome in the 1940s when biologists demonstrated that mutations in *E. coli* do arise stochastically. Another key advance was the discovery that bacteria have evolved a variety of ways to exchange genetic material, even though they do not engage in sexual reproduction as it is typically defined (i.e., two cells joining to combine their DNA). Most influential was the discovery of bacterial conjugation, during which one bacterium inserts a piece of its DNA directly into another bacterium (Lederberg, 1947). In hindsight, it was fortuitous that this research was carried out with the K-12 strain of *E. coli*, because only about 5% of all *E. coli* strains are capable of conjugation (Davis, 2003, p. 91).

Once researchers were able to create *E. coli* mutants and crossbreed different strains, the use of *E. coli* as a model species gathered speed. It accelerated further after the discovery of DNA's double-helical structure in 1953, which immediately suggested many new lines of investigation. Researchers created thousands of *E. coli* mutants and developed ever more powerful techniques for manipulating *E. coli*'s genome. Along the way, they uncovered numerous molecular and cellular processes, ranging from gene transcription and regulation to metabolic pathways. Most of these processes were thought to be broadly conserved, prompting Monod's famous dictum that anything true of *E. coli* must also be true of elephants (see this chapter's opening quotation). A few dissenters argued that "the steadily growing recognition of the existence of alternate pathways, of qualitative and quantitative differences in enzymatic patterns, of differences in submicroscopic cell structure, permeability and rate of cell division [argue for] a 'disunity in biochemistry'" (Racker, 1954; quoted in Friedmann, 2004), but such comments were largely ignored at the time.

Research on *E. coli* slowed down after the 1970s, as biologists turned to new questions and other model systems, notably yeast and mice. In the words of François Jacob, who had performed Nobel Prize–winning work on *E. coli* with Monod (see the appendix):

At the end of the 1960s, it was clear that the center of gravity in biology was shifting. Although the study of bacteria and viruses still had much to teach us, it was slipping to second place. If we didn't want to stand around rehashing the same questions, we needed the courage to abandon old lines of research and old models, to turn to new problems and study them with more suitable organisms. (1998, p. 6)

Jacob went on to study mice. However, *E. coli* has continued to be important for diverse forms of research, including comparative genomics, experimental evolution, and the study of toxic strains (Fux et al., 2005; Ferens & Hovde, 2011; Tenaillon et al., 2012). *E. coli* has also been used to synthesize therapeutic compounds, such as insulin or human growth hormone, as well as perfume fragrances and biofuels (Liu & Khosla, 2010; Koppolu & Vasigala, 2016; Idalia & Bernardo, 2017).

4.1.2 Viruses

Biologists in the early 1900s realized that some infectious diseases were caused by particles much smaller than bacteria. These particles, which we now call viruses, range in diameter from 5 to 300 nanometers. The first photographs of viruses were taken with an electron microscope in 1939. Since those early days, biologists have worked with many different viruses, including for example the Rous sarcoma virus, which causes cancer in chickens (see chapter 5). One of the first viruses to be studied in detail was the tobacco mosaic virus, which causes mosaic disease in tobacco and some other plants. It served as the subject for early studies on the structure and chemical composition of viruses and is still widely used in plant biology (Creager, 2002). However, the viruses that have had the greatest, broadest impact on biology are viruses that attack bacteria; they are known as bacteriophages or, more simply, phages.

With an estimated 1 million species and a total of 10^{31} individuals (Keen, 2014), phages are the most diverse and numerous organisms on earth. They infect specific strains or species of bacteria and then use the host's intracellular machinery to make additional phage particles. Their rate of replication is extremely rapid, with infection by a single virus generating up to 100 viral copies in 45 minutes (Davis, 2003). Eventually, the infected bacterium breaks down and releases the new virus particles, which are then ready to infect additional bacteria. In addition to these so-called virulent phages, evolution has fashioned temperate phages, which integrate their genome into that of the infected bacterium. The viral genes then replicate when the bacterium divides but do not create additional virus particles within the host until some triggering event causes the bacteria to enter the "lytic phase," during which the bacteria produce large numbers of phage particles. Either way, viral replication is so rapid that most of the classic phage experiments could be conducted within a single day.

Early studies with phage focused on the T series of virulent phage. These viruses were selected mainly because they infect *E. coli*, which (as reviewed earlier) is an experimentally convenient bacterium. Coupling the high replication rate of *E. coli* with that of the T phages, researchers were able to detect phage mutants as rare as 1 in 100 million. Such experiments allowed high-resolution mapping of mutations inside of single genes (Benzer, 1959), which in turn facilitated the discovery of the triplet nature of the genetic code (figure 4.2) that converts DNA sequences to a succession of amino acids (Crick et al., 1961; Yanofsky, 2007). Work with T phages also led to other important findings, such as the discovery of messenger RNA (mRNA) being an intermediary between DNA and proteins, and of phages injecting their DNA into host cells while leaving their protein coat outside (Stent, 1963; Brock, 1990; Summers, 2004). An interesting fact related to the topic of this book is that Max Delbrück, who led much of the earliest phage research, pressured his colleagues to work with the same T phages and specific *E. coli* strain (strain B) that he had been using because he felt that standardization of the research organisms would allow studies to build on one another more reliably and, hence, more rapidly.

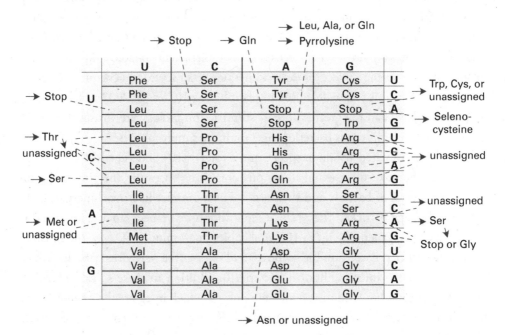

Figure 4.2
Variation in the standard genetic code. This table shows how triplets of the RNA nucleotides uracil (U), cytosine (C), adenine (A), and guanine (G) specify which amino acids should be translated or whether translation should stop. The first, second, and third positions within each triplet (i.e., codon) correspond to the letters along the left, top, and right edges of the table. Although this "universal" genetic code is widely conserved, modifications are found (as indicated) in mitochondria and in a handful of bacteria, fungi, ciliates, green algae, and diplomonads. Adapted from Sengupta & Higgs (2015); Koonin & Novozhilov (2017).

Despite Delbrück's efforts, several very influential biologists preferred to work with temperate phages, especially one called *lambda* (Casjens & Hendrix, 2015). This phage infects the K-12 strain of *E. coli*, and, together, these two microbes became the preferred model system for a number of research areas. Especially important was research on how phage DNA becomes integrated into the host cell's DNA. This work led to the discovery of restriction enzymes and eventually to techniques for inserting foreign genes into host DNA—and thus to gene cloning and recombinant DNA technology more generally. Research with the *lambda* phage also led to the discovery of a repressor protein that prevents viral DNA from being transcribed, which ultimately led to detailed models of transcriptional regulation (i.e., the lac operon model). Overall, research on *E. coli* and its viruses revealed, within a few decades, a long list of molecular mechanisms underlying gene replication, recombination, regulation, and function (Davis, 2003). Many of the principles embodied by those mechanisms have turned out to be conserved across a wide array of species and cell types. In that sense, what is true for *E. coli* really is true for elephants.

However, not all the processes and mechanisms found in *E. coli* and its phages are universally conserved. For example, the *lambda* genome is linear while that of many other viruses is circular; some viruses inject only their DNA into host cells, while others inject some of their protein as well; and most animal viruses do not inject their DNA into host cells at all but are instead endocytosed. Similarly, the lac operon model of gene regulation tends not to hold within eukaryotes. Even the triplet genetic code is not as universal as most biologists assume (figure 4.2). The molecular techniques that emerged from research on *E. coli* and its viruses also had to be fine-tuned. In words of one author, "the technical reasons that initially led to the choice of lambda as experimental material have long since been superseded either by new techniques or by advances that allowed old techniques to be applied to a wider range of material" (Campbell, 1986, p. 280).

None of this imperils the idea that research on bacteria and phage served as "crucibles of the molecular revolution" (Davis, 2003, p. 129). Nor has such research ceased to be fruitful. It has led, for example, to the discovery of the CRISPR/Cas9 gene editing system, which probably evolved as a bacterial defense against phage infections (Stern & Sorek, 2011) and is now used by biologists to edit the genomes of many different organisms. I mostly wish to make the point that not everything that is true of elephants is true of *E. coli* or phage. Indeed, some of those very differences are what made *E. coli* such an ideal organism for the research questions that occupied biologists in the second half of the 20th century.

4.1.3 Yeast

Yeasts are unicellular fungi and, as such, belong to the eukaryotes (see figure 4.1). Thus, in contrast to bacteria, yeasts have distinct cell nuclei, a complex cytoskeleton, intracellular organelles (notably mitochondria), and intracellular membrane systems (e.g., an endoplasmic reticulum). They also divide by means of mitosis and meiosis, rather than the simple binary fission found in prokaryotes.

The most intensively studied species of yeast is *Saccharomyces cerevisiae*, also known as budding yeast, brewer's yeast, or baker's yeast (Greig & Leu, 2009). Like *E. coli*, it is not pathogenic, can live on simple media, and does not form clumps, which makes it relatively easy to estimate the concentration of cultured cells. Budding yeast cells tend to be only 4–6 microns in diameter and can divide every 90 minutes, roughly half as fast as *E. coli*. Importantly, *S. cerevisiae* has a complex life cycle that includes an asexual phase, during which haploid cells divide by budding, and a sexual phase in which cells of different "mating types" fuse their genetic material and then divide by two successive rounds of meiosis (Barnett, 2007). Over the years, biologists have learned to control this process so that they can isolate recessive mutations in haploid, asexually dividing strains and then perform planned crosses using strains of different mating types. They have also figured out how to freeze and store mutant strains for long periods of time, thus creating immense strain libraries for later use. Another major technical breakthrough was the creation of the first transgenic yeast strain in 1978, using an *E. coli* plasmid (a small extrachromosomal piece of DNA) to insert DNA from one yeast strain into another (Hinnen et al., 1978). Soon it became easier to move genes in and out of yeast than any other species.

An especially powerful approach was gene cloning by complementation, which allows researchers to identify the genes underlying any particular mutation in yeast (Elledge et al., 1993). In essence, the approach involves transfecting a mutant yeast strain with a library of many different plasmids that each carry a different snippet of normal yeast DNA and then asking which plasmid is able to convert the mutant to a wild-type phenotype. The successful plasmid can then be replicated in *E. coli* to produce large amounts of the gene and protein of interest. Ultimately, researchers can use restriction enzymes to determine the gene's nucleotide sequence and the protein product to generate useful antibodies. Using these and related techniques, researchers identified many genes that function in critical aspects of yeast biology, including cell cycle control, metabolic pathways, DNA repair, protein targeting and degradation, and cell death (Botstein & Fink, 2011).

Surprisingly, many yeast genes were found to have homologs in the genomes of humans or other mammals. Even more surprising was that many of the human genes could substitute for their yeast homologs in the previously described complementation assays (Tugendreich et al., 1994). These findings supported the idea that most of the "basic" molecular functions of genes are broadly conserved across species. Based on this assumption and the ease with which gene functions could be studied in yeast, findings obtained in yeast came to dominate gene ontology databases, which were originally conceived as species-independent summaries of gene-function associations for "a generic eukaryotic cell" (Ashburner et al., 2000, p. 26). One should note, however, that one of the few studies to explore systematically the conservation of eukaryotic gene-function associations revealed that fewer than one-third of 621 examined yeast genes can be functionally replaced by their human homologs (after accounting for the likelihood of false-negative results) (Hamza et al., 2015). These data indicate a substantial amount of functional divergence between yeast genes and their human homologs.

Considerable divergence is also apparent when one compares the entire genome of *S. cerevisiae* to that of humans or other mammals. The yeast's nuclear genome is 12.8 megabase pairs in size and contains only about 5,600 genes. Roughly 30% of these genes have clearly identifiable mammalian homologs (Botstein et al., 1997); conversely, approximately 30% of human disease genes have yeast homologs with very similar nucleotide sequences (Foury, 1997). These data indicate both a significant amount of genetic conservation and substantial divergence. In this context, it is interesting to observe that the fission yeast *Schizosaccharomyces pombe* shares about 80% of its protein-coding genes with *S. cerevisiae* (Wood et al., 2002), from which it diverged roughly 1 billion years ago (Hedges, 2002). Similarly, it is intriguing that the network of genetic interactions in these two species exhibits substantial differences, even when one considers only homologous genes (Roguev et al., 2008). Whether these genetic differences are surprisingly large or surprisingly small depends on one's expectations, as it does for the yeast-human comparisons.

In any case, most yeast biologists surely agree that "yeast offers invaluable guidance for approaching human disease-associated gene functions. In contrast to humans, the *S. cerevisiae* genes can be easily deleted, mutated and reintroduced into yeast cells, overexpressed, tagged and thoroughly studied, very quickly providing a considerable amount of information useful for understanding the molecular basis of diseases" (Foury, 1997, p. 8). That said, most would also agree that "every major conclusion [reached in yeast] . . . will have eventually to be tested directly in higher organisms" (Botstein & Fink, 1988, p. 1442).

4.2 CELL AND TISSUE CULTURE MODELS

Many in vitro models do not involve microbes but instead entail cells or tissues that were originally isolated from multicellular organisms. The following sections briefly review the major types of models based on animal cells, focusing on their principal advantages and limitations.

4.2.1 Hanging Drop Cultures

The first cultures of explanted animal tissue were created by Ross Harrison in 1910, using tissue from the peripheral nervous system of frogs. Adapting a method previously used to observe living bacteria under the microscope, Harrison attached a fragment of frog tissue to a glass coverslip by means of coagulated lymph, which also provided the explanted cells with nutrients. The coverslip was then inverted and laid over an indentation that had been ground into a standard microscope slide so the "drop" of tissue could hang suspended from the coverslip (Millet & Gillette, 2012) (figure 4.3A). After sealing the contraption to prevent desiccation, Harrison was able to watch individual cells under the microscope. Most importantly, he observed individual neuronal cell bodies extend axons that grew longer over time and were tipped with distinctive growth cones. These observations helped to establish the neuron doctrine, which holds that neurons are discrete cells connected via synapses (Guillery, 2007).

Harrison's hanging drop technique was later modified by switching from lymph to clotted plasma, which was easier to work with, and extended to tissues from warm-blooded animals (Carrel & Burrows, 1911). The main advantage of the hanging drop technique remained the ability to observe the behavior of living cells, but the cultures also facilitated the design of elegant experiments. For example, Levi-Montalcini et al. (1954) explanted pieces of spinal ganglia together with bits of mouse tumors and found that the tumor cells promoted the outgrowth of axons from the explanted neurons. Subsequent research then revealed the molecular identity of the diffusible axon growth-promoting substance, which we now call nerve growth factor (Cohen, 1960).

A major limitation of the hanging drop technique was the tissue's relative inaccessibility, which made it difficult to supply the tissue with fresh nutrients and oxygen, or to create subcultures (i.e., starting new cultures from portions of the original culture). These issues were addressed by the development of special culture flasks (figure 4.3A) that were not as convenient for direct microscopic observation of the cultured cells but made it easier to change culture media, create subcultures, and generally maintain cultures for long periods of time.

4.2.2 Organotypic Slice Cultures

In early tissue culture experiments the explants consisted of tissue chunks that had been cut rather haphazardly. However, experimenters in the 1970s and 1980s began to make more deliberate cuts, sectioning the tissue into thin slices along carefully selected planes so that each slice would optimally reflect the tissue's internal organization. These organotypic slices were then cultured in clotted serum "drops" using a variety of

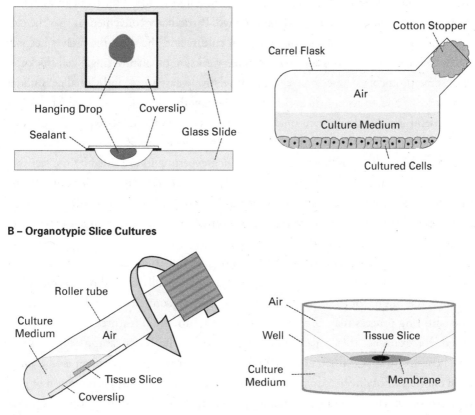

A – Hanging Drop and Carrel Flask Cultures

Cotton Stopper

Carrel Flask

Air

Culture Medium

Cultured Cells

Hanging Drop   Coverslip

Sealant

Glass Slide

B – Organotypic Slice Cultures

Roller tube

Culture Medium   Air

Tissue Slice

Coverslip

Air

Well   Tissue Slice

Culture Medium   Membrane

Figure 4.3

Cell and tissue culture methods. (*A*) Hanging drop cultures (top and side views on the left) are convenient for observing cells under a microscope, but so-called Carrel flasks (right) are better for long-term cultures because they facilitate the exchange of culture medium. (*B*) In organotypic slice cultures, thin slices of tissue can be cultured in rotating test tubes (left), where they are exposed alternately to culture medium and air, or on a semipermeable membrane that floats on the surface of culture medium (right). Adapted from (*A*) Wellbourne-Wood & Chatton (2018); (*B*) Cavaliere & Matute (2011).

protocols that exposed the slices to both culture medium and air, either alternately or at the interface between the two (figure 4.3B) (LaVail & Wolf, 1973; Gähwiler, 1981; Stoppini et al., 1991; Humpel, 2015).

Using these methods, and keeping slices to less than 500 micrometers in thickness, it is possible to maintain the cultures for several weeks or even months. By now, biologists have developed protocols for maintaining organotypic slice cultures of many different organs, including the kidney, lung, liver, and brain. They have also created organotypic cultures of structures like the retina that do not need to be sliced because they are inherently planar and thin.

The principal advantage of organotypic cultures is that they maintain much of the tissue's natural organization while allowing experimenters to access the tissue in ways that are not possible or are extremely difficult in vivo. Particularly convenient is that the clot surrounding the tissue gradually disappears in culture and the slices themselves become much thinner over the course of two to three weeks, approaching single-cell thickness with some protocols. These changes make it relatively easy to see individual cells under a microscope and to record their electrical activity with microelectrodes, which is especially useful for elucidating neuronal functions. The use of voltage-sensitive molecules and fluorescence microscopy further allows investigators to monitor patterns of electrical activity within a slice (Phillips et al., 2015). Such experiments are aided by the fact that organotypic slices tend to be relatively healthy because, after one to two weeks in culture, most of the slicing-induced tissue damage has been cleared (Gähwiler, 1981, 1988). It is even possible to monitor some aspects of normal development in organotypic slices made from immature organs (LaVail & Wolf, 1973).

Of course, tissue slices cannot retain an organ's full complexity whenever important structural features extend over long distances in all three dimensions. Slices through the brain, for example, are bound to destroy some long-range connections. The culturing process may also induce some new, unnatural features. It is well established, for example, that organotypic slices through the hippocampus sprout some novel connections (Coltman et al., 1995). Yet another potential problem is that long-term cultures of brain slices may be distorted by the uncontrolled proliferation of glial cells. This problem is usually prevented by blocking all cell divisions in the slice either with radiation or pharmacologically.

Although brain slices can be cultured for many weeks, researchers often study brain slices in the first few hours after they have been cut. Such acute brain slices are maintained in elaborate slice chambers that continuously perfuse highly oxygenated medium across the slice, thereby minimizing cell death (Schwartzkroin, 1975). Such preparations have been used extensively in neurobiology to study intracellular and circuit-level processes. Their main advantage is that they do not depart from the natural, in vivo anatomy as much as the long-term slice cultures do. However, acute brain slices do exhibit significant damage near the cut surfaces, as well as extensive molecular and synaptic changes that can profoundly affect neuronal function (Kirov et al., 1999; Taubenfeld et al., 2002). Thus, they clearly are imperfect, albeit very useful, models.

4.2.3 Dissociated Cell Cultures

One major limitation of explanted tissue cultures is that the tissue tends to contain multiple cell types, which can make it difficult to gather detailed information about specific cells. Biologists have solved this problem by using mechanical forces and enzymes that destroy connections between cells to "dissociate" the explanted tissue. They then use cell type–specific antibodies to isolate the desired cells and start new, more homogeneous cultures. Many of the cells in such cultures will divide repeatedly (neurons being a major exception, since most mammalian neurons become postmitotic after embryogenesis). At some point the original (aka primary) cultures become overpopulated and risk deterioration, unless the experimenters take a subset of the cells to seed new subcultures.

Many cell types can be subcultured repeatedly, sometimes for years, but it is now generally agreed that normal somatic cells eventually become senescent, which means that they stop dividing and die. Fibroblasts (i.e., connective tissue cells) isolated from normal human fetuses, for example, can only divide approximately 50 times in ideal culture conditions (Hayflick & Moorhead, 1961; Witkowski, 1980). A major mechanism underlying cellular senescence involves the telomeres, which cap individual chromosomes and shorten with each cell division; when those telomeres become too short, DNA damage results. An interesting (albeit tentative) observation is that the doubling potential of primary cultures tends to increase with the life span and size of the species from which those cells originate (Hayflick, 1975), which in turn correlate with the average number of divisions those cells undergo in vivo (see the discussion of Peto's paradox in chapter 7, section 7.1.4).

Although explanted cells of many tissues can divide in vitro, explanted neurons from postembryonic individuals are usually postmitotic; the study of such primary neuronal cultures can nonetheless provide important results. In particular, dissociated neurons may be diluted to such an extent that individual neurons can be observed in relative isolation, independently of interactions with other cells. Such studies have shown, for example, that different types of neurons tend (to a large degree) to adopt their distinctive in vivo morphologies as they extend new processes in vitro (Bray, 1973; Banker & Goslin, 1998). Other studies have revealed the adhesion- and tension-based mechanisms through which axons elongate (Lamoureux et al., 1989), as well as some of the molecular cues that they use for guidance (Zheng et al., 1994).

4.2.4 Continuous Cell Lines

Because primary cell cultures and their subcultures have a limited life span, working with them requires taking samples from animals that likely suffer some distress as a result. This is true especially for cultured neurons, which become postmitotic as they differentiate. Another issue is that primary cell cultures tend to vary with the genotype of the donor animal (or human), and this variability can reduce the replicability of the experiments. All these problems are overcome by the creation of continuous cell lines, which are also known as immortal or immortalized cell lines. Many of these continuous cell lines are derived from tumor biopsies. HeLa cells, for example, were taken from the cervical tumor of a woman called Henrietta Lacks in 1951 (without informed consent, which was not standard at the time), and her cells have been dividing since then in laboratories across the world (Skloot, 2010). More than 60,000 published papers describe work on these cells (Bhatia et al., 2019), and the cells themselves are so hardy and proliferate so rapidly that they have contaminated or completely displaced numerous other cell lines (Gartler, 1968; Hughes et al., 2007; Capes-Davis et al., 2010). Indeed, some biologists have suggested that HeLa cells form a new species that is highly adapted to the laboratory niche (Van Valen & Maiorana, 1991).

Continuous cell lines can also arise spontaneously from primary cultures that have been subcultured repeatedly; presumably this kind of transformation occurs via mutations that would cause cancer if they occurred in vivo. Intriguingly, spontaneous immortalization appears to be extremely rare for chicken cells, more common for human cells, and most common for cells from mice (Macieira-Coelho et al., 1977). A third way to obtain continuous cell lines is to transform cultured cells with viruses that can insert cancer-causing genes into the genomes of their hosts (Cepko, 1989). Some of these genetically engineered cell lines proliferate in one set of culture conditions, but cease dividing and differentiate (e.g., into neurons) when the conditions (e.g., temperature) are changed; that is, they are conditionally transformed. Altogether, biologists now have at their disposal more than 4,000 continuous cell lines that they can order from various cell line repositories (e.g., the American Type Culture Collection at ATCC.org).

The principal advantage of dissociated cell cultures, especially continuous cell lines, is that researchers can examine the properties of specific cell types in tightly controlled contexts and do so repeatedly, using virtually identical (isogenic) cells. Thus, they can examine the effects of various genetic mutations or environmental manipulations and compare them across cell types. Continuous cell lines can also be used to synthesize therapeutic compounds. For example, a cell line derived from a hamster's ovary has long been used to synthesize tissue plasminogen activator, which helps dissolve blood clots (Rahimpour et al., 2016). Similarly, mouse and human cancer cells have been used to synthesize therapeutic antibodies by fusing them to antibody-producing spleen cells to create "hybridoma" cell lines (Köhler & Milstein, 1975; Olsson & Kaplan, 1980) (see chapter 5). Although mammalian cell lines tend to be less efficient at high-volume synthesis than *E. coli*, they contain more of the cellular machinery required for proper folding and post-translational modification of the desired, mammalian proteins (Wurm, 2004).

A key limitation of dissociated cell cultures is that cells tend to change their phenotype as they acclimate to culture conditions. For example, cultured lung endothelial cells express only about 42% of the membrane-associated proteins that they express in intact rats (Durr et al., 2004) (figure 4.4A). Similarly, placing microglia from human or mouse brains into primary cultures causes almost 4,000 genes in the explanted cells to be up- or down-regulated more than fourfold, relative to the in vivo condition (Gosselin et al., 2017). Transplantation experiments (with microglia derived from stem cells) suggest that many of these culture-induced changes in gene expression can be reversed when the cultured microglia are brought into their natural, in vivo environment (Hasselmann et al., 2019) (figure 4.4B), but this merely underscores that the culture environment can have profound effects on cellular phenotypes (Matarese et al., 2012). Because culture conditions may vary across laboratories and with different batches of culture medium ingredients (e.g., animal-derived serum), this form of phenotypic plasticity reduces replicability. A related problem plaguing primary cultures is that the more rapidly proliferating cell types may come to dominate the cultures, thereby causing experimental results to vary with time in culture.

A major problem with continuous cell lines is that they often exhibit significant genetic mutations. The aforementioned HeLa cells, for example, exhibit genetic abnormalities on 20 chromosomes, and four of their chromosomes are shattered into multiple pieces (Mittelman & Wilson, 2013). These abnormalities are probably extreme, but tumor-derived cell lines often exhibit genetic mutations and copy number variations that exceed those observed with in vivo tumors (Domcke et al., 2013). Spontaneously transformed cells are subject to analogous problems because the senescence-induced telomere shortening tends to cause chromosomal aberrations (Hornsby, 2007). In addition, continuous cell lines that are maintained as separate populations tend to diverge over time, as their cells acquire random mutations (via genetic drift) and adapt differentially to the cell culture environment. This has been well documented in a comparative analysis of various cancer cell lines (Ben-David et al., 2018). In the words of one commentator, "different stocks of widely used cancer cell lines—a staple of cancer research over many decades—are highly heterogeneous in terms of their genetics, transcriptomics and responses to therapies" (Hynds et al., 2018, p. 1). Many of these molecular changes probably relate to the previously mentioned selective pressure for increased proliferation rates in culture conditions (Auman, 2010).

For all of these reasons, many biologists agree that key experiments performed on continuous cell lines should be replicated in primary cells, at least if the intent is to obtain knowledge that applies also to in vivo conditions (Kaur & Dufour, 2012). Nonetheless, continuous cell lines are often selected as "model cells" for their availability, hardiness in culture, experimental tractability, and accumulated knowledge of their properties. PC12 cells, for example, have been widely used as a model for neurons and, in particular, for the study of transmitter release. This cell line was originally derived

A – Fresh *Ex Vivo* vs. *In Vitro*

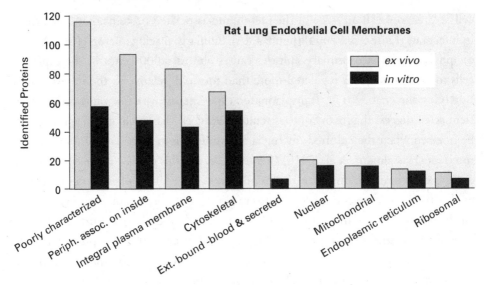

B – iPSc-derived Microglia Implanted into Mice

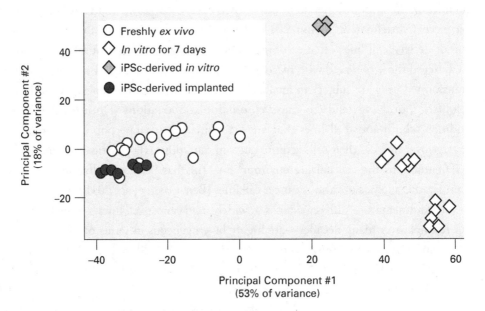

Figure 4.4

Cells are different in vitro versus in vivo. (*A*) Fresh ex vivo versus in vitro. Durr et al. (2004) isolated microvascular endothelial cells from rat lungs and then examined the proteins associated with their cell membranes, either right after isolating the cells (ex vivo) or after they had been cultured for some time. They found that many proteins were expressed ex vivo but not in vitro (varying somewhat with the presumed location of the proteins; shown along the *x*-axis). (*B*) Hasselmann and colleagues derived microglia from human induced pluripotent stem cells (iPSCs) and transplanted them into mouse brains. They found that the gene expression profile of the implanted microglia was quite similar to that of freshly isolated human microglia, but significantly different from that of the cells before implantation and from isolated human microglia cultured for seven days (as visualized in a principal components analysis). Adapted from (A) Durr et al. (2004); (B) Gosselin et al. (2017); Hasselmann et al. (2019).

from a rat's adrenal gland tumor (Greene & Tischler, 1976), but its cells synthesize and release several neurotransmitters that are also found in postganglionic sympathetic neurons, which are developmentally related to the adrenaline-secreting cells of the adrenal gland. Treatment with nerve growth factor also causes PC12 cells to grow long processes that resemble axons. Therefore, these cells are clearly neuron-like. However, they do not develop dendrites or proper synapses, and their phenotype does not precisely match that of any in vivo neurons (Greene et al., 1998; Banker & Goslin 1998; Westerink & Ewing, 2008).

Thus, PC12 cells are also an example of "*in vivo* veritas—*in vitro* artificia" (Matarese et al., 2012). Fortunately, most cell biologists are keenly aware of this issue and usually perform critical experiments on multiple cell types and, not infrequently, on primary cells and intact animals as well.

4.2.5 Stem Cells

In the last two to three decades, some of the most commonly cultured cell types have been stem cells, which we can define as self-renewing proliferating cells that, given various specific treatments, can differentiate into a broad range of cell types. Stem cells vary in the range of potential cell fates they may pursue, but most of the in vitro research has focused embryonic stem cells (ESCs) and induced pluripotent stem cells (iPSCs). The former are derived from very early (preimplantation) embryos and can develop into virtually all types of adult cells. Mouse ESCs were first created in 1981, but efforts to culture human ESCs were unsuccessful until researchers began to "feed" them with factors derived from cultured mouse cells (Thomson et al., 1998). It also helped to recognize that mouse and human ESCs respond differently to a few critical molecules (Yu & Thomson, 2008). For example, the transcription factor c-Myc is used to promote self-renewal in mouse ESCs, but it causes human ESCs to differentiate or die (Sumi et al., 2007; Watt et al., 2008).

In contrast to ESCs, iPSCs are derived from adult somatic cells (e.g., skin cells) that have been engineered to express, or otherwise treated with, several transcription factors that collectively "induce" the cells to dedifferentiate and then become capable of redifferentiating into a wide variety of different cell types. Essentially, iPSCs are adult somatic cells that have been reprogrammed in vitro to become similar to ESCs. They even resemble ESCs in terms of their "landscape" of epigenetic modifications (Guenther et al., 2010), which change systematically across development and regulate cellular differentiation (Atlasi & Stunnenberg, 2017). However, closer analyses have revealed some significant differences between the epigenomes of iPSCs and ESCs, as well as between different iPSC lines (Lister et al., 2011). These differences imply that the reprogramming of the iPSCs is often incomplete, which causes the cells to have a "memory" of their somatic origin (Efrat, 2021). The reprogramming also tends to be somewhat aberrant, which causes iPSC lines derived using identical protocols to be notoriously variable (Ortmann & Vallier, 2017; Bar & Benvenisty, 2019). The first human iPSCs were created in 2007, one year after the discovery of mouse iPSCs (Takahashi et al., 2007).

Mouse ESCs and iPSCs can be used in place of mouse primary cell cultures so long as protocols have been developed to differentiate the stem cells into the desired cell types (McQuade et al., 2018); their principal advantage is the reduced animal sacrifice. Similarly, human ESCs can be derived from unwanted preimplantation embryos containing just 200 to 300 cells, and human iPSCs can be derived from easily harvested tissue (e.g., skin or connective tissue). Especially useful is that these stem cells can be derived from human subjects with known genetic diseases, thus creating cellular disease models that are "genetically precise" (Kaye & Finkbeiner, 2013). The patient-derived cells can then be studied in vitro to gain a better understanding of the molecular disease mechanisms or to test which of several possible therapies is most effective on a patient's own cells. Patient-derived stem cells can also be differentiated into a desired cell type and transplanted back into the patient without risking the immune rejection that usually accompanies transplantation of foreign cells. The implanted cells may be able to replace cells that were lost as part of the disease or they may secrete various factors that can slow or reverse disease progression (Drago et al., 2013; Grade & Götz, 2017; Sharma et al., 2019). If necessary, scientists can "correct" the mutations of the stem cells before the transplantation by use of genome editing techniques (An et al., 2012). Genetically corrected cells can also serve as an ideal control group in experiments on disease mechanisms because they are isogenic with the patient's cells except for the disease-causing defects.

Despite their enormous promise, stem cells have significant limitations. For human ESCs, the principal concern is that their use requires the destruction of human embryos. Although those embryos are usually less than two weeks old and left over after in vitro fertilization (IVF) (Hyun et al., 2016), they have the potential to develop into full-fledged human beings. Many people—including many with pro-life views (Lo & Parham, 2009)—find such a use of human IVF embryos acceptable, but public attitudes on this subject are varied and complex, as illustrated by the extended debates over the tight restrictions on research with human ESCs that were imposed on US researchers while George W. Bush was president (Connolly, 2005). Fortunately, those ethical concerns can be sidestepped entirely by working with human iPSCs because they are derived from tissue humans can spare. Nonetheless, the donors of the original cells must be fully informed and then consent to how their cells are going to be used (Lo & Parham, 2009).

A second limitation of stem cells is that they tend to acquire significant genetic abnormalities during their time in culture, ranging from extra chromosomes to insertions, deletions, and point mutations (Draper et al., 2004; Gaztelumendi & Nogués, 2014; Martin, 2017). Some of these aberrations may increase the fitness of the cells in the culture environment, and not all of them would necessarily be harmful in vivo, but they clearly make the behavior of extensively cultured stem cells less predictable for purposes of experimentation or therapy. The fact that stem cell cultures vary with the genotype of the organism from which they were derived, together with the "aberrant" epigenetic reprogramming mentioned previously, further increases this unpredictability. Given this variability, one wonders whether regulators charged with ensuring the safety and efficacy of clinical products might have to approve each batch of stem cells separately (Anderson & Cummings, 2016). A related problem is that the differentiation of stem cells before transplantation is rarely complete, which means that some of the transplanted cells may remain proliferative and, hence, potentially tumorigenic (Kaye & Finkbeiner, 2013; Burns & Verfaillie, 2015).

A third significant complication is that human stem cells intended for transplantation into humans must first be tested in animals. However, transplanting human cells into mice tends to trigger a strong immune response that kills the foreign cells unless the mice are genetically modified to lack crucial parts of the immune system (e.g., Marsh et al., 2017). This is potentially problematic because immune responses are integral to many diseases, either as part of the pathology or in response to it (Doty et al., 2015) (see chapter 5). Therefore, modeling such diseases in immunodeficient mice may not accurately predict the human response. A potential solution to this dilemma is to work with mice that have a "humanized" immune system (L. D. Shultz et al., 2007; Yong et al., 2018), but this humanization can only be partial. As a result of all these uncertainties, it currently remains unclear to what extent stem cell therapies will ultimately live up to their enormous potential (Marcheque et al., 2019). The most promising result to date is that the transplantation iPSC-derived cells into patients with macular degeneration stabilized the vision of one patient for at least one year (Mandai et al., 2017). However, the US Food and Drug Administration has not yet approved any human stem cell therapies other than the transplantation of hematopoietic (i.e., blood cell producing) stem cells from human umbilical cords (FDA Commissioner, 2020).

Thus, despite their promise, stem cell therapies should be regarded cautiously (Singh et al., 2020). Unfortunately, this caution is often drowned out by hope and greed, which is why poorly regulated for-profit centers around the world are offering a wide variety of unapproved stem cell therapies, many of which are probably ineffective, and some of which can cause real harm (Cyranoski, 2013; Turner & Knoepfler,

2016). Debates about how best to regulate these therapies and then enforce those regulations are ongoing (Marks et al., 2017; Laurencin & McClinton, 2020). It is not an easy problem to solve, especially because the diversity of stem cell types and treatment procedures is still expanding rapidly.

4.2.6 Complex Culture Systems

Dissociated and stem cell cultures are usually grown with the goal of culturing a single cell type, but sometimes it is useful to co-culture two or more cell types that interact. For example, the previously mentioned discovery of nerve growth factor involved the co-culture of explanted neurons and a separate set of growth factor–secreting cells (Levi-Montalcini et al., 1954). In those experiments the different cell types were placed next to one another within a single hanging drop, but later co-culture experiments often employed culture chambers with multiple compartments that were interconnected in ways that allow cells in different compartments to interact, either through the culture medium or, in some cases, via axons that grow from one compartment into another (Campenot, 1977).

Starting in the early 2000s, culture chambers with multiple compartments were miniaturized and outfitted with pumps that cause the culture medium to flow along channels that are only a few hundred microns wide (Beebe et al., 2002; Whitesides, 2006). These "microfluidic" devices minimize the quantities of cells and culture medium, thereby reducing costs. They can also be augmented by depositing molecular cues, growth factors, or extracellular matrix onto specific surfaces in one or more of the compartments. Furthermore, they may be equipped with miniature motors that create dynamic physical forces similar to those found in vivo, such as heart muscle contractions or gut peristalsis (Huh et al., 2010; Balijepalli & Sivaramakrishan, 2017; Haring et al., 2017). Because microfluidic devices typically contain multiple cell types and are manufactured using methods similar to those used in the production of microchips, they are often called organs-on-a-chip. Compared with traditional culture chambers, the on-chip systems provide the cultured cells with a far more physiological, in vivo–like environment (Li et al., 2005). This verisimilitude clearly facilitates the modeling of both normal and disease-related processes. However, organs-on-a-chip are commonly populated with cells from continuous cell lines (Marx et al., 2012), which keeps costs low but brings with it the genetic abnormalities and other problems (as mentioned in section 4.2.4).

Another way to generate complex cell cultures is to let the cells grow in three dimensions (3D), rather than the two dimensions (2D) of traditional cell culture plates. Except for epithelia, most cells of the body grow in complex 3D environments that are filled with other cells and extracellular matrix. These conditions can be mimicked in vitro by culturing cells in biodegradable scaffolds that resemble extracellular matrix. Such scaffolds may be created in diverse ways, but they typically involve natural or synthetic polymer networks that contain a lot of water. Cells can be mixed into these hydrogels prior to hardening, or they can deposited on top of the matrix and migrate into it (Merceron & Murphy 2015; Funaki & Janmey 2017; Lovett et al., 2020). Such 3D cultures can also be integrated with the on-chip devices described previously, and perfusable channels can be designed into the 3D matrices using micromolds or 3D printing (Xie et al., 2020). Overall, work on these systems has shown that many cells behave differently when grown in 3D versus 2D environments and that they are responsive to the physical and chemical properties of the surrounding matrix. In general, the 3D cultures more closely mimic the in vivo condition (e.g., Pickl and Ries, 2009). This is true especially for the culture of stem cells, which tend to exhibit some "spontaneous morphogenesis" (Paşca, 2018) when cultured in 3D systems, often forming an organoid (Lancaster & Knoblich, 2014).

Organoids result when stem cells self-aggregate in culture and then begin to differentiate into several different cell types, which in turn self-organize to create morphologies that bear some similarity to in vivo organs. By now, scientists have managed to create organoids of many different tissues, including blood vessels, kidneys, intestines, retina, and brain (Kretzschmar & Clevers, 2016; Boutin et al., 2019). The latter are particularly impressive, as they exhibit multiple types of neurons arranged in layers that resemble those of the cerebral cortex (Lancaster et al., 2013). These cerebral organoids, affectionately known as mini-brains, make it possible to study human brain development in vitro. They can also be used to create and study models of various diseases. For example, cerebral organoids have been used to develop potential treatments against Zika virus infection (Y.-P. Xu et al., 2019) and to create a model of Alzheimer's disease that resembles the human disease more precisely than any other in vitro system (Choi et al., 2014). However, mini-brains and other types of organoids typically lack a functioning vasculature, which means that they must be relatively small and frequently contain dead or dying cells in their center. Furthermore, an analysis of many different mini-brains revealed that their constituent neurons do not fully mimic the in vivo cell types and, in general, exhibit symptoms of significant cellular stress (Bhaduri et al., 2020).

An interesting approach to overcoming some of these problems is to implant organoids into live animals. Kidney organoids derived from human iPSCs, for example, have been transplanted into mouse kidneys, where they survive for at least six weeks and become at least partially vascularized (Nam et al., 2019). Similarly, cerebral organoids become vascularized upon transplantation into mouse brains; they even develop active neuronal connections and are infiltrated by microglia, which were not present in the original organoids (Mansour et al., 2018). Ultimately, researchers aim to transplant organoids into humans to replace or repair damaged organs. For now, however, this goal remains a distant one. The aforementioned kidney organoids, for example, remain immature after transplantation and develop some abnormal features, such as cartilage and cysts. Similarly distant are dreams (or nightmares) of mini-brains attaining consciousness (Reardon, 2018) or becoming capable of suffering. Still, the mere existence of mini-brains would have been considered science fiction a few decades ago. Indeed, when one recalls the humble beginnings of cell culture as hanging drops under Harrison's microscope, one must marvel at how far this field has come.

4.3 ALTERNATIVE MODELS IN TOXICOLOGY

Humans generally want to know which pesticides, cosmetics, pharmaceutical products, or other substances are likely to make them sick. Historically, this need was met by examining the effects of these chemicals on a variety of mammals, especially rodents, rabbits, and dogs. Such toxicity testing was mandated by governments in response to major disasters in which humans were exposed to highly toxic compounds. The prime example is elixir of sulfanilamide (i.e., sulfanilamide dissolved in diethylene glycol), which had been marketed as an antibiotic (see chapter 5) but not been tested for toxicity; it killed at least 100 people in 1937 before it was pulled from the market (Ballentine, 1981). An even more influential disaster was the drug thalidomide, which was widely taken by pregnant women in the 1950s to prevent morning sickness. It ended up causing at least 10,000 stillbirths and birth defects worldwide because, sadly, it had not been tested for detrimental effects on fetal development (Rehman et al., 2011).

By now, scientists and regulatory agencies have developed a broad array of animal toxicity tests that include diverse forms of chemical exposure (e.g., inhalation versus ingestion) and look for various kinds of toxicity (National Research Council, 2006). In 2019 roughly a quarter of all experimental procedures conducted on vertebrates in the United Kingdom were performed to satisfy regulatory requirements. However, the ability of these animal tests to predict human toxicity is far from perfect, and differences among test species are relatively common (Calabrese, 1988; Bailey et al., 2005) (figure 4.5A). Aspirin, for example, is substantially more to toxic to cats than to dogs or humans, mainly because cats are slower to clear it from their circulation (Court, 2013). The sensitivity to various carcinogens likewise varies across species, with the concordance (i.e., rate of agreement on both positives and negatives) between rats and mice being just 74% (Haseman, 2000; Smith & Anderson, 2017; Smith & Perfetti, 2018). Given this prevalence of species differences, regulatory agencies usually request toxicity testing in two different species, one of which should be a nonrodent; the latter

A – Mammalian Toxicity Testing

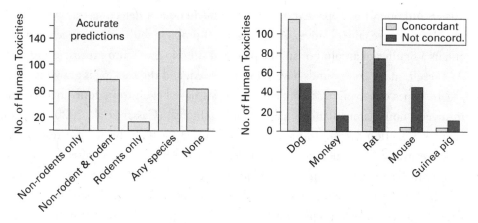

B – Non-mammalian and *in vitro* Toxicity Screens

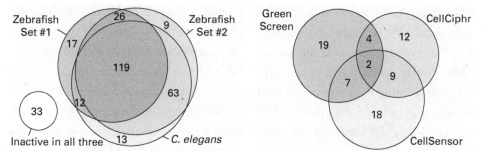

Figure 4.5

Concordance and discordance in toxicity testing. (*A*) Mammalian toxicity testing. Olson et al. (2000) gathered human and animal toxicity data for 150 compounds and asked how frequently the data from various species were concordant with the human data. *Left:* Tests on nonrodents (mainly dogs) are better than tests on rodents (mainly rats) at predicting human toxicity, and combining the results from rodents and nonrodents increases test sensitivity. *Right:* The ratio of correct positives to false predictions (i.e., nonconcordance) is higher in dogs and monkeys than in rodents. (*B*) Non-mammalian and in vitro toxicity screens. *Left:* Results of developmental toxicity tests for 292 compounds in two sets of zebrafish and one set of nematodes (*Caenorhabditis elegans*). Roughly half of the compounds were correctly identified as being either toxic or inactive in all three assays. The results from the two zebrafish assays were 65% concordant. Additional analyses (not shown) revealed that the zebrafish and *C. elegans* data predict rabbit and rat toxicity with a balanced accuracy of approximately 50% (averaging the true-positive and true-negative rates). *Right:* Results of geno- and cytotoxicity testing for 309 compounds in three different in vitro assays based on human cell lines. The overlap between the three data sets is low. Additional analyses indicated that the concordance rates between the individual assays and the Ames test or rodent tumorigenicity data lie between 50% and 60%. Adapted from (A) Olson et al., 2000, in part by recategorizing rabbits as nonrodents; (B) Boyd et al. (2016); Knight et al. (2009).

is usually a dog, a monkey, or, increasingly, a pig (Brown et al., 2013; Monticello et al., 2017). Although this approach increases the likelihood of detecting true positives, it can also promote false positives. A more general problem with this approach is that it entails a significant amount of animal pain and suffering; it is also expensive and slow. As a result, many compounds have yet to be tested, and the backlog is growing.

In hopes of solving this problem, biologists have begun to perform toxicity tests in diverse non-mammalian animals, notably zebrafish (Cassar et al., 2020) (see also chapter 3, section 3.7.3). This species is well suited for high-throughput testing and raises relatively few ethical concerns, at least for the first five days after fertilization (before the larvae begin hunting for food; Strähle et al., 2012). Moreover, test results in zebrafish are positively correlated with at least some of the analogous mammalian data (Milan et al., 2003; Sipes et al., 2011). However, two different zebrafish toxicity studies examining the same large set of substances yielded a concordance rate of just 65% (figure 4.5B), and both studies predicted the analogous rabbit and rat data with only about 50% accuracy (calculated as the average of true-positive and true-negative rates); analogous screens on the roundworm *Caenorhabditis elegans* yielded quantitatively similar results (Boyd et al., 2016). Whether such concordance/accuracy rates are disappointing or promising depends on your perspective and interests, but they do appear to be lower than those usually reported for nonhuman primates or rodents. Nonetheless, a high degree of discordance could result from various factors other than species differences. One might note, for example, that exposing zebrafish to potential toxins dissolved in ambient water is very different than having rats inhale or ingest the same compounds. This might explain, at least in part, why "the zebrafish was on average 180 times more sensitive than the mammalian system for inhalation [of the same compounds]" (Ducharme et al., 2015).

A promising alternative to toxicity testing in animals is to perform it in vitro. A significant advance in this direction was the development of the Ames mutagenesis test, which uses a mutant strain of the bacterium *Salmonella typhimurium* to quantify a test compound's ability to reverse the effect of the original mutation, indicating that the substance is a mutagen and, by implication, a likely carcinogen. Early studies suggested that the Ames test identifies 90% of known carcinogens (McCann et al., 1975), but later research with larger libraries of test compounds found its rate of true-positive results to range from just 45% to 60%, with a false-positive rate of 14% to 25%, respectively (Tennant et al., 1987; Kirkland et al., 2005). Moreover, roughly 60% of 331 compounds that induce tumors in mice or rats fail to indicate mutagenicity in the Ames test, indicating either that the rodent assays overestimate carcinogenicity (Smith & Perfetti, 2018) or that many carcinogens induce tumors by something other than mutagenesis, such as increasing cellular proliferation or impairing the body's ability to fight incipient tumors (Shaw & Jones, 1994). Both possibilities are likely true, and species differences are certainly involved as well. Thus, although the Ames test is often hailed as the gold standard for in vitro testing, its ability to predict mammalian carcinogenesis is not as strong as researchers initially believed (Walmsley & Billinton, 2011).

A more recent development is the push for a "new paradigm" in toxicology that calls for in vitro testing on human cells (National Research Council, 2007). This idea is appealing because the use of human cells removes the specter of species differences; it also eliminates most ethical concerns, reduces costs, and in all likelihood accelerates testing. Although such tests do not allow researchers to quantify toxicity at the whole organism level, signs of cellular toxicity (e.g., DNA damage) can certainly be quantified. Researchers can also use modern molecular techniques to examine whether a test compound perturbs specific molecular and cellular pathways, especially those deemed to be "adverse outcome pathways" (Villeneuve et al., 2014). Although this strategy has been widely hailed and adopted (Kleinstreuer et al., 2013), serious questions remain. Particularly concerning is that the use of many different tests, assembled into test batteries, may lead to numerous false positives (figure 4.5B). In the words of Cox et al. (2016) "the predictive power of the in vitro test batteries is not much better than would be achieved by simply assuming that all chemicals are rodent carcinogens, thus creating excellent sensitivity (no false negatives) but poor specificity (many false positives)" (p. 55).

Another issue with the in vitro tests is that one must decide which human cells ought to be used and how they are to be cultured. The cells most frequently used for in vitro toxicity testing are primary human hepatocytes. These cells are a good choice because most toxins affect the liver, and hepatocytes are the dominant liver cell type, performing most liver functions. In addition, many substances do not become toxic until they are metabolized by liver cells, which would lead to false-negative results if cells from other organs were used. It can be difficult to obtain healthy human hepatocytes for primary cultures, but samples obtained from liver surgeries are sometimes available.

Sadly, explanted hepatocytes tend not to proliferate in vitro, even though liver cells in vivo can regenerate. An additional problem is that hepatocytes rapidly dedifferentiate in culture and undergo profound metabolic alterations (Cassim et al., 2017). Indeed, a proteomic comparison between fresh livers and hepatocytes that were cultured under standard conditions for two days found 457 proteins to be differentially expressed (Bell et al., 2016). The expression changes were less dramatic when the hepatocytes were cultured as small 3D aggregates (i.e., spheroids), but they were nonetheless significant. Scientists are trying to overcome these limitations by making the culture conditions (e.g., oxygen concentrations and substrate stiffness) more naturalistic, but obstacles remain (Ruoß et al., 2020).

Aside from primary hepatocytes, toxicologists have used a variety of other human-derived cells. Some have used hepatocyte-derived continuous cell lines, which solves the challenge of obtaining the cells but decreases the model's similarity to the in vivo condition. Others use stem cell–derived hepatocyte-like cells (Vinken et al., 2012). Yet other groups combine primary hepatocytes with nonliver cells, such as iPSC-derived heart muscle and motor neurons derived from fetal stem cells, in multicompartment body-on-a-chip cultures (Oleaga et al., 2019). Such systems have enormous potential, but their ability to predict in vivo toxicities is just starting to be examined (Herland et al., 2020). They are also neither simple nor cheap.

Many in vitro toxicity tests are designed specifically to replace the Draize eye test, which uses rabbits to test compounds for their tendency to cause eye irritation or damage. These Draize tests, which had been used extensively in the cosmetics industry, predict the human reaction roughly 85% of the time (Wilhelmus, 2001). However, they have long been criticized for the distress they cause rabbits. In response, many scientists have sought suitable in vitro alternatives. An exciting recent advance is the use of human corneal epithelium cells to create a 3D in vitro epithelium that closely resembles the human cornea. These cutting-edge in vitro tests can predict eye irritation with 80% to 86% accuracy (Organisation for Economic Co-operation and Development [OECD], 2019; Lim et al., 2019). Unfortunately, they cannot distinguish between eye irritation and serious eye damage, and regulators admit that "in the foreseeable future, no single in vitro test method will be able to fully replace the *in vivo* Draize eye test" (OECD, 2019, p. 2). Still progress on this front has been remarkable, which is timely considering that the European Union in 2013 banned the marketing of all new cosmetic products developed with animal testing.

In general, we can conclude that no single test, be it in vitro or in vivo, can fully predict human toxicity. Cell culture artifacts and species differences stand in the way. The most widely accepted solution to this dilemma is to employ multiple tests, such as a rodent and a nonrodent species, or a whole battery of in vitro tests. But what are we to do when test results are discordant? A positive in vitro result is usually used to prioritize the test compound for further testing, including in vivo tests; alternatively, it might halt product development. This seems reasonable and safety conscious, but how many products are prematurely abandoned on the basis of false-positive results? Nobody knows. By contrast, false negatives seem easier to guard against so long as a sufficient number of different tests are conducted and any positive results trigger warnings—but how many different tests are sufficient? Moreover, as the number of tests goes up, so will the number of false positives. To address at least some of these issues, scientists are using computer algorithms and machine learning techniques to analyze many different kinds of toxicity-related data, ranging from chemical structure analysis to in vivo testing. Not surprisingly, they find that integrating all of the available data provides the best results (Guan et al., 2018). Unfortunately, this does not tell us which tests are "best." Perhaps that question is ill-posed.

Instead of seeking a single test that perfectly predicts human toxicity, we should probably employ a variety of tests and then seek to understand the variation between them so that we can weigh the evidence accordingly (Adriaens et al., 2018). In addition, it will be important to collect more data on what is toxic to humans; we currently seem to have more detailed information on toxicity in rodents than humans. In that context, it is interesting to note that toxicity testing historically developed after scientists discovered that humans tended to get sick after engaging in certain activities, such as chimney sweeping, or being exposed to certain chemicals (Adami et al., 2011). The putative toxins were then tested in animals to confirm the hypothesized effects and elucidate the toxic mechanisms of action. This kind of epidemiological research still plays an important role in toxicology, and it is likely to persist.

4.4 MODEL ECOLOGY REVISITED

Some animal and in vitro models became widely used but then waned in popularity; others are still ascendant. It is tempting to view these fluctuations as simple fads. As Erwin Chargaff, who made significant contributions to our understanding of DNA, put it in 1976,

> At one time, you could get money only for work on animal organs; bacteria were out. Then suddenly bacteria were in, everything else was taboo. A little later, all this changed again. Bacteriophage workers, once the kings of the coven, now populate the debtors' prisons. Procaryotes are proscribed, eucaryotes acclaimed. Animal viruses, long an object of suspicion, commiseration, and neglect, are highly quoted and oversubscribed. . . . It is the style of a period that conditions, or even compels, us to accept and adopt its various fashions or fads. (Gest, 1995)

Extending Chargaff's view, we might say that mice now rule the day, together with stem cells. It is simplistic, however, to view the historical fluctuations in model popularity as mere fashions. It is more instructive to view them as the outcome of a competition between models, creating a sort of ecology in which only the fittest thrive (see chapter 3, section 3.8). In addition, it is important to realize that the models themselves are hierarchically organized, with the major types having subtypes (e.g., different mouse strains or cell types), which in turn may be composed of multiple variants (e.g., different housing or culture conditions). From this perspective, model systems can be viewed as evolving at multiple levels, so that major model types may well be popular for an extended period of time while also changing in their specifics. The fascinating question then becomes: What factors govern the success or demise of the individual, competing models?

One critical factor, enshrined in Krogh's principle (see chapter 2), is experimental convenience. This term embraces many dimensions, including animal/materials costs and availability, housing requirements, and experimental tractability. For genetic studies and many in vitro experiments, the rate of organismal reproduction and cell proliferation, respectively, are also key variables—the faster the better, other things being equal. The availability of highly standardized animals and cell types is another significant aspect of convenience, as variability among research subjects makes it more difficult to obtain statistically significant results. In general, researchers tend to prefer to study models that allow them to obtain statistically significant, replicable results in relatively short amounts of time. Given the importance of research productivity in academic promotions and grant reviews, this makes good sense.

A second major factor is that researchers often want their findings to be generalizable, or, more formally, they want their model to have broad "representational scope" (Ankeny & Leonelli, 2011). A model's scope is difficult to determine ahead of time, however, since even closely related species and cell types may differ in some unexpected respects. Still researchers tend to be aware that some of the experimentally most-convenient organisms—dubbed "Krogh organisms" by Green et al. (2018)—possess at least a few features that are unlikely to be found in other species. Owls, for example, are exceptionally good at localizing sounds, and the study of the mechanisms underlying this capacity has yielded beautiful results (Carr & Konishi, 1988), but some of the specific mechanisms employed by owls do not generalize to mammals (Grothe et al., 2010). Therefore, research on sound localization with owls has now been supplemented with extensive studies in mammalian species (Grothe & Pecka, 2014). Similarly, paramecium had been a heavily studied microbe in the 1950s and 1960s, but it fell out of favor because many aspects of its genetics did not generalize to other species (Preer, 1997).

In short, for a model system to attract practitioners, it should be "experimentally tractable yet at the same time typical enough so that lessons learned in the model have a good likelihood of still being true in many other organisms" (Botstein and Fink, 1988, p. 1440). This does not mean that the study of unusual features cannot be justified. For one thing, it can reveal or confirm general principles, which is how August Krogh himself approached the study of extreme organisms (see chapters 2 and 7). For another, discoveries made in unusual species can be very useful for specific applications. The polymerase chain reaction (PCR) technique employed in DNA sequencing, for example, was made possible by the discovery of an exceptionally heat-resistant DNA polymerase only found in a bacterium adapted to hot springs (Chien et al., 1976).

The third principal factor influencing model selection is the concern for animal welfare. As reviewed in chapter 3, the history of animal models demonstrates an overall movement away from nonhuman primates and pets toward food animals, pests, and other organisms that elicit few ethical concerns. This movement peaked in the early days of molecular biology, when bacteria, viruses, and yeast reigned supreme. Then, as investigators turned their attention to more complex aspects of biology, multicellular animals were once again ascendant, although worms, fish, rats, and mice took center stage. Nonhuman primates were generally used only for studies where similarity to human is paramount, such as late-stage safety testing. The general calculus is this: "What animals are enough like us to make laboratory results obtained from them generalizable to humans, but not so much like us that we ethically prohibit their being the subjects of experiments?" (Rader, 2004, p. 22). As discussed further in chapter 7, different people tend to answer this question differently, based largely on the research questions that interest them and on their rather subjective notion of how similar the various animals are to humans.

One of the attractions of in vitro biology is that it circumvents the ethical concerns over animal welfare almost entirely, as long as no animals or embryos are killed to harvest cells. Historically, in vitro models experienced a shift in prevalence from tissue explants to continuous cell lines and, more recently, to human cells, especially human stem cells and their derivatives. Clearly, the principal attraction of working with human cells is that they eliminate the concern over species differences that plagues animal research. However, in its place we find the challenge of making the in vitro models complex enough to answer questions about intact humans. To paraphrase Rader (from the previous paragraph), the question is, Which in vitro systems are enough like us to make laboratory results obtained from them generalizable to humans in vivo? The question is clearly troublesome, as demonstrated by the efforts of so many researchers to create ever more complex in vitro systems, be they co-cultures, organs-on-a-chip, or organoids. Unfortunately, as the complexity of these systems goes up, their experimental convenience (in terms of costs and required experimental expertise) tends to decrease.

The fourth major factor governing model selection is sociological. As many science historians have noted, the success of a particular model can often be attributed to one or a few particularly brilliant and charismatic leaders who fostered the development of a collaborative community centered around their model of choice (Kohler, 1994; Rader, 2004; Endersby, 2007). These communities share animals, technical resources, and the latest data, and they generally make their members feel that their research has an eager audience. By comparison, developing a new model is lonely, often frustrating work. Indeed, the mere fact that a model has attracted a number of fans and resources is likely to increase its popularity—and grant funding—further. It is an example of the Matthew effect, according to which "to him who has will more be given, and he will have abundance; but from him who has not, even what he has will be taken away" (Matthew 13; quoted in Merton, 1968). Similarly, the accumulation of knowledge about a model system stimulates further research on that model, as researchers can build on the existing information (Matthews & Vosshall, 2020). In the words of Joshua Lederberg, who pioneered much of the early work on *E. coli*, "The very accumulation of knowledge, mostly concentrated on a single strain, 'K-12,' made it more likely that it would be a prototype for still further studies" (Lederberg, 1998, p. 231). That said, scientists sometimes do develop new models (e.g., consider Benzer's switch from *E. coli* and phage to flies), regardless of the obstacles and personal investments they may have already made in established models (i.e., sunk costs). It is a risky but often effective way to open new frontiers and make one's name.

Finally, models wax and wane in popularity because the questions of interest to scientists change over time. Some questions in biology lose their fascination before they are fully resolved, but many do get answered to the satisfaction of the scientific community. Either way, new questions arise, and often they require new models to answer them. The fundamentals of the genetic code, for instance, were worked out in bacteria, but scientists then turned to eukaryotes for the study of more complex processes. Similarly, fruit fly genetics was largely replaced by mouse and zebrafish genetics as research interests shifted from the basics of Mendelian inheritance to the control of vertebrate development and cancer biology.

These observations are consistent with the commonly expressed idea that the selection of a model system should be driven by the question that is asked. This is certainly true insofar as scientific progress generally depends on selecting an appropriate model. However, as knowledge about a model system accumulates and the associated research community grows, it becomes increasingly likely that the model itself will determine the research questions rather than the other way around. When this happens, scientists risk learning more and more about the model but losing sight of how the model relates to other models and its presumed targets (Horrobin, 2003). Such concerns should not be overblown, as model-specific knowledge is often necessary to answer general questions. However, the fact that one observes a relatively high degree of discordance among the various models and their target (e.g., in toxicology) suggests that some caution is warranted.

In particular, it is good to remember that generality should never be assumed. Fortunately, having detailed information about one model makes it easier to examine other, related models and, thus, to ascertain how general the findings are (Striedter et al., 2014). That is, once a model has been studied in detail, the next step is—or ought to be—determining its representational scope. The narrower a model's scope, the less likely it is to remain popular for long.

5 MODELS AND THERAPIES: INFECTIOUS DISEASES, CARDIOVASCULAR DISEASE, AND CANCER

From all the above tests it was clear that this substance possessed qualities which made it suitable for trial as a chemotherapeutic agent. Therapeutic tests were therefore done on mice infected with streptococci, staphylococci and *Cl.* [Clostridium] *septique.* . . . The results are clear cut, and show that penicillin is active in vivo against at least three of the organisms inhibited in vitro. It would seem a reasonable hope that all organisms inhibited in high dilution in vitro will be found to be dealt with in vivo.

—CHAIN ET AL. (1940), P. 228

The two previous chapters introduced the principal in vivo and in vitro models used in biomedical research. By contrast, the present and following chapters will focus on several major human diseases, exploring how the various models were used to understand and develop treatments for those maladies. In many cases the research advanced from in vitro to in vivo, as it did in the work on penicillin (see the chapter's opening quote). Much of the research focused on mouse models, but other species were also used, especially in the historically older studies and in the development of cardiovascular surgery. Human tissues and cells have featured heavily in recent work, especially as transplants into mice. We return to these and other trends at the end of this chapter.

Given that it is neither possible nor desirable to cover all diseases equally, the present chapter focuses on three types of disease that, between them, account for the majority of human deaths: infectious diseases, cardiovascular disease, and cancer. Chapter 6 will address neurological disorders, which represent a huge drain on humanity because they tend to be long-lasting and difficult to treat. Both chapters focus on therapies that were historically very important or are particularly relevant to the topic of model use.

5.1 INFECTIOUS DISEASES

Infectious diseases are caused by parasites (e.g., malaria), bacteria, or viruses; I here focus only on the latter two. Both bacterial and viral infections can be prevented (and in some cases treated) with vaccines, but it has proven difficult to develop effective vaccines against some major viral diseases, notably human immunodeficiency virus/acquired immunodeficiency syndrome (HIV/AIDS) and hepatitis C. Bacterial and viral diseases can also be treated with antibiotics and antiviral drugs, respectively.

5.1.1 Vaccines

Vaccine development was pursued most famously by Louis Pasteur, whose efforts were already mentioned in chapter 3. Briefly, Pasteur created his vaccines by weakening the original infectious particles by physical means, such as heat or exposure to dry air. He began with vaccines against several animal diseases, such as chicken cholera, swine flu, and cattle anthrax, but then developed a vaccine against the rabies virus, which can infect humans. In 1885, Pasteur gave his rabies vaccine to a boy who had been bitten by a rabid dog, and the boy survived (Rappuoli, 2014). Some say that Pasteur had previously tested this vaccine on many dogs, but others are not so sure and have uncovered doubts about several of Pasteur's vaccine-related claims (Anderson, 1993). In any case, Pasteur's research on vaccines brought him worldwide acclaim. Chapter 3 also included a brief review of polio vaccine development. As we discussed, this research initially required many nonhuman primates, both for virus production and vaccine testing. This changed with the development of in vitro systems, which significantly reduced the need for nonhuman primates in polio research.

The very first vaccines—predating Pasteur's—were directed against the smallpox virus, which had long been a global scourge and killed approximately 400,000 people a year in Europe during the 18th century (Riedel, 2005). The effort was spearheaded by Edward Jenner in the 1790s. He had heard stories of milkmaids becoming immune to smallpox after becoming infected with cowpox—a related but much milder disease. Acting on this idea, Jenner inoculated a boy with fluid from the sores of someone who had been infected with cowpox and then challenged the boy with an injection of smallpox. The boy was fine, but it took some time for Jenner to find additional volunteers for this somewhat daring procedure, which he named vaccination (*vacca* being Latin for "related to cows"). Still, Jenner's vaccine turned out to be quite effective and quickly gained worldwide recognition. After additional vaccine development and worldwide vaccination campaigns, the World Health Assembly declared the planet free of smallpox in 1980. For our purposes, it is interesting that Jenner did not have a detailed, mechanistic understanding of what causes smallpox and did not, apparently, test his vaccine on animals before experimenting with the boy.

Tuberculosis is another highly infectious disease, but it is caused by a bacterium rather than a virus. The bacterium causing tuberculosis was identified by Robert Koch in 1882 as *Myobacterium tuberculosis*. Koch managed to grow this bacterium in vitro and used these cultures to infect guinea pigs, which then developed the disease. Koch had also experimented with mice and dogs but found the guinea pigs to be more susceptible to tuberculosis. In addition, Koch developed a histological stain for the tuberculosis bacterium that he then used to demonstrate the bacterium's presence in infected guinea pigs as well as humans (Cambau & Drancourt, 2014). Koch attempted to develop a treatment for tuberculosis, but, despite some exaggerated claims, he ultimately failed. Emil von Behring, who in 1892 pioneered the development of serum therapy (which entails treating infected organisms with serum from others who previously conquered the infection), likewise failed in his attempts to develop a human tuberculosis vaccine (Grundman, 2001).

The first successful vaccine against tuberculosis was developed between 1908 and 1921. The researchers discovered, somewhat fortuitously, that cultivating the tuberculosis bacterium in a medium containing ox bile reduced its virulence. Exhibiting remarkable persistence, they subcultured their bacterial colony 230 times over the course of 11 years and, eventually, obtained a bacterium that did not cause progressive tuberculosis when injected into various animals, including horses, cattle, and guinea pigs (Luca & Mihaescu, 2013). This vaccine, commonly referred to as Bacillus Calmette–Guérin (or BCG), was first given to humans in 1921 and has been administered to more than 1 billion people since then (Behar & Sassetti, 2020). Because this vaccine's effectiveness wanes several years after the immunization, researchers are eagerly seeking improved vaccines or administration schemes. It is exciting, therefore, to learn that the BCG vaccine is more effective in rhesus monkeys when it is given intravenously, rather than injected into the skin, which is the traditional route (Darrah et al., 2019). Whether this will also be the case in humans remains to be seen.

Human papilloma virus (HPV) can cause anogenital warts and cervical cancer. Scientists found it difficult to prove this causal relationship, however, because HPV does not infect nonhuman species, and intentionally infecting humans with HPV would be unethical. To get around this problem, researchers in the 1980s exposed human cervical tissue to HPV in vitro and then implanted it into immunodeficient mice; several weeks later the grafted tissue exhibited several features reminiscent of the anogenital warts typically observed in HPV patients (Kreider et al., 1985). This experiment helped to prove that HPV causes the warts; together with analogous experiments using grafts of human foreskin from circumcised infants (Kreider et al., 1987), it also gave researchers a way to cultivate the virus in animals. This rather cumbersome approach was necessary at the time because HPV was extremely difficult to cultivate in traditional two-dimensional cell cultures, even with human cells. However, studies in the 1990s revealed that HPV can go through its complete life cycle when it is cultivated in 3D human cell cultures that closely resemble epidermis (Doorbar, 2016; De Gregorio et al., 2020).

The early 1990s also saw the development of the first HPV vaccine. In contrast to traditional vaccines, which use weakened or killed viruses, this HPV vaccine employed viruslike particles. Specifically, researchers expressed one of the two main HPV proteins in *Escherichia coli* or an insect cell line and found that the viral proteins self-assembled into large viruslike particles. When injected into animals, these particles trigger a strong immune response that then results in good protection against future HPV infections (Roldão et al., 2010). Because the vaccine contains no viral genes, there is no risk of the viruslike particles replicating and becoming infectious. This novel approach was pioneered using rabbits, dogs, and viruses closely related to HPV (Kirnbauer et al., 1992; Suzich et al., 1995). The human vaccine, called Gardasil, was approved for the prevention of human HPV in 2006. Other, similar vaccines were developed later.

Novel approaches to vaccine development have also been employed to fight the COVID-19 pandemic (van Riel & de Wit, 2020). In addition to traditional vaccines that use weakened versions of the responsible virus—severe acute respiratory syndrome coronavirus 2 (SARS-CoV-2)—some researchers developed vaccines that consist of viral messenger RNA (mRNA) encoding just a single SARS-CoV-2 protein. Encapsulated within special nanoparticles, this mRNA can enter host cells and then be translated into the viral protein, which triggers the desired immune response. Other promising COVID-19 vaccines use weakened versions of relatively harmless viruses that have been engineered to express one of the SARS-CoV-2 proteins. When those viruses infect host cells, they cause them to make the SARS-CoV-2 protein, transport it to the cell surface, and trigger an immune response. Both of these vaccine types are presumed to be relatively safe because the SARS-CoV-2 protein cannot form complete SARS-CoV-2 particles. That said, the new vaccines might well cause some unexpected adverse reactions, which is why they are tested in animals, including monkeys, for both safety and efficacy (Corbett et al., 2020; Doremalen et al., 2020). An intriguing aspect of COVID-19 vaccine development is that the animal tests are often run concurrently with early phase human trials, rather than sequentially. This extraordinary step seems justified, given the great urgency created by this lethal pandemic.

Overall, we can conclude from this brief sampling of vaccine development history that the field has become much less reliant on animal models for basic virus research and production. Advances in cell culture technique and genomics now make it possible to study most bacteria and viruses without infecting animals. However, animal models continue to play a role in the development of all vaccines, especially when it comes to testing for safety and efficacy (Pardi et al., 2018). Given the regrettably widespread skepticism toward vaccines that already exists, humanity can ill afford vaccines that cause substantial harm or do not work.

5.1.2 Antibiotics

Antibiotics are used to fight bacterial infections. They are what Paul Ehrlich called "magic bullets" because, unlike real bullets, they kill the pathogen without harming the host (Bosch & Rosich, 2008; Strebhardt & Ullrich, 2008). Some antibiotics are synthesized by scientists, but many others are naturally produced by other microbes, which is consistent with the idea that different microbial species often compete aggressively with one another.

The first widely used antibiotic was discovered in the quest to conquer syphilis, a sexually transmitted disease whose symptoms begin with genital ulcers, followed by painful rashes and abscesses; in the long run, syphilis often leads to cardiovascular and neurological problems as well. This disease, sometimes referred to as "the great pox" (Bowater, 2016), erupted into devastating epidemics throughout much of Europe in the late 1400s. Diverse treatments for syphilis had been proposed over the years, including ointments containing potentially toxic levels of mercury (Abraham, 1948).

Significant progress came only after 1905, when scientists discovered that syphilis is caused by a spirochete bacterium called *Treponema pallidum*. Inspired by this discovery, Paul Ehrlich and his Japanese student Sahachiro Hata in 1909 infected rabbits with syphilis and then tested hundreds of arsenic-related compounds that Ehrlich and his collaborators had already synthesized in the quest to treat a different disease (African sleeping sickness). One of these compounds, arsphenamine, turned out to cure the rabbits of syphilis very effectively. Curiously, two of Ehrlich's assistants had previously tested this compound and failed to see an effect (Williams, 2009), thereby illustrating how false negatives and issues with replicability (see chapter 1) have frustrated scientists for a long time.

By the end of 1910, 65,000 doses of arsphenamine (trademarked as Salvarsan) had been administered to more than 20,000 patients with syphilis; these were unprecedented numbers at the time (Williams, 2009). Indeed, arsphenamine was more effective than any previous syphilis treatment. Unfortunately, treatment with arsphenamine involved a complex, protracted regimen, and unpleasant side effects were commonplace. Researchers tried to avoid these problems by slightly modifying the drug's chemical structure, but problems persisted (Bosch & Rosich, 2008). In the long run, arsphenamine was eclipsed by penicillin, which proved to be a simpler, even more effective treatment for syphilis.

Penicillin was discovered by Alexander Fleming in 1928, when he noticed a bacteria-free area around a bit of mold that had contaminated one of his bacterial culture dishes

(Fleming, 1929). The mold turned out to be of the genus *Penicillium,* and Fleming showed that it secreted a compound capable of killing *Staphylococcus* and several other kinds of bacteria. Fleming also showed that the mold's secretions were not toxic to rabbits or mice, but he did not examine whether they could halt ongoing infections. Part of the problem was that Fleming could not produce large amounts of the mold's secretions, which he called penicillin; nor was he able to isolate its active component. These problems were solved 11 years later by Howard Florey and Ernst Chain, who then demonstrated penicillin's antibacterial effectiveness in mice and, one year later, in humans (Chain et al., 1940; Science History Institute, 2016). After penicillin production had been scaled up (Bentley, 2009), the drug was used widely to treat innumerable patients, including many soldiers wounded in World War II.

Although the development of penicillin clearly involved extensive animal research, animal rights advocates sometimes point out that the drug is toxic to guinea pigs. This observation supposedly undermines the notion that biologists can extrapolate findings from animals to humans. It is important to note, however, that penicillin is not toxic to germ-free guinea pigs (Formal et al., 1963) (figure 5.1). Indeed, penicillin and some related antibiotics alter the gut microbiome of ordinary (i.e., not germ-free) guinea pigs in such a way that highly toxic coliform bacteria come to dominate, thereby killing the guinea pigs indirectly (Farrar et al., 1966). Apparently, this shift toward a toxic gut microbiome does not happen in other mammalian species. Therefore, we can conclude

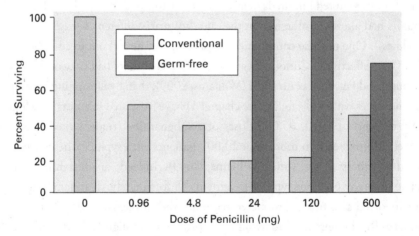

Figure 5.1

The microbiome's effect on penicillin toxicity in guinea pigs. Formal et al. (1963) gave varying doses of penicillin to guinea pigs that were either raised in a germ-free environment (and thus without microbes in their gut) or raised conventionally. Considering the three highest doses penicillin together, only 1 out of 20 animals died within seven days in the germ-free group, whereas 41 out of 54 animals succumbed in the conventional group. These data demonstrated that it is something about the microbiome that makes penicillin toxic to guinea pigs. Adapted from Botting (2015).

that species differences are real and important—indeed, antibiotics can only work as magic bullets because of species differences between hosts and their pathogens—but that those differences can be explained (at least sometimes). Nor do they negate the utility of all animal research. Indeed, the species differences in penicillin sensitivity argue in favor of working with multiple animal models, rather than just one.

Roughly coincident with the discovery of penicillin, Gerhard Domagk tested a variety of sulfur-containing azo dyes for their ability to kill highly virulent streptococcal bacteria. One of these compounds, eventually marketed as Prontosil, did not kill the bacteria in vitro but was able to cure streptococcal and other bacterial infections in mice and rabbits (Domagk, 1947). Domagk and his collaborators then tested Prontosil on humans, including Domagk's own daughter. The results were positive, with minimal side effects, and were published in 1935 (Bentley, 2009; Bailey, 2010). Although Prontosil's antibacterial effectiveness was discovered before that of penicillin, penicillin kills a broader range of bacteria and has, therefore, been more widely used. Another interesting twist in this story is the relatively late discovery that Prontosil is metabolized to sulfanilamide, which turns out to be the active antibacterial agent. That discovery explains why Prontosil did not kill the bacteria in vitro, where the requisite metabolic enzymes are lacking. It also led to the development of second-generation sulfa drugs based on slight modifications of sulfanilamide. Also of interest is that the sulfanilamide tragedy of 1937, which killed more than 100 people (see chapter 4, section 4.3), involved sulfanilamide dissolved in diethylene glycol; it was the latter molecule that killed, not the sulfanilamide!

5.1.3 Antiviral Drugs

AIDS was first identified in 1981 and was characterized by a progressive decrease in the number of helper T cells, which are type of white blood cell. AIDS is transmitted primarily through sexual intercourse; left untreated, it leads to a slow and painful death by other diseases such as tuberculosis or cancer. Researchers in the early 1980s determined that AIDS is caused by a complex retrovirus called human immunodeficiency virus (HIV) (Vahlne, 2009). This virus probably evolved from simian immunodeficiency virus (SIV), which is quite similar to HIV but infects a variety of nonhuman primates, rather than humans. Indeed, HIV easily infects only humans and chimpanzees, and infected chimpanzees rarely develop AIDS. In the words of Varki et al. (2011), "more than 100 chimpanzees in the United States and Europe were experimentally infected with HIV. Surprisingly, after more than 10 years, only one chimpanzee progressed to a full-blown acquired immune deficiency syndrome (AIDS)-like syndrome" (p. 376).

Given the ethical problems inherent in chimpanzee research and the fact that these animals rarely develop AIDS, researchers have tried to understand the human disease by studying SIV-infected macaques, many of which do develop AIDS-like symptoms (Letvin et al., 1983). Efforts are also underway to modify HIV's genetic sequence so that the virus can infect monkeys, but this modification is nontrivial (Thippeshappa et al., 2020). In addition, AIDS researchers frequently study immunodeficient mice that have been implanted with HIV-infected human tissue (Namikawa et al., 1988; Hatziioannou & Evans, 2012). These humanized mouse models are ethically less troubling than the nonhuman primate models, but they are expensive, difficult to work with, and limited insofar as the virus will infect only the transplanted human cells, not the rest of the mouse. Finally, scientists have long been able to cultivate the HIV virus in cell lines derived from human T cells (Mitsuya et al., 1985), but those in vitro models cannot reveal the more complex aspects of AIDS.

In short, most models of HIV/AIDS are rather limited. This constraint largely explains why, despite long-standing efforts to develop an AIDS vaccine, none has been forthcoming. At least as important is that, as a retrovirus, HIV can "hide" from antibodies in the host's genome and then re-emerge later. Yet another factor is that HIV can spread not only through the intercellular fluid but also via direct cell-to-cell contact, which can make it inaccessible to most antibodies (Agosto et al., 2015).

Although it is frustrating to have no AIDS vaccine, scientists have had considerable success with antiviral drugs that target HIV. The first and most influential of these antiviral drugs is azidothymidine (AZT). It was discovered in the mid-1980s by screening several related drugs in an in vitro assay where AZT prevented HIV from infecting and killing T cells (Mitsuya et al., 1985). We now know that this protective effect results from AZT's ability to inhibit an enzyme (a reverse transcriptase) that HIV requires for replication. We also know that AZT must be phosphorylated before it can exert its antiviral activity, but fortunately the cells in the original in vitro assay possessed the requisite enzymes; otherwise, the discovery of AZT might well have been delayed. Once the in vitro data were in hand, AZT moved very quickly to clinical trials. Because of intense societal pressure to find a treatment for AIDS, which by 1985 had killed 20,000 people worldwide, human trials of AZT were begun in 1985 without prior testing in animals (Yarchoan & Broder, 1987; Mitsuya et al., 1990; Wyand, 1992). However, some animal testing was conducted concurrently (Ruprecht et al., 1990). High doses of AZT turned out to have substantial adverse effects, but lower doses were reasonably safe and effective. Nowadays, AZT is usually taken in combination with other antiviral drugs.

Hepatitis C is another major disease that has resisted vaccine development but can now be treated with antiviral drugs. Plaguing roughly 3% of the world's population, the hepatitis C virus (HCV) mutates at a high rate in human liver cells and thus tends to evade the host's immune response. The resulting chronic infection often leads to liver cancer, cirrhosis, or other forms of liver failure. Like HIV, HCV naturally infects only chimpanzees and humans. Tree shrews (which are closely related to primates) can be infected with HCV, but only if they are severely immunosuppressed, and mice are normally resistant to HCV. To overcome the latter limitation, biologists sometimes study HCV infection in immunodeficient mice implanted with human hepatocytes. However, humanized mouse models of hepatitis C are (like the mouse models of HIV/AIDS) expensive and limited insofar as only the transplanted cells are infected. Perhaps the biggest obstacle to hepatitis C research was, for many years, that HCV is extremely difficult to propagate in cell culture.

This in vitro cultivation problem was overcome in 1999 when researchers developed a "replicon system" that allows most of the viral genome to replicate efficiently in a human liver cell line. Early versions of this model did not produce complete HCV particles, but this was later rectified (Woerz et al., 2009). Moreover, even the early versions allowed researchers to screen numerous drugs for their ability to interfere with many (though not all) aspects of HCV's life cycle. Most notably, these screens yielded drugs like sofosbuvir and simeprevir, which inhibit enzymes needed to replicate the viral genome and cleave a viral protein precursor, respectively (Horscroft et al., 2005; Eltahla et al., 2015). These and several other drugs that directly interfere with the life cycle of HCV have been approved for the treatment of hepatitis C since 2011 (Horsley-Silva & Vargas, 2017). They are relatively simple to administer and have fewer negative side effects than earlier treatments. Although they were initially discovered through in vitro drug screens, they were subsequently tested in rats and monkeys for safety and to estimate their effective tissue concentrations and excretion/degradation rates (i.e., pharmacokinetic parameters) (Rosenquist et al., 2014; Spera et al., 2016). Similarly, newer generations of anti-HCV drugs usually undergo safety and pharmacokinetic tests in diverse animal models but are tested for efficacy in humanized mouse models and in vitro assays (e.g., Rajagopalan et al., 2016).

A third important infectious disease that can be treated with antiviral drugs is COVID-19. Specifically, a randomized placebo-controlled clinical trial with 1,063 COVID-19 patients indicated that the drug remdesivir can reduce the duration of hospital stays (Beigel et al., 2020). Earlier research had already shown that remdesivir reduces lung damage in monkeys infected with Middle East respiratory syndrome coronavirus (MERS-CoV), which is similar to SARS-CoV-2, and that it inhibits SARS-CoV-2 replication in a monkey kidney cell line (M. Wang et al., 2020). Rodents, it turns out, do not make good animal models for testing remdesivir (and some other antiviral drugs, such as Tamiflu), because they express high levels of an enzyme that interferes with remdesivir's bioavailability (Bahar et al., 2012; Warren et al., 2016). In primates, however, remdesivir causes replication of the viral genome to be aborted prematurely; that is, it interferes with the ability of SARS-CoV-2 to replicate.

5.1.4 Sepsis

Sepsis is an extreme inflammatory response to bacterial or viral infections that harms internal organs; it often develops in trauma patients because their immune system is weakened. In the US, sepsis is responsible for more than 250,000 deaths per year, and approximately 6 million people die from sepsis every year worldwide (Korneev, 2019). Attempts to treat sepsis have a poor track record, including an apparently successful clinical trial that could not be replicated and at least one drug that was withdrawn after having been approved (Fink, 2014).

One potential explanation for this woeful approval history is that the most commonly used animal model of sepsis may not reflect the human condition very well. Specifically, much of the research on sepsis is based on mice that have been injected with lipopolysaccharides (LPS), which are found in the outer membranes of toxic bacteria. Injections of LPS do trigger strong, systemic immune responses in both humans and mice, but the median lethal dose of LPS in mice is "about 1000-fold to 10,000-fold greater than the dose of LPS that is required to induce severe illness and hypotension in humans" (Fink, 2014, p. 148). Moreover, the pattern of up- and down-regulated genes observed in white blood cells after LPS injection in humans differs substantially from that observed in LPS-injected mice (Seok et al., 2013) (figure 5.2). Major burn injury or blunt trauma, which also cause intense inflammatory responses, likewise elicit surprisingly different responses in mice and humans. In the words of the study authors, "Although acute inflammatory stresses from different etiologies result in highly similar genomic responses in humans, the responses in corresponding mouse models correlate poorly with the human conditions and also, one another" (Seok et al., 2013, p. 3507). In short, mice appear to make poor models for human sepsis.

This conclusion caused quite a stir in the scientific community, and several objections were raised. Some critics argued, for example, that one cannot compare a single inbred strain of mice to a highly heterogeneous human population (Osuchowski et al., 2014). However, this criticism could be leveled at most of the existing mouse models and is, therefore, a poor defense of mouse models. More interesting is the suggestion that Seok and colleagues should have excluded from their analysis any genes that are up- or down-regulated only in humans, not in mice (Takao & Miyakawa, 2015). The rationale for this recommendation was that mouse models are always merely partial models of the human condition (see chapter 6, section 6.5.3) and thus would never mimic all the human genomic responses. The critics argued that excluding the human-only genomic responses from the comparative analysis is standard practice in the field, but others disagree (Shay et al., 2015).

I, too, believe that it is more appropriate to compare all the available data between a model and its target, at least as a first step. If the model system is then found to

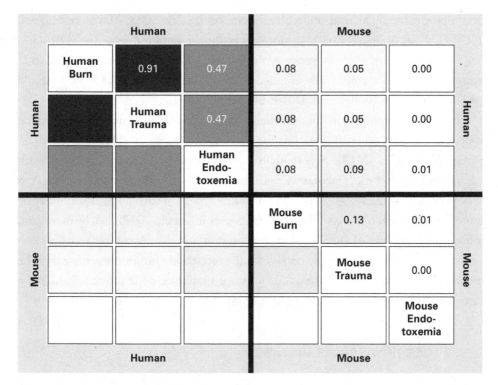

	Human			Mouse		
Human Burn	0.91	0.47	0.08	0.05	0.00
	Human Trauma	0.47	0.08	0.05	0.00
		Human Endo-toxemia	0.08	0.09	0.01
			Mouse Burn	0.13	0.01
				Mouse Trauma	0.00
					Mouse Endo-toxemia

Figure 5.2

Sepsis-related gene expression changes reveal species and model differences. Seok et al. (2013) compared how gene expression in white blood cells was altered in human patients suffering from serious burns, major blunt trauma, and bacterial toxins (endotoxemia), all of which cause serious systemic inflammation (aka sepsis). They found that gene-expression changes in these three conditions are positively correlated with one another (Pearson's correlation coefficients shown as numbers and reflected in gray levels). However, the changes observed in these human conditions correlated poorly with corresponding mouse models. Nor did the mouse models correlate well with one another. The analyses covered almost 5,000 genes. Adapted from Seok et al. (2013).

mimic only restricted aspects of the target system, this finding can be pursued and specified in more detail. In other words, it is better to identify and, if possible, explain specific differences between a model and its target than to bias one's analyses a priori in favor of similarities. That said, an unbiased reanalysis of the data published by Seok et al. (2013) supposedly suggests that the similarities between mice and humans with regard to sepsis are greater than originally reported (Shay et al., 2015).

Whether one sees the mouse models of sepsis as being similar to or different from their human target may well be a matter of seeing the glass half-full or half-empty. However, the record of translational failures for potential therapies emerging from these models suggests that the species differences are meaningful. Indeed, the immune systems of mice and humans are already known to exhibit a great number of morphological, cellular, and molecular differences (Haley, 2003; Mestas & Hughes, 2004; Gibbons & Spencer, 2011). More generally, the immune system of vertebrates has undergone extensive divergent evolution, which is not surprising given that immune systems are continually engaged in an evolutionary arms race with rapidly evolving pathogens (Bailey et al., 2013; Bitar et al., 2019).

In order to minimize the species differences, one can try to humanize the immune system of mice (or other species), but the transformations are likely always to remain partial (Yong et al., 2018). A potentially complementary approach is to work with animals that, unlike most laboratory mice, have faced multiple immune challenges as they grew up (Tao & Reese, 2017). The immune system of such animals will likely be more similar to that of their wild counterparts (Abolins et al., 2017) and, by extension, more similar to that of humans. However, significant differences between humans and mouse models will probably remain for the foreseeable future, suggesting that the modeling of human immune responses in mice should be done cautiously. In some cases, nonhuman primate models may be appropriate (Messaoudi et al., 2011).

5.2 CARDIOVASCULAR DISEASES

Heart attacks and strokes (defined as disruptions of the brain's blood supply) are the world's top two killers of humans, accounting for a combined 15.2 million deaths in 2016 (World Health Organization, 2018a, 2018b). Cardiovascular death rates have been dropping in high-income countries but, globally, cardiovascular disease still kills more people than cancer (Dagenais et al., 2020) (figure 5.3).

5.2.1 Heart Medications

The principal cardiac health problems are heart failure, defined as insufficient cardiac pumping ability and an irregular heartbeat (arrhythmia). These problems are sometimes addressed with coronary bypass surgery and the implantation of cardiac pacemakers, both of which were developed in the 1950s with heavy reliance on dogs as test subjects (Callaghan & Bigelow, 1951; Zoll, 1973; Konstantinov, 2000). However, heart failures and arrhythmias are most commonly treated with medication. For our purposes, therefore, it makes sense to focus on four very influential heart medications, namely digoxin/digitalis, nitroglycerin, propranolol, and angiotensin-converting enzyme (ACE) inhibitors.

Digoxin and digitalis belong to a class of plant-derived drugs that has been used to treat heart conditions for more than 200 years. William Withering (1785) first extracted the drug digitalis (aka digitoxin) from the common foxglove plant (*Digitalis purpurea*) and recommended it for the treatment of many conditions, including what we now call

A – Top 10 Global Causes of Death, 2016

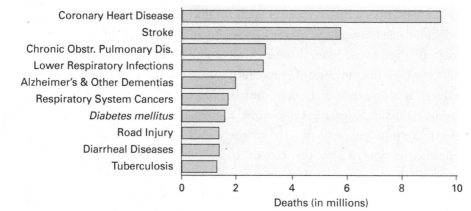

B – Causes of Death in the United Kingdom, 2012

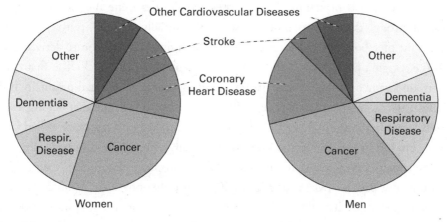

Figure 5.3

Statistics on causes of death. (*A*) Top 10 global causes of death, 2016. According to the World Health Organization, coronary heart disease and stroke were the two top causes of death globally. The respiratory cancers in this chart include lung, bronchial, and tracheal cancers. (*B*) Causes of death in the United Kingdom, 2012. Cancer was the most common cause of death, and women were more likely than men to die from Alzheimer's disease or other dementias (12% versus 6%). Cardiovascular diseases were equally common causes of death in men and women, but within that category men were more likely to die of coronary heart disease (16% versus 10%). Based on (A) World Health Organization (2018a); (B) adapted from Bhatnagar et al. (2015).

congestive heart failure (Silverman, 1989). A very similar drug, called digoxin, was isolated in 1930 from a different species in the foxglove genus (Hollman, 1996); it is sometimes prescribed instead of digitalis because it has a longer half-life in the body. Both drugs are cardiac glycosides that block the cellular sodium-potassium pump, which indirectly leads to an increase of intracellular calcium, which then leads to more forceful and regular contractions of the heart muscle. Human use of these drugs predates animal testing, but both clinical trials and animal testing later revealed potentially serious side effects. The principal problem is the relatively small difference between the therapeutically effective dose and one that is toxic. Indeed, the world's most prolific serial killer, Charles E. Cullen, used his position as a nurse to kill an estimated 400 patients with digoxin in the 1980s and 1990s (Harrington, 2018). Nowadays, digoxin and digitalis are used infrequently, mainly in combination with other heart medications.

In contrast to digoxin and digitalis, other heart medications tend not to increase the force of cardiac contractions. Instead, they decrease heart rate and increase the flow of blood through the coronary blood vessels. This (initially counterintuitive) approach reduces the metabolic stress placed on the heart and thus decreases heart-derived chest pain (i.e., angina pectoris) and the risk of a subsequent heart attack. The first drugs of this kind were organic nitrates, including nitroglycerin, which had originally been synthesized as a novel explosive (Fye, 1986). After becoming aware that licking a spot of freshly synthesized nitroglycerin induces violent headaches, a physician called Constantin Hering in the 1840s supposed that nitroglycerin might cure headaches (on the homeopathic principle that "like cures like") and proceeded to test it on himself and numerous patients (Marsh & Marsh, 2000). Another physician, William Murrell, noticed in the 1870s that self-administering nitroglycerin not only gave him headaches but also affected his heart, which prompted him to test the drug as a possible treatment for angina pectoris (Fye, 1995). The drug worked well for this purpose, and we now know that it does so, at least in part, by dilating the coronary veins. Later studies showed that nitroglycerin can have some serious side effects and that prolonged treatment leads to drug tolerance (Parker & Parker, 1998; Thadani & Rodgers, 2006). Experiments on rats further suggest that prolonged treatment with nitroglycerin may make the heart more vulnerable to subsequent attacks (Sun et al., 2011). Still, the drug remains in use for the temporary treatment of angina pectoris.

Propranolol is another influential drug that reduces heart stress. Because it selectively inhibits beta-adrenergic receptors, it is referred to as a beta-blocker. Propranolol was discovered in 1964 after an impressively systematic search for a novel drug that would selectively decrease heart rate and blood pressure (Black et al., 1964). This effort was carried out in a variety of material models, ranging from isolated, spontaneously beating guinea pig hearts to anesthetized dogs and cats (Black, 1989). The first clinical trial of propranolol was conducted in 1968 and provided positive results (Hebb et al., 1968). Over the years, several additional beta-blockers have been developed, but propranolol is still often prescribed for angina pectoris and several other conditions (e.g., migraines and stage fright).

An important breakthrough in the treatment of heart conditions was the discovery of drugs that inhibit angiotensin-converting enzyme (i.e., ACE inhibitors). These drugs lower blood pressure by dilating blood vessels, both directly and indirectly via the hormone aldosterone. They were originally discovered in 1967 by researchers who tested extracts of pit viper venom—already known to kill prey by causing its blood pressure to plummet (Péterfi et al., 2019)—for their ability to inhibit angiotensin II activity. The assays used in these early experiments relied on cell-free extracts of dog lung to obtain the enzyme, and on in vitro pieces of rat colon to test for its inhibition (Ng & Vane, 1968; Bakhle, 1968). The first ACE inhibitors were tested in animals and even a few humans, but they could not be taken orally and, therefore, attracted little commercial interest (Cushman & Ondetti, 1991). However, the scientists then tested a strategically synthesized series of peptides in an in vitro assay (based on guinea pig intestine) to optimize the drug; in 1975, they ended up with a compound that could be taken orally. This drug, called captopril, was subsequently tested in intact rats and approved for human use in 1980. Additional refinements to reduce unwanted side effects led to enalapril (aka Vasotec), which became the first billion-dollar drug in 1988 (Bryan, 2009). Enalapril has now been eclipsed by even newer ACE inhibitors, but as a class the ACE inhibitors remain a mainstay of cardiovascular therapy.

5.2.2 Atherosclerosis

Atherosclerosis (derived from the ancient Greek words *athéra* and *sklérōsis*, meaning gruel and hardening, respectively) is a progressive inflammatory disease characterized by the buildup of lipids and cells inside of arteries. The resulting accumulations, called plaques, narrow the arterial cavity and thus obstruct blood flow. Plaques can also rupture, which then creates a blood clot that tends to obstruct the vessel completely. If prolonged, the blockage will cause extensive cell death in the supplied tissues. Most seriously, the obstructions can lead to heart attacks (myocardial infarcts) or brain damage (i.e., ischemic stroke). The treatment of atherosclerosis typically involves changes in diet, cholesterol-lowering drugs, and, in advanced cases, endovascular surgery. In the following paragraphs, I discuss those therapies in turn.

Cholesterol is a lipid that helps to keep cell membranes pliable. It is synthesized mainly in the liver but can also be obtained by eating animal products (plants do not make cholesterol). Cholesterol is not water soluble and must, therefore, be transported in the blood by special carrier proteins (i.e., lipoproteins). Most of the cholesterol in the blood is carried from the liver to other tissues by low-density lipoprotein (LDL), which is often referred to as the "bad cholesterol." Some cholesterol is carried from the tissues back to the liver by high-density lipoprotein (HDL).

The link between atherosclerosis and cholesterol was first suggested in 1908–1910, when scientists discovered that rabbits fed a diet rich in meat, fats, and milk developed atherosclerosis and that human arterial plaques contain high levels of cholesterol. An influential study in 1912 then showed that high levels of dietary cholesterol correlate with plaque formation in rabbits (Kritchevsky, 1995). As it happens, the choice of rabbits in these early experiments was fortuitous because rabbits are far more sensitive to dietary cholesterol than rodents or dogs (Getz & Reardon, 2017). The link between cholesterol and atherosclerosis in humans was firmly established by the Framingham Heart Study, which followed 5,209 men and women over many years (Castelli et al., 1992). It showed that, indeed, high levels of LDL-bound cholesterol are associated with coronary heart disease (i.e., constriction of the coronary arteries that supply the heart). It also revealed that having low levels of HDL augments the risk.

Aside from changing one's diet, the principal treatment for atherosclerosis consists of medications that lower cholesterol (i.e., statins). The initial discovery of these drugs dates back to the 1960s, when Akira Endo hypothesized that some *Penicillium* fungi might secrete compounds that inhibit cholesterol synthesis in other organisms (but do not harm the fungi because they rely on ergosterol rather than cholesterol to accomplish similar functions). After screening several thousand "fungal broths" in in vitro rat liver assays, Endo discovered compactin, the first statin (Endo, 2010). Surprisingly, this compound does not reduce cholesterol synthesis in intact rats—a finding later confirmed by other researchers—but Endo persisted. Eventually, he found that compactin does reduce plasma cholesterol levels in chickens, dogs, and monkeys. The drug also showed promising results in humans, but at very high doses it caused cancer in dogs and was, therefore, abandoned.

Nonetheless, work on statins continued. In the late 1970s two groups of researchers independently discovered lovastatin, which is structurally quite similar to compactin but derived from a different fungus. This second major statin was shown to inhibit cholesterol synthesis in human cells and intact rats; it also reduced plasma levels of cholesterol, including LDL cholesterol, in dogs (Alberts et al., 1980, 1989). Importantly, long-term toxicity in dogs and several other animal species was relatively low. Clinical trials of lovastatin began in 1982 using healthy volunteers as well as subjects with congenitally high levels of cholesterol. Overall, lovastatin was found to reduce LDL cholesterol reliably with few negative side effects. It was approved by the US Food and Drug Administration (FDA) in 1999 and remains available, even though several additional statins have been developed in the intervening years. An important, relatively recent addition to the arsenal of cholesterol-lowering drugs is the use of synthetic antibodies that lower LDL cholesterol by inhibiting PCSK9, an enzyme that helps degrade the LDL receptors. These therapeutic antibodies were shown to reduce LDL cholesterol in mice and monkeys (Chan et al., 2009). They have been approved for human use since 2015 and are usually taken in conjunction with statins.

When atherosclerotic plaques threaten the blood supply to vital organs, surgeons may try to dilate the obstructed vessels. The surgical technique for this procedure, called angioplasty, was originally developed in the 1960s and 1970s and involves the insertion of long catheters through the relevant blood vessels into the constriction (Payne, 2001). In early versions of angioplasty the surgeon inflated a small balloon at the tip of the inserted catheter. This inflation temporarily expands the vessel's interior, stretching its walls. Because the constriction often returns after such balloon angioplasties, surgeons invented stents, which are wire mesh cages that could be left in place to keep the blood vessel open. Unfortunately, material tends to accumulate inside regular stents, causing renewed constriction (i.e., restenosis). In response, scientists created drug-eluting stents, which release molecules (e.g., sirolimus) that reduce inflammatory responses and reduce restenosis rates substantially. Some of these drug-eluting stents appear to be effective for at least five years (Morice et al., 2007; Kastrati et al., 2007). However, there is some risk of blood clots forming inside of stents. One should also note that, in cases of stable coronary heart disease, stents appear to work no better than less invasive treatments (Maron et al., 2020).

The surgical techniques and devices involved in angioplasty were extensively tested and refined in large animal models, such as dogs, pigs, and cattle (Feng & Jing, 2018). This makes practical sense because the hearts and blood vessels of these animals are similar to ours in size, even if their anatomical details differ in some respects. It is worth noting, however, that the very first angioplasty was performed on patients without prior animal tests (Payne, 2001). Indeed, surgeons are generally given more latitude than other medical professionals in developing new therapies without prior testing in animals (Darrow, 2017). The reasons for this difference are unclear, but variations in surgical skill level are probably involved. It is also generally accepted that surgeons need to "learn in practice," which is to say that they must, from time to time, try new ways of doing things (Morlacchi & Nelson, 2011; Gardner, 2013). In any case, surgical devices, including stents, are typically subjected to more extensive animal tests than the surgical procedures themselves.

Although angioplasty was developed mainly in humans and large animal models, mouse models of atherosclerosis have contributed significantly to our understanding of the underlying disease mechanisms (Shen et al., 2017). Nonetheless, it is important to note that mice tend to develop plaques in different locations than humans, and that these tend to rupture only in response to additional mechanical disturbance (Emini Veseli et al., 2017). Moreover, many treatments that are effective in the mouse models of atherosclerosis have not done well in clinical trials. One possible explanation for this discrepancy is that the manipulations performed on blood vessels in the mouse models do not mimic human angioplasty very precisely (Libby, 2015). Even the large animal models differ from the human condition in the time course of disease progression and response to therapies. In addition, the animal models are generally young and otherwise healthy, which tends not to be the case for humans with cardiovascular disease. In part because it is so difficult to develop good animal models of human cardiovascular disease, there is considerable interest in using human data to construct in vitro and in silico (i.e., computational) models of the human heart and blood vessels (Z. Li et al., 2019; Savoji et al., 2019). Although such models hold significant promise, their ability to deliver novel types of therapies remains to be seen.

5.2.3 Stroke and Neuroprotection

Because strokes cause damage to the brain, it is justifiable to classify stroke as a neurological disorder (Shakir, 2018). However, I here discuss stroke as a cardiovascular disease because it results from problems with cerebral blood vessels. Specifically, hemorrhagic stroke involves the rupture of cerebral blood vessels, and ischemic stroke occurs when the brain's blood supply is blocked, depriving the downstream neurons of the oxygen and nutrients they need for survival.

Some strains of rats, a few transgenic mouse lines, and even some dogs develop ischemic strokes spontaneously (O'Collins et al., 2017; Hermann et al., 2019), but these natural stroke models are difficult to work with because their strokes are unpredictable in time and location. Therefore, researchers generally induce strokes experimentally—for instance, by tying off a cerebral artery. Gerbils were often used for such experiments because they have an incomplete circle of Willis (a set of small vessels that interconnect the major cerebral arteries at the base of the brain), which means that one can deprive an entire cerebral hemisphere by blocking one of the carotid arteries (Graham et al., 2004). However, ischemic strokes in humans usually affect only parts of one cerebral hemisphere. Therefore, researchers nowadays prefer to block only select branches of one carotid artery (e.g., the middle cerebral artery), and they do so mostly in rats and mice (van der Worp et al., 2010).

Another relevant consideration is that the arteries running across the surface of the brain are interconnected via collateral branches, so blocking one artery may cause blood flow to be rerouted around the obstruction. Because the anatomy of this collateral system differs across species, analogous blockages in different species may lead to different outcomes (Howells et al., 2010; Sommer, 2017; Hancock & Frostig, 2017).

Given such species differences, one might expect that therapies derived from animal models of ischemic stroke would translate poorly to humans. Indeed, "in animal models of acute ischemic stroke, about 500 'neuroprotective' treatment strategies have been reported to improve outcome, but only aspirin and very early intravenous thrombolysis with alteplase (recombinant tissue-plasminogen activator) have proved effective in patients, despite numerous clinical trials of other treatment strategies" (van der Worp et al., 2010, p. 1).

One major problem, aside from the already mentioned species differences, is that treatments in animals are usually administered very soon after the induced stroke (or even before), whereas human stroke patients often remain untreated for several hours after the incident. In addition, the animal research is usually performed on young and otherwise healthy individuals, whereas human stroke victims tend to be elderly and suffer from a variety of other health problems (Hermann et al., 2019). Two additional concerns are that the existing animal and clinical studies on stroke both tend to have insufficient sample sizes (i.e., are underpowered), and that the animal research suffers from a positive publication bias and other methodological weaknesses (Philip et al., 2009; O'Collins et al., 2017). On a more positive note, recent clinical trials indicate that surgical extraction of blood clots through endovascular surgery is relatively effective against ischemic stroke in humans (Goyal et al., 2016).

5.3 CANCERS

Physicians have known about cancers since ancient times, defining them as abnormal growths that tend to grow back even if surgically removed. More recently, biologists have characterized malignant, cancerous cells as exhibiting six essential features, namely (1) the ability to proliferate in the absence of external growth signals, (2) an insensitivity to external growth-inhibiting signals, (3) the ability to evade programmed cell death, (4) a limitless proliferative potential, (5) the ability to attract new blood vessels, and (6) the potential to invade other tissues and metastasize (Hanahan & Weinberg, 2000). Although these attributes characterize most, if not all, malignant cancers, they tend to be acquired over time, with cells becoming progressively more cancerous. In a quasi-Darwinian competition among cell lineages, the cells with the greatest ability to survive will proliferate, spread, and come to predominate, which is how precancerous growths become malignant cancers. A second important point is that the details of how cancerous cells acquire their hallmark attributes can vary significantly between different types of cancer. Indeed, biologists have identified more than 100 types of cancers (National Cancer Institute, 2007). These may differ not only in the tissues where they initially arise but also in their molecular characteristics and responses to therapy.

As scientists have struggled to understand and treat cancer, they have used a variety of model systems. They started in 1915 with rabbits, smearing coal tar on their ears to prove that it is a carcinogen; this hypothesis had emerged from the much earlier observation that chimney sweeps often got cancer (Thomas et al., 2016). Along a very different track, C. C. Little (the founder of the Jackson Laboratory; see chapter 3) in the 1930s promoted the study of inbred mouse strains with varying rates of cancer incidence as the best strategy for uncovering cancer's genetic basis (Little, 1935); he considered nonheritable factors to play little or no role in cancer. It is ironic, therefore, that some of the biggest breakthroughs in cancer research came through the use of carcinogenic viruses (Vogt, 2012; Bister, 2015).

The first of these carcinogenic viruses was Rous sarcoma virus, which infects chicken cells and makes them cancerous (Rous, 1959, 1967). Studies of this and other cancer-causing viruses led to the identification of specific genes that, when inserted into the genome of host cells, transform those cells (into tumor cells). Subsequent research revealed that many of these cancer-causing genes, called oncogenes, closely resemble genes natively present in the host genomes, suggesting that the viruses had acquired them from a host at some point in the evolutionary past. Importantly, this discovery strongly suggested that animals possess potential oncogenes (aka proto-oncogenes) that when malfunctioning can cause cancer even in the absence of a viral infection. Later studies confirmed this hypothesis and showed that tumor formation usually requires a whole series of malfunctioning proto-oncogenes (as well as malfunctioning tumor repressor genes). It also showed, as Little had argued, that some individuals have heritable cancer predispositions (H. T. Lynch et al., 2004; Pomerantz & Freedman, 2011).

Complementing this research, which was largely in vitro, were studies on mice implanted with cancerous cell lines or explanted tumors (see section 5.3.4). Many recent studies of this kind have examined, for example, the microenvironment of developing tumors (Watnick, 2012; Jiang et al., 2020), focusing especially on the conditions that allow tumors to thrive (e.g., the ingrowth of blood vessels). It is impossible here to discuss these and other fundamental aspects of cancer biology in any depth. Instead, let us focus on a few key developments in cancer therapy and on the models that led to them.

5.3.1 Radiation Therapy

Because cancer cells divide more rapidly than normal cells, it makes sense to treat cancer with therapies that interfere with cell division. Historically, the first such method was X-ray irradiation. Within a few years of Wilhelm Röntgen's 1895 discovery that X-rays could be used for bioimaging, it became apparent that X-rays can also damage dividing cells. In particular, early experiments showed that irradiation of the testes in various mammals triggered a massive degeneration of sperm progenitor cells but not of mature sperm (del Regato, 1976). Extensive studies on rams, which have conveniently large testicles, then showed in the 1920s that dividing the total dose of radiation into multiple smaller doses, given at regular intervals, made the treatment even more selective for the rapidly dividing cells. Henri Coutard and others subsequently extended this insight to the treatment of humans with cancer (Coutard, 1934; Thames, 1988).

Radiation therapy was widely regarded as a breakthrough in cancer treatment during the 1920s and 1930s, and it has been refined considerably in recent years. Especially important has been the development of computer-assisted methods for three-dimensional (3D) tumor imaging and the use of multiple X-ray beams to deliver high doses of radiation to regions where the beams intersect (Thariat et al., 2013). The "gamma knife" technique, for example, allows physicians to target tumors within the brain without damaging overlying areas. These technical advances involved extensive animal testing, much of it in laboratory mice (Verhaegen et al., 2018; Butterworth, 2019).

5.3.2 Hormone Therapy

A causal link between cancer and male sex hormones was firmly established in 1940, when Charles Huggins and his collaborators discovered that castration causes the enlarged prostate glands of old, senile dogs to shrink; injections of estrogen had similar effects (Huggins & Stevens, 1940). The following year, these scientists performed analogous experiments on dozens of men with prostate cancer and observed a similar degree of tumor regression (Huggins & Hodges, 1941). Nowadays prostate cancer can be treated in diverse ways, including surgical removal and targeted radiotherapy or chemotherapy. However androgen deprivation therapy via castration or antitestosterone drugs continues to be widely used in the treatment of prostate cancer (Gunner et al., 2016); even injections of estrogen continue to be employed in select instances (Reis et al., 2018). Not surprisingly, both treatments can have adverse effects on libido and hormone-related sexual characteristics.

Breast cancer used to be treated mainly by surgical removal of both breasts as well as the adjacent pectoral muscles and lymph nodes (Cotlar et al., 2003), but such radical mastectomies are rarely done today because other, less invasive therapies have become available. Particularly important was the realization that breast cancer, like prostate cancer, is often hormone dependent. The initial stimulus for this concept came from the 1896 discovery that removal of the ovaries causes degeneration in the mammary glands of lactating rabbits. Furthermore, ovariectomy diminished breast cancer in a small trial of three human subjects (Stockwell, 1983). However, this line of inquiry was not pursued in earnest until the 1950s, when Charles Huggins began to study breast cancer in female Wistar rats, which reliably develop mammary cancers when fed high doses of a carcinogen (Bashyam, 2007). Using this model, Huggins et al. (1959) found that mammary cancers decrease in size when the rats have their ovaries removed or are injected with testosterone. Soon thereafter, a large clinical trial revealed that surgical removal of the ovaries (or the adrenal glands, which produce significant amounts of estrogen in older women) reduces breast cancer; importantly, it only works on tumors that express high levels of estrogen receptors (Jensen et al., 1977).

Also in the 1970s, scientists discovered that the drug tamoxifen is an effective breast cancer treatment with few side effects. Tamoxifen was initially developed as an antiestrogen that might serve as a morning-after pill, and it does act as a contraceptive in rats. However, and rather paradoxically, tamoxifen induces ovulation in women (Quirke, 2017). Despite this initial disappointment, studies in 1970s showed that tamoxifen could prevent the development of mammary cancer in carcinogen-fed rats (Jordan, 1976, 2008). A series of clinical trials subsequently showed that tamoxifen is quite effective against breast cancers and has fewer side effects than the earlier therapies. Particularly important was that tamoxifen can prevent breast cancer in at-risk women and avert metastasis after surgical breast cancer removal. It was approved by the FDA in 1977, and millions of women (and some men) have benefited from tamoxifen worldwide. Some newer antiestrogen cancer drugs have been developed in the last couple of decades, but tamoxifen continues to be prescribed, often in conjunction with other therapies.

5.3.3 General Chemotherapy

In parallel with the discovery of hormone therapy, scientists in the 1940s developed drugs that selectively kill rapidly dividing cells and could, therefore, be used as a broad anticancer treatment. The first of these general chemotherapy drugs was nitrogen mustard, which is closely related to sulfur mustard (aka mustard gas). These mustard agents, which are unrelated to the mustard plant but smell like it, had been synthesized as toxins for use in chemical warfare. They were rarely used on the battlefield, but mustard gas did kill hundreds of sailors when it was accidentally released during the bombing of an Allied cargo ship carrying a secret load of this toxin during World War II (Conant, 2020b). Examination of the deceased, as well as subsequent experiments on rabbits and several other mammalian species, indicated that mustard agents selectively kill the fast-dividing bone marrow cells that generate white blood cells (DeVita & Chu, 2008; Conant, 2020a). We now know that they do so, at least in part, by forging cross-links between DNA strands, which then interfere with DNA transcription and replication.

The idea of using mustard agents to fight cancer began to be pursued in the early 1940s. Extensive early studies in rabbits and then mice showed that mustard agents, like X-rays, destroy primarily the rapidly dividing cells. More importantly, when nitrogen mustard was given to mice that had been implanted with lymphomas, the tumors regressed. These "experiments in mice were sufficiently encouraging to consider a therapeutic trial in man" (Gilman, 1963, p. 576). Sure enough, the first clinical trial, published in 1946, showed that nitrogen mustard reduces tumors in patients with lymphosarcoma and Hodgkin's disease, both of which are cancers of the lymphatic system (Goodman & Wintrobe, 1946). It is much less effective against leukemias, both in humans and in mice. This is interesting because the entire project might have been dropped if the scientists had focused their early animal studies on leukemia rather than lymphomas (Gilman, 1963). It also shows that even general chemotherapy agents are more effective against some types of cancer than others.

Although the development of nitrogen mustard as an anticancer agent was a breakthrough in cancer therapy, the drug has a variety of adverse side effects, including a dramatic decrease in the number of white blood cells. Scientists therefore developed a variety of closely related compounds, such as cyclophosphamide, that are similarly effective but less toxic. They also developed a variety of other drugs that are not related to mustard agents but likewise interfere with various aspects of cancer cell proliferation. Particularly interesting is the history of platinum-containing drugs, which remain widely used against some cancer types. These drugs derive from a serendipitous observation made by Barnett Rosenberg in the 1960s. He had been examining the effect of electric fields on cell division in *E. coli* and discovered that passing current through the platinum electrodes he had been using created a platinum salt, later called cisplatin, that stopped the bacteria from dividing (even though they continued to grow in size). Rosenberg and colleagues (1969) then showed that cisplatin could be used to treat two types of cancer in mice. The drug was approved for use in humans in 1978 and has spawned a wide variety of other platinum-related drugs (Johnstone et al., 2014). One must note, however, that cisplatin all too frequently causes permanent hearing loss in children (Knight et al., 2005; Sheth et al., 2017), a finding that has been replicated in monkeys (Stadnicki et al., 1975) but could not have been obtained in *E. coli*.

Another class of general chemotherapy agents blocks DNA synthesis, which cancer cells require to proliferate. First to be discovered were drugs that inhibit folic acid, an enzyme that is involved in the synthesis of some DNA nucleotides. The initial experiments showing that inhibitors of folic acid can counteract cancer were performed in mice with implanted tumors (Leuchtenberger et al., 1945; Sneader, 2005). Related compounds were then synthesized and tested in humans (Farber et al., 1948). The two historically most impactful antifolate compounds are aminopterin and methotrexate, which have been used for the treatment of leukemia and several other malignancies (Goldin et al., 1955). A second class of compounds that interfere with DNA synthesis consists of modified pyrimidines, including 5-fluorouracil. This drug was created specifically with the aim of fighting cancer, and studies in the late 1950s showed it to be effective in mice with transplanted tumors (Heidelberger et al., 1957). The first clinical trials were already promising (Curreri et al., 1958), although tumor regression was typically associated with toxic side effects. Nonetheless, 5-fluorouracil remains widely used to treat colorectal and related cancers, early-stage breast cancer, and several cancer-related skin conditions. It remains on the World Health Organization's List of Essential Medicines (as do cisplatin and methotrexate) (WHO, 2020).

The chemotherapy agents we have discussed in this section interfere with cell division generally and, therefore, kill not only cancer cells but also rapidly dividing cells in other tissues, such as hair follicles, the lining of the gastrointestinal tract, and white blood cell precursors. Because of these negative side effects, general chemotherapy drugs were widely regarded as "poisons" in the 1960s and used reluctantly by many physicians who, after all, are sworn to "do no harm." It took another decade or so before general chemotherapy became more widely accepted, mainly because, in combination with anticancer surgery, it did effectively cure some types of cancer (DeVita & Chu, 2008).

5.3.4 Targeted Chemotherapy

As discussed in section 5.3.2, the existence of hormone-sensitive cancers revealed that tumor cells may differ in the genes and proteins they express and, therefore, may require different types of therapy. The resulting quest for compounds that target specific, molecularly defined types of cancer accelerated dramatically in the 1970s, when biologists began to document the molecular genetic profiles of diverse cancers (National Cancer Institute, 2018). This work revealed that cancers vary not only between the sexes and in the tissues where they originate, but in their cancer-causing mutations.

Indeed, as mentioned earlier, scientists have by now discovered many different oncogenes and tumor repressor genes, defined as genes that cause cancer when they become abnormally active or are disrupted, respectively. To test the causal role of these genes in producing the hallmarks of cancer, researchers created hundreds of genetically modified mouse strains that express altered versions of the putative cancer-related genes. Study of these "oncomice" (Hanahan et al., 2007) revealed an enormous amount of information about the intricacies of cell cycle regulation and tumor growth, including the observation that it often takes multiple cancer-related mutations to trigger tumor formation (Hutchinson & Muller, 2000). However, using these genetically modified mice to develop anticancer therapies is difficult because one cannot know precisely when and where the cancer-prone mice will develop their tumors, which in turn makes it difficult to conduct large-scale screens for anticancer drugs.

To overcome this problem, researchers began to transplant tumor cells from mice with different types of cancer into normal mice of the same strain. Because the two sets of mice share the same genetic background, immune rejection of the graft is unlikely. Left untreated, the transplants reliably develop into tumors. Because the transplanted tissue is typically placed just under the skin on the side of the animal, tumor growth is easy to spot and quantify. Thus, these "isograft" models are well suited to drug screening. Unfortunately, many of the drugs that work in these models have not succeeded in human clinical trials (Rangarajan & Weinberg 2003; Anisimov et al., 2005). The most likely reason for this failure is the existence of species differences in cancer biology. For example, a major type of breast cancer that expresses high levels of estrogen receptor in humans does not do so in mice (Herschkowitz et al., 2007). Similarly, disruption of the retinoblastoma gene causes retinal cancer in humans but pituitary cancer in mice (Robanus-Maandag et al., 1998). In addition, researchers have noted that it commonly takes four to six distinct mutations to trigger cancer in humans but fewer in rodents (Hahn & Weinberg, 2002).

In light of these species differences, cancer researchers increasingly focused their efforts on human cancer cells. In particular, they created large libraries of 60 or more human cancer cell lines that can be tested in vitro for their responses to diverse drugs. Because cell culture conditions differ substantially from the in vivo context (see chapter 4), the human cancer cells are often transplanted into immunodeficient mice, thus converting the aforementioned isograft models into "xenograft" models (*iso-* and *xeno-* meaning "same" and "foreign," respectively, in ancient Greek). The latter models are experimentally convenient, but the use of immunodeficient mice is potentially problematic because the immune system is known to be involved in combating cancers (see section 5.3.5). One can try to get around this problem by using mice with partially humanized immune systems (Wege, 2018), but such mice are difficult to use and expensive. Another problem is that the human cancer cell lines tend to acquire additional genetic modifications while they are maintained in vitro and adapt to those new conditions (Ben-David et al., 2018). Many cancer cell lines do retain some similarity to the original tumors, but the deviations tend to be significant (Domcke et al., 2013). Consequently, xenograft models using cancer cell lines have fallen out of favor with many cancer researchers (Auman, 2010).

Currently more popular are xenograft models that use patient-derived live tumor tissue rather than cancer cell lines. These patient-derived xenograft models are less convenient because patient-derived tumors can be difficult to procure. However, carefully curated "tumor banks" have been established and now offer a wide selection of different human tumors for research. A second challenge is that each patient-derived tumor can support only a limited number of transplant experiments. Researchers attempt to minimize this problem by extracting the tumors growing in previously xenografted mice and implanting pieces of those extracted tumors into additional mice (effectively passaging the tumors). Unfortunately, some mouse cells tend to invade, and thus contaminate, the transplanted tumors, and this problem gets worse with each round of extraction-transplantation. Moreover, after several rounds of tumor passaging, the human tumor cells acquire genetic abnormalities that differ from those they would acquire in humans. Indeed, some evidence suggests that tumor cells face different selection pressures in patients versus mice, resulting in divergent tumor evolution (Ben-David et al., 2017). Yet another problem is that implanting tumor cells under the skin, rather than in the location of the original tumor, can affect how the resulting tumor spreads through the body (i.e., metastasis) (Day et al., 2015; Ireson et al., 2019). Of course, this last limitation is shared with the cell-derived xenograft models, as are the aforementioned problems associated with the use of immunodeficient mice.

Given all these limitations, it is not surprising that merely 5% to 11% of the anticancer drugs showing promise in mouse models have passed their clinical trials (Kola & Landis 2004; Ocana et al., 2011; Begley & Ellis 2012; Hay et al., 2014). Despite this high failure rate, the work has yielded major successes. Especially useful have been small molecules that inhibit one or more kinases that regulate cell proliferation. More than three dozen of these kinase inhibitors had been approved by 2018 (Bhullar et al., 2018); one of them is imatinib (aka Gleevec). This drug was strategically designed using in vitro assays but was then tested in a mouse isograft model before moving to clinical trials (Druker et al., 1996, 2001). It is used to treat chronic myelocytic leukemia, a rare type of bone marrow cancer featuring a unique chromosomal abnormality, as well as several other cancer types.

In addition to small-molecule drugs, synthetic antibodies are often used to target specific cancer-related genes. Trastuzumab (aka Herceptin), for example, is a humanized antibody that blocks a growth factor receptor called Her2, which is required for cell proliferation in breast cancers that express high levels of this particular protein (Harwerth et al., 1993; Mokbel & Hassanally, 2001; Pegram & Ngo, 2006). This very successful drug was derived from a xenograft model using a cancer cell line.

In short, extensive research on both animal and in vitro systems has led to the development of several highly specific anticancer drugs that each target a relatively small, highly specific subset of cancers. Importantly, the side effects of these treatments tend to be relatively mild, at least compared with general chemotherapy drugs.

The realization that cancers come in many different forms, each requiring a different treatment (Sánchez et al., 2017), ushered in the era of personalized medicine (aka precision medicine; see chapter 2). Of course, researchers had long realized that cancers found in different tissues require different treatments (e.g., breast cancer versus prostate cancer), but the ability to characterize cancers by their genetic mutations and expression profiles greatly expanded the number of different cancer types. This abundance, in turn, stimulated the development of many different targeted chemotherapy drugs. It also required a shift in how clinical trials are run. Instead of testing a single drug on a heterogeneous population of cancer patients, scientists had to enroll subjects with the specific type of cancer targeted by the trial drug. As more and more drugs required testing, scientists designed very large trials that match different subject to different trial therapies based on the molecular details of their tumors. Such trials have yielded notable successes (Sánchez et al., 2017); but the vast majority of enrolled subjects end up without a match (Letai, 2017; Flaherty et al., 2020). Thus, the currently available set of targeted chemotherapy drugs addresses only a relatively small subset of cancers.

A related problem is that developing novel drugs for small groups of patients makes it difficult for pharmaceutical companies to recoup their investments, which helps explain the currently high costs of chemotherapy. Yet another troubling issue is that many of the drugs targeting specific cancer-related genes appear to work in mice even when the supposed targets are deleted (Lin et al., 2019), raising doubts about their supposed specificity.

5.3.5 Cancer Immunotherapy

A recent breakthrough in oncology has been the development of immune checkpoint therapy. This form of cancer treatment is based on the discovery, made in the 1990s, that the T cells of the body's immune system can be suppressed, notably by tumor cells (Pardoll, 2012). Some degree of T cell suppression is normal and adaptive, as it prevents the immune systems from attacking its own normal cells (i.e., autoimmunity). However, reducing the level of T cell suppression with immune checkpoint inhibitors can prompt the immune system to attack cancer cells while leaving normal cells in peace. This works because cancer cells accumulate so many mutations that the disinhibited T cells can distinguish them from normal cells.

The first checkpoint inhibitors were antibodies that partially inhibit CTLA-4, a receptor on the surface of T cells. After showing that these antibodies increase T-cell activation in primary cell cultures (Krummel & Allison, 1995), researchers tested them on mice injected with mouse tumor cells. The principal finding was that anti-CTLA-4 can prevent tumor growth and cause even established tumors to regress (figure 5.4). Although high doses of anti-CTLA-4 can lead to autoimmune problems, the drug's benefits can be quite long-lasting, presumably because the body's immune system develops a "memory" for the tumor cells and quickly attacks them if tumors recur. Based on these benefits, a humanized antibody against CTLA-4, called ipilimumab, was approved in 2010 for the treatment of metastatic melanoma.

The second immune checkpoint inhibitor to be developed targets a receptor called PD-L1 (programmed death ligand 1). When molecules that bind the PD-L1 receptor

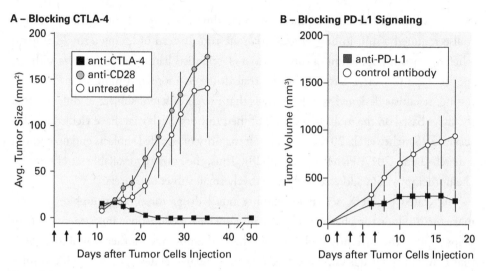

Figure 5.4

CTLA-4 and PD-L1 suppress the immune system's antitumor response. (*A*) Blocking CTLA-4. Leach et al. (1996) injected mice with cells from a mouse tumor cell line and then treated the mice with function-blocking antibodies against either CTLA-4 or CD28 (three sequential antibody injections indicated with arrows). Untreated controls and the anti-CD28 treated animals developed large tumors, whereas the anti-CTLA-4 treated mice exhibited tumor regression and then remained tumor free for at least 90 days. This finding suggests that CTLA-4 normally suppresses the immune system's ability to combat tumor growth. (*B*) Blocking PD-L1 signaling. Imai et al. (2002) injected mice with cells from a mouse tumor cell line express-ing PD-L1. When these cells were co-injected with a normal rat immunoglobulin (control antibody), all the mice developed large tumors. By contrast, co-injecting anti-PD-L1 significantly inhibited tumor growth. Adapted from (A) Leach et al. (1996); (B) Iwai et al. (2002).

are expressed by cancer cells implanted into mice, tumor growth is accelerated (relative to tumor cells that lack this expression). More importantly, tumor growth is prevented when the implanted mice are treated with antibodies that block PD-L1 signaling (Fig 5.4B). Clinical trials ultimately led to the approval of several human antibodies that target PD-L1 for the treatment of several types of cancer (note that not all cancer types up-regulate CTLA-4 or PD-L1). In 2018, the Nobel Prize in Medicine and Physiology was awarded to James Allison and Tasuku Honjo for their work on CTLA-4 and PD-L1, respectively (see the appendix).

An interesting distinction between immune checkpoint therapy and the targeted chemotherapy we discussed in section 5.3.4 is that the former targets receptors on T cells of the immune system, whereas the latter targets the cancer cells directly. None-theless, a major similarity is that both types of cancer therapy work only for a rather limited set of cancer subtypes (Gay & Pradad, 2017). This narrowness of scope is expected for the tumor-targeting drugs but somewhat surprising for the immune checkpoint inhibitors. One likely explanation is that cancers vary in how effectively they can trigger an immune response, even when the brakes on T-cell activation are removed. It is also true that not all tumors suppress the immune cells in the first place. Regardless of the details, the bottom line is that all cancer therapies are limited in scope and, therefore, cancer treatment generally varies with cancer type.

5.4 PATTERNS OF MODEL USE

Despite the ongoing crisis of translation in biomedical research (see chapter 1), medical progress over the last 150 years has been remarkable. Major epidemics have been controlled, and cancer patients now have a much better chance of survival than in the past. Many deaths have been delayed by early intervention (especially in cardiovascular diseases), and modern medicines tend to have fewer negative side effects than their predecessors. An excellent example is the increased selectivity of modern anticancer drugs, compared to radiation therapy and general chemotherapy.

Thus, Paul Ehrlich's dream of discovering a magic bullet that kills the disease-causing agent but does no harm to the patient has been realized for many diseases (Strebhardt & Ullrich, 2008). However, we now realize that many diseases, notably cancer, are in fact a collection of many different diseases that each require a different treatment. Moreover, many diseases do not yet have a magic bullet of their own, and different patients may respond differently to the same therapies even when they have the same disease (see chapter 7). In short, personalized medicine is both imperative and imperfect.

Animal and in vitro models have featured prominently in virtually all the medical advances we have reviewed (although it is hard to prove that they were *necessary* for the advances; see Matthews, 2008). In general, biomedical researchers have managed to discover or create model systems that reliably develop the disease of interest, or at least some symptoms reminiscent of the human disease (for more discussion of "partial models," see chapter 6). The scientists then expose their model to a variety of compounds or other manipulations (e.g., radiation) to test whether any of these treatments can prevent or reverse one or more of the disease-related attributes.

Historically, researchers have tested relatively large and semirandom sets of compounds that were either synthesized by chemists or isolated from microbes and plants. Beginning in the mid-1980s, the pattern changed. Biologists began to gather enormous amounts of information about the genetic and molecular aspects of the various diseases and then targeted specific disease-linked molecules and pathways for therapy development. Often they were able to design potential therapies "rationally" by creating antibodies against specific molecules or designing compounds likely to interact with putative binding sites.

Some observers have suggested that the adoption of more target-based screens has contributed to the current translational crisis, but this remains debatable (Swinney & Anthony, 2011). In any case, even target-based drug development typically requires multiple rounds of trial and error. As Paul Ehrlich used to say, success in medical research requires patience, skill, money, and luck (i.e., Geduld, Geschick, Geld, und Glück)—that is still true.

In vitro models have made significant contributions to many medical advances. The development of polio vaccines, for example, progressed much faster after scientists discovered how to grow the virus in cultured cells (see chapter 4). The discovery of statins and ACE inhibitors also owes much to in vitro systems, ranging from cell-free assays to isolated animal organs. Major developments in cancer therapy likewise built on knowledge gained from in vitro systems, including the carcinogenic viruses, the microbial systems that provided early insights into cell cycle control (i.e., yeast and *E. coli*), and the cultured T cells used to develop immune checkpoint therapy. Discoveries made in these in vitro models were always followed by in vivo animal studies, except in cases of exceptional urgency (i.e., the AIDS and COVID-19 pandemics).

For the therapies we reviewed here, the in vivo results were generally concordant with the in vitro data, but this is not surprising, given our focus on the most successful therapies. Even so, a few compounds were effective in vivo but not in vitro because they had to be metabolized by liver cells before they could exert the desired function (e.g., Prontosil). In addition, immortalized cell lines often produced dubious results because those cells tend to harbor serious chromosomal aberrations and, like all extensively cultured cells, acquire mutations that adapt them to the in vitro environment. Therefore, abandoning a research avenue after negative in vitro results might sometimes be ill-advised.

A major challenge of working with in vivo animal models is the existence of species differences. It is unfortunate, for example, that the viruses responsible for AIDS and hepatitis C naturally infect only a limited number of nonhuman primate species, or that the antiviral drug remdesivir is metabolized differently in rodents and primates. Similarly, the observation that penicillin is far more toxic to guinea pigs than other species had long been puzzling. Another odd aspect of rodent biology is that rats do not respond to compactin, which lowers cholesterol in many other species, including humans (Endo, 2010). In reaction to all these species differences, some observers have concluded that animal experiments are inherently misleading and should be abandoned (Knight, 2007; Langley et al., 2015).

In my view, a more productive response is to work with multiple species so that the species differences become more apparent and research avenues are not abandoned too early. In the words of Abraham Kaplan (1973), "The dangers are not in working with models, but in working with too few, and those too much alike" (p. 263; quoted in van der Staay, 2006). Of course, this more comparative approach raises new questions of its own: for instance, how many species is enough? We will return to such questions in chapter 7. For now, it is useful merely to note that regulatory agencies have traditionally required preclinical results to be presented for at least two species, one of which should be a nonrodent. In toxicology, this dual-species approach seems to optimize the prediction of human results (Monticello et al., 2017) (see also chapter 4, section 4.3).

As we discussed in chapter 2, biologists often select their model systems for experimental convenience (i.e., Krogh's principle). In biomedical research, it is common to think that one should start with the most convenient model, which is often in vitro, and then work one's way up through "lower animals" (e.g., fruit flies) to "higher animals" (e.g., dogs or nonhuman primates) before beginning clinical trials. Indeed, many successful therapies began with in vitro research, and nonhuman primates are usually tested only after other models have been examined.

However, the model sequence is rarely linear. This is most apparent when in vitro models are literally combined with in vivo models, as in the transplant models for cancer and hepatitis C research. It is also unclear where human cultured cells fit into the supposed sequence because they are high on the species spectrum but lack in vivo complexity. Even more important is that "convenience" has many dimensions (e.g., cost, ease of breeding or experimental manipulation, ethical concerns), and what is most convenient depends on the experimental aims. For the last few decades, mice have been regarded as most convenient for nearly all in vivo experiments, while rabbits, guinea pigs, and other mammals have receded in popularity (see chapter 3). However, nonhuman primates have remained critical for determining safe and effective doses for human use.

All of which raises an interesting question: What about the non-mammalian in vivo models that, according to the 3Rs of Russell and Burch (1959), should replace mammalian models whenever possible (see chapter 2)? Although fruit flies, *Caenorhabditis elegans,* and zebrafish are all embraced by sizeable research communities and regularly used for high-throughput screening of drugs (see chapter 3), their contributions to medical breakthroughs have thus far been meager.

One possible explanation for the strong focus of the biomedical research community on mammalian models—and within mammals, the persistence of nonhuman primate models—is that even broadly conserved biological principles are often implemented differently in different species. That is, the principles may be conserved, but the molecular details often vary. Crucially, those species differences can impact how a system reacts to pathogens, toxins, genetic mutations, and potential therapies.

Biomedical researchers therefore seek (at least subconsciously) to study systems that tend to minimize those species differences; some even experiment on themselves, as in the case of Hering and Murrell tasting the nitroglycerin (see section 5.2.1). Although species differences do not always increase predictably with phylogenetic distance, they do increase on average (section 2.3.1; Striedter, 2019). Therefore, nonhuman primates tend to be more similar to humans than rodents or other mammals—so long as one averages across a multitude of traits. Similarly, nonprimate mammals tend to be more similar to humans than non-mammalian vertebrates or, for that matter, invertebrates.

We will return to these ideas in chapter 7. For now, it is interesting simply to point out that the two organ systems that have diverged most dramatically in evolution are the immune system, which featured heavily in this chapter, and the nervous system, which shall be our focus in the next.

6 NEUROLOGICAL DISORDERS: TRIALS AND TRIBULATIONS

> In the 50-odd years since the introduction of clinically effective medications for the treatment of behavioral disorders such as depression, anxiety or schizophrenia there has recently been growing unease with a seeming lack of substantive progress in the development of truly innovative and effective drugs for behavioral disorders; an unease indicated by escalating research and development expenditure associated with diminishing returns.
>
> —MCARTHUR AND BORSINI (2008), P. XV

I here define neurological disorders very broadly to include the effects of neuronal injury (e.g., spinal cord injury) as well as neurodegenerative and psychiatric disorders. Other authors may distinguish between neurological and behavioral/mental/psychiatric disorders, but I here prefer a single, comprehensive term (Insel et al., 2010). It is certainly reasonable to include stroke among the neurological disorders because it involves some brain damage, but we already discussed stroke in chapter 5, in the context of cardiovascular disorders (section 5.2.3).

Collectively, neurological disorders cause enormous suffering and inflict huge losses on society in terms of premature death and disability. By some measures, the costs are greater than those caused by cancer or cardiovascular disease. Part of the problem is that neurological disorders often afflict relatively young people who then, in the absence of cures, must live with their problems for many years (figure 6.1). As mentioned in the chapter's opening quote, some effective treatments for neurological disorders do exist, but they are generally not cures, and the rate at which new therapies are being developed is painfully slow. Many promising therapies have emerged from animal and in vitro models, but most of the clinical trials have failed.

In the following sections, I review a representative selection of neurological disorders, focusing on the major models biologists have used to study them and on the

A – Disability Costs of Neurological Disorders

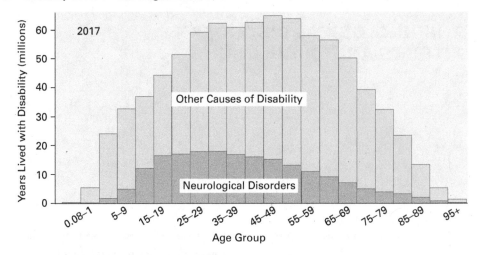

B – Sex Differences in Neurological Disorders

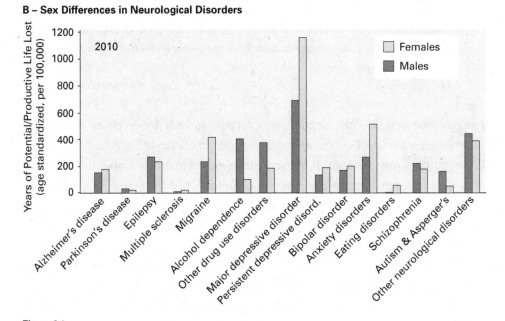

Figure 6.1

The impact of neurological disorders varies across the life span and between the sexes. (*A*) Global burden of neurological disorders (here grouped with "mental disorders" but excluding cases of CNS injury) in terms of total years lived in less-than-ideal health. The burden falls heavily on people of working age. (*B*) Histogram showing that many neurological disorders (again excluding injuries) impact the two sexes differently. The metric here is years lost to disability, either as a result of premature death or in terms of years of productive life lost due to disability (aka disability-adjusted life years, DALY). Neurological and mental disorders are combined. Adapted from (*A*) Institute for Health Metrics and Evaluation (2018); (*B*) Whiteford et al. (2015). Data reported in (*A*) Murray et al. (2012) and (*B*) James et al. (2018).

breakthrough therapies that have emerged. In addition, we will discuss some instances in which the therapies that worked in animal models have failed in human clinical trials. The chapter ends with some possible explanations for why progress on treating neurological disorders has been so slow.

6.1 BRAIN AND SPINAL CORD INJURIES

Damage to the central nervous system (CNS) typically results from trauma to the head or spine. It often involves the death of neurons at the site of injury as well as damage to the axons that course through the area of injury. In addition, CNS injury typically involves inflammatory responses that help to contain the damage in the short term but may have a variety of long-term negative effects. Because the initial neuronal losses are generally irreversible, potential therapies aim mainly to reduce the deleterious follow-on effects and to restore some of the damaged connections. Not discussed here are mechanisms of compensatory plasticity that contribute to successful rehabilitation after CNS injuries.

6.1.1 Traumatic Brain Injury

Traumatic brain injury (TBI) affects over 25 million people annually (James et al., 2019). It comes in many different forms, ranging from closed-head injury (e.g., when the head hits the windshield during an automobile accident) to penetrating injury (e.g., a bullet wound). The type of damage is also quite varied, ranging from outright neuron death to the disruption of axonal connections and widespread neuroinflammation. It is no surprise, therefore, that biologists have created a wide variety of techniques for modeling such injuries (Ma et al., 2019).

Descriptions of these methods are difficult to read without triggering strong emotional reactions because they emphasize the importance of inflicting serious damage with great precision and then measuring the behavioral sequelae as objectively as possible. Such rigor is essential to the modeling, but it nonetheless elicits ethical concerns, especially when the test animals are sentient. In addition, one must ponder whether the animals should be anesthetized during trauma induction, given that humans in traumatic accidents are usually awake. Some anesthetics are already known to modulate the severity of TBI (Wojnarowicz et al., 2017), but, as the controversy surrounding the University of Pennsylvania monkey experiments revealed (see chapter 3, section 3.6.2), the idea of intentionally damaging the brains of awake monkeys is repugnant to most. To solve these problems, researchers have explored the use of various anesthetics, and some have modeled TBI in flies (Shah et al., 2019) and other presumably "less sentient" species, or even in vitro.

Although numerous preclinical studies on TBI have yielded successful neuroprotection, clinical trials of the putative remedies have frequently failed, just as they did in the case of neuroprotection from stroke (see section 5.2.3). A systematic analysis of 191 clinical TBI trials revealed that only 20% of the interventions resulted in a positive clinical outcome, and even the most promising interventions showed poor consistency across trials (Perel et al., 2007; Bragge et al., 2016). In fact, there currently are no drug treatments approved by the US Food and Drug Administration (FDA) for TBI.

One of the most promising drugs for TBI treatment had been the steroid hormone progesterone. More than 300 studies in various rodent models of CNS injury showed progesterone to be neuroprotective, and the drug passed phase II clinical trials for safety (Stein, 2015). These positive results led to two large phase III trials, but one of them was "stopped for futility," and the other found no difference from placebo after six months. There are many possible explanations for these expensive and frustrating failures, including the possibility of suboptimal drug doses or insufficient treatment duration in the clinical trials (Howard et al., 2017). However, it is also possible that the animal models differ from the human condition in critical ways. Similar uncertainties surround the hormone erythropoietin, which has been reported to be neuroprotective after various forms of TBI in rats (Mammis et al., 2009). A meta-analysis of five clinical trials examining erythropoietin in 2016 indicated some benefits in humans, but the difference from placebo was not statistically significant (Liu et al., 2017). More recent meta-analyses that included additional trials are similarly disappointing (Lee et al., 2019). More promising are efforts to reduce brain damage by cooling the CNS soon after injury, but even those trials have yielded equivocal results (Bragge et al., 2016).

One interesting response to these frustrations has been the formation of a research consortium in which multiple laboratories coordinate their efforts while working with multiple TBI models in rats and pigs (Kochanek et al., 2018). The work with pigs is included mainly because the large brains of these animals have proportionately more white matter and more highly folded cortices than smaller brains, two features that likely influence how brains respond to dangerous impacts (Sorby-Adams et al., 2018). The hope is that this coordinated effort will lead to treatments that are effective in both rodents and pigs and thus are likely to work in humans as well. Sadly, the consortium thus far has found that their tested therapies "have produced less robust benefit than expected from the published literature upon which many of the therapies were chosen and treatment regimens designed" (Kochanek et al., 2020, p. 2357).

6.1.2 Spinal Cord Injury

Traumatic spinal cord injury (SCI) is less common than TBI, but it still affects almost 1 million new patients per year worldwide, leading to a general prevalence of 27 million people in 2016 (James et al., 2019). It typically kills neurons whose cell body is located in the damaged portion of the spinal cord, and, at least as important, it disrupts the axons passing through that site. The loss of axons that ascend to the brain accounts for the loss of sensations below the site of injury, and the loss of descending axons causes a corresponding loss of voluntary movements. Unfortunately, the vast majority of neurons in the CNS of mammals do not divide in adulthood and, thus, cannot be replaced once they have been lost. Moreover, the axons damaged by SCI in adult mammals also do not regenerate (Huebner & Strittmatter, 2009). Therefore, a major goal for SCI therapy development is to induce axon regrowth experimentally.

In pursuit of this aim, neurobiologists have studied SCI in a variety of species. Some of the earliest research on SCI was done with dogs and cats, but since 1946 approximately 87% of the work has been performed on rodents, especially rats (Sharif-Alhoseini et al., 2017; Filipp et al., 2019). These animal models typically involve surgical exposure of the spinal cord followed by contusion (i.e., interruption of the blood supply), compression, or transection. In most animals, the damaged tissue is quickly removed by phagocytic cells, leaving a fluid-filled cavity that presents an obstacle to axon regrowth. In mice, however, the site of injury fills with a matrix of connective tissue that then contracts longitudinally so that the two ends of the injured spinal cord draw close together (Inman & Steward, 2003); this should facilitate axon regrowth in mice.

Another potentially important species difference is that the projections from the neocortex to the spinal cord are less extensive in rodents than in large-brained primates, where they extend far down the spinal cord and deep into its ventral horn (Bortoff & Strick, 1993; see Striedter, 2005). This probably explains why rodents with damage to the corticospinal tract are generally less impaired than primates with similar injuries (Nardone et al., 2017), and why mice, in stark contrast to primates, can stand within a week or two after SCI, even in the absence of axonal regeneration (Khan et al., 1999; Filipp et al., 2019). Based on these species differences, recovery from SCI should a priori be easier to accomplish in mice than in primates, but this hypothesis has not been tested directly. Be that as it may, biologists have developed several effective therapies for SCI in rodents, and some of them show promise in humans.

One form of SCI therapy is based on the discovery of molecules that inhibit axon growth and, thus, regeneration in the adult CNS. Most of the research has focused on Nogo-A, a protein that is produced by the glial cells that myelinate axons in the CNS (i.e., oligodendrocytes). Antibodies against this protein promote the sprouting and regeneration of axons after SCI in adult rats and macaque monkeys (Schnell & Schwab 1990; Freund et al., 2007). Although the degree of axonal regeneration is limited, especially in primates, the treated animals do recover at least some of their lost motor abilities (Bregman et al., 1995; Freund et al., 2006). Based on these findings, clinical trials of anti-Nogo therapy have begun and are now in phase II.

One significant concern is that the treatments inhibit Nogo globally and may, therefore, lead to undesirable axon sprouting elsewhere in the brain (Mohammed et al., 2020). It is also becoming clear that Nogo is just one of several molecules that inhibits CNS axon growth and that its inhibition may have modest or even insignificant effects (Zheng et al., 2003). Deletion of the molecule PTEN (phosphatase and tensin homolog) in cortical neurons, for example, enhances axon sprouting after SCI much more than does deletion of Nogo (Liu et al., 2010; Geoffroy et al., 2015). Additional uncertainty is created by the observation that genetic deletion of the Nogo receptor in mice has very little effect on axon outgrowth (Zheng et al., 2005).

A second type of SCI therapy is the implantation of stem cells into or near the site of injury. The implanted cells generally do not replace the damaged cells but instead facilitate recovery by secreting various growth factors, reducing inflammation, or providing a scaffold for axon growth (Drago et al., 2013). Numerous studies supporting the general safety and efficacy of this approach have been conducted in animals, primarily rodents (Cummings et al., 2005; Keirstead et al., 2005; Iwanami et al., 2005; Ahuja et al., 2020). More than a dozen clinical trials have also been performed, although most were relatively small and none have completed phase III (Donovan & Kirshblum 2018; Zhao et al., 2019; Platt et al., 2020).

Overall, stem cell therapy does show significant promise for the treatment of SCI, but questions remain about the optimal timing of the implantation and the optimal number of implanted cells. Arguably the biggest challenge is that the various studies have used a variety of different stem cells, derived from various cellular lineages and treated with different protocols. In light of this variation, it remains unclear which types of cells are optimal, especially since even superficially similar cell types have sometimes yielded very different results in these studies (Anderson et al., 2017). Immune rejection of the implanted cells also remains a concern. Patient-derived stem cells are a potential solution to this problem, but they vary across patients (by definition). Collectively, all this variation makes one wonder whether regulatory agencies should approve induced pluripotent stem cells (iPSCs) for each patient separately, or perhaps approve a limited set of iPSC stock lines that are likely to be "safe" for different patient groups (Tsuji et al., 2019). A third possibility is to approve specific procedures for obtaining high-quality patient-derived stem cells rather than the cells themselves, but this strategy could end up being undermined by lapses in quality control (even if they are relatively rare).

Other treatments for SCI continue to be explored as well, but progress has been unsteady. For example, the drug methylprednisolone (a synthetic corticosteroid) was approved as an anti-inflammatory drug in the 1950s, and a large clinical trial in 1990 showed that a high dose of methylprednisolone injected intravenously within 8 hours of SCI had some beneficial effects (Bracken, 1992). Based on this finding, methylprednisolone became widely accepted as the standard of care for SCI. However, several additional trials (Donovan & Kirshblum, 2018), as well as analogous studies in rats (Rabchevsky et al., 2002), have raised serious questions about this therapy's safety and efficacy (Hugenholtz, 2003; Hall & Springer, 2004).

Some of the neuroprotective treatments we discussed in the context of stroke (chapter 5) are also being tried on patients with SCI (Ulndreaj et al., 2017). A very recent but also very promising approach to treating SCI is a form of gene therapy that induces expression of a "designer cytokine" in neurons of the motor cortex, which then promotes regeneration of the corticospinal tract and behavioral recovery, at least in mice (Leibinger et al., 2021).

6.2 NEURODEGENERATIVE DISORDERS

The neurodegenerative diseases are those in which various types of neurons die progressively, unrelentingly; they include Alzheimer's, Parkinson's, and Huntington's diseases, as well as amyotrophic lateral sclerosis (ALS). Collectively, neurodegenerative diseases place an enormous burden on individuals and on society because the patients often live for many years while their condition deteriorates. Individual cases are typically characterized as being familial or sporadic, with the former being "inherited" and the latter having no obvious genetic basis, as inferred from family history or genetic screening. The familial cases are sometimes referred to as early onset, because they tend to emerge at younger ages than the sporadic cases, but logically this need not be the case. It is possible, for example, for someone to inherit the mutation that causes Huntington's disease yet develop the disease late in life.

A common feature of neurodegenerative diseases is that a subset of neurons develops intracellular aggregates of mutated and misfolded proteins, with the identity or mix of those proteins varying across the disorders. These aggregates are often considered to be harmful to the affected neurons, although some of the aggregates may be protective, and soluble forms of the mutated proteins can also be neurotoxic (Saudou et al., 1998; Krstic & Knuesel, 2013). Another common element of neurodegenerative diseases is extensive inflammation of the brain, which involves the activation of glial cells, especially microglia. These cells scan the brain for cellular damage and then respond by releasing inflammatory molecules and engulfing cellular debris. These activities are usually beneficial, but microglial cells may destroy reasonably healthy neurons if the inflammation is prolonged. Studying these complex processes in diseased human brains is difficult because human brain tissue usually becomes available only at autopsy, after many neurons have already died. Therefore, biologists have studied neurodegeneration in diverse animal and in vitro models.

In the following sections, I review this research for the four most intensively studied neurodegenerative diseases, focusing on the models that have been used and the translational efforts. I begin with Huntington's disease, because scientists understand it better than any other neurodegenerative disorder (especially at the genetic level). Huntington's disease therefore provides a good backdrop against which to discuss the other, more mysterious forms of neurodegeneration.

6.2.1 Huntington's Disease

Huntington's disease (HD) is relatively rare, affecting 2 to 3 of every 100,000 people worldwide (Pringsheim et al., 2012). It is characterized by erratic involuntary movements, called chorea, as well as cognitive decline. Sadly, the disease is relentlessly progressive and fatal. Despite its rarity, HD is worth discussing in some depth because it is arguably "the most curable incurable brain disorder" (Wild, 2016). The reason for this optimism is that HD results from a well-described mutation in a gene called *huntingtin*, which encodes a protein called Huntingtin.

Discovered in 1993, the *huntingtin* gene contains a sequence of the nucleotides C, A, and G that is normally repeated 14 to 29 times (MacDonald et al., 1993; Lee et al., 2012). Unfortunately, this CAG repeat has a tendency to expand across generations, and when it is repeated more than 35 times, the person is very likely to develop HD. The longer the repeat, the earlier symptoms tend to appear, although other factors clearly play a role as well (figure 6.2). Importantly, the CAG expansion is inherited in an autosomal dominant fashion, meaning that a single copy of the mutant gene suffices to cause HD in the offspring.

In the brain, HD is associated with the death of neurons in the striatum, which plays a major role in action selection and movement control, but other types of neurons also die as the disease progresses (Han et al., 2010; Rüb et al., 2014). Many of the remaining neurons exhibit proteinaceous aggregates that consist mainly of mutant Huntingtin (DiFiglia et al., 1997), but the extent to which these aggregates are neurotoxic or protective remains a matter of debate (Arrasate et al., 2004; Hoffner & Djian, 2015).

Most of what we know about mutant and wild-type *huntingtin* and the Huntingtin protein comes from animal and in vitro research. By now biologists have mutated or otherwise manipulated *huntingtin* in a wide variety of cell types from many different species, including yeast, fruit flies, pigs, marmosets, and macaques (Li & Li, 2012; Marsh et al., 2012; Hofer et al., 2018). However, the vast majority of studies on mutant *huntingtin* have involved genetically modified mice (Menalled et al., 2014).

A – Huntington's Disease and CAG Repeat Length

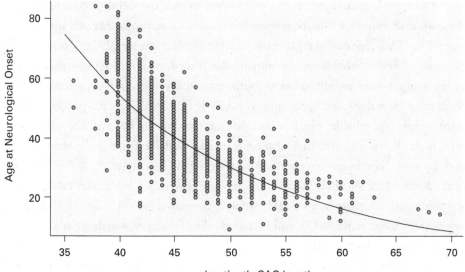

B – Distribution of Huntington's Disease *in vivo* Models

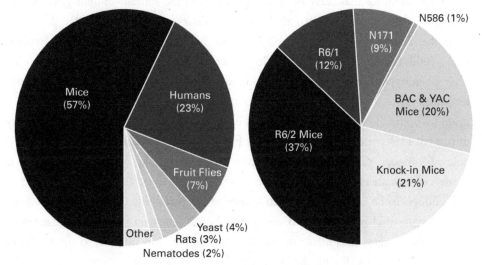

Figure 6.2

Huntington's disease (HD) in humans and in vivo models. (*A*) HD and CAG repeat length. The greater the length of the CAG repeat in the *huntingtin* gene (*htt*), the earlier humans develop the neurological symptoms of HD. Overall, CAG repeat length accounts for 67% of the variation in age at neurological onset; the remaining variation is poorly understood. (*B*) The distribution of species used to model HD in vivo (*left*) and the major genetically modified mouse models of HD (*right*), based on the author's analysis of roughly 1,000 data papers published between 1993 and 2017, retrieved through a PubMed search using the keywords "Huntington's disease gene" or "huntingtin protein" (conducted by the author in October 2018 and searching the full text for species or model mentions). Virtually all the species listed as "other" are vertebrates. The R6/1 and R6/2 mice are transgenic for a mutant form of human *huntingtin* (*mHtt*) exon-1. The BAC and YAC mouse lines express a full-length human *mHtt*, in addition to their endogenous wild-type *htt*. In contrast, the knock-in mice have one or both of their endogenous *htt* alleles modified to include an expanded CAG repeat or, in some cases, a mutant human *htt* exon-1. (A) Adapted from Gusella & MacDonald (2009).

One of the very first and most frequently studied animal models of HD is the R6/2 line of transgenic mice (figure 6.2B). In addition to their own, endogenous *huntingtin*, these animals express a mutant fragment of human *huntingtin* that was modified to carry 150 CAG repeats (Mangiarini et al., 1996). This degree of CAG expansion is extreme, relative to what is seen in humans, but it is effective for HD modeling insofar as the mutant mice rapidly develop motor symptoms and die prematurely. Indeed, these mice are widely used by researchers precisely because disease progression in these mice is rapid and reliable, which means that sample sizes can be small and publication rates high. The R6/2 mice have also been readily available from commercial sources and, by now, have been characterized in great detail (Hockly et al., 2003; Dragatsis et al., 2009; Menalled et al., 2012). Despite these advantages, R6/2 mice exhibit much less neuron death than human HD patients, develop different types of Huntingtin aggregates (André et al., 2017), and frequently die of epilepsy, which is not a common cause of death in human HD.

Since 1996, when the R6/2 mice were first produced, biologists have created many additional HD model mice that are less divergent from the human condition. These include mice that carry full-length versions of the mutant human gene instead of (rather than in addition to) their endogenous *huntingtin* and various knock-in mice that feature an expanded CAG sequence in their endogenous gene (Menalled et al., 2002; Southwell et al., 2013). An interesting aspect of these later models is that their phenotypes are generally mild and slow to develop. To counteract this problem researchers tend to use homozygous mice with CAG repeat lengths of 150 or more (Woodman et al., 2007; Southwell et al., 2017). In contrast, the vast majority of human HD patients are heterozygous for the mutation and possess CAG lengths below 65 (figure 6.2A). Thus, the efforts to create genetically "more accurate" mouse models of human HD have failed in the sense that the most accurate models are not sick enough to be convenient for laboratory research. In this context it makes sense that many researchers continue to use the R6/2 mice. The data from the more accurate models also suggest that mice are simply less sensitive than humans to the effects of mutant *huntingtin*, although this possibility is rarely discussed.

Some of the discoveries first made in HD models were later found to apply to human patients as well. Most notably, Huntingtin-containing aggregates were described in R6/2 mice before they were identified as such in HD patients (Roizin et al., 1979; DiFiglia et al., 1997). Overall, however, the data from the various HD models present a rather complex picture, replete with species, strain, and cell type differences. For instance, fragments of mutant Huntingtin tend to form insoluble aggregates, but the pattern and extent of aggregate formation vary across species, strains, and cell types (Tallaksen-Greene et al., 2005; Sawada et al., 2007; Malinovska et al.,

2015; Carty et al., 2015; Chongtham et al., 2020). Similarly, the tendency for the CAG repeats to expand somatically (i.e., in nongonadal tissues as the body develops and matures) varies across organs and within the brain across cell types; it also varies with the genetic background of the HD model mice (Telenius et al., 1994; Shelbourne et al., 2007; Gonitel et al., 2008) (figure 6.3). Unfortunately, such variation is rarely studied explicitly (Han et al., 2010; Ooi et al., 2019). In general, scientists have learned a great deal about what *can happen* when *huntingtin* is mutated, but what *actually happens* depends on the genomic and cellular context and is, therefore, not easily predictable (Bard et al., 2014).

If the preceding statement is true, then drugs that ameliorate Huntingtin toxicity in one model system might not work was well in a different model. The extent to which this is the case remains unclear because many therapies that appear promising in one model have not been tested in others. However, in the few cases where such tests have been carried out, they reveal only modest agreement (i.e., concordance) across models. For example, Varma et al. (2007) used a high-throughput screen to identify 29 compounds (out of around 43,000) that selectively reduce cell death in an immortalized striatal cell line expressing a mutant fragment of human *huntingtin*. They then tested these "hit" compounds in a second cellular HD model and in HD model worms (*Caenorhabditis elegans*); only four compounds were effective in multiple models, and only two were effective in all three. Additional testing in a rat model of HD then reduced the number of concordant hits to one, and this remaining drug was not effective in a yeast model of HD. Given this low concordance, it is not surprising that none of the four initially so promising therapies have, as far as I can tell, advanced to clinical trials. In fact, only a handful of drugs have entered clinical trials for HD (Hersch & Ferrante, 2004; Kumar et al., 2015), and virtually none have advanced to regulatory approval.

The principal exception is tetrabenazine, a drug that reversibly inhibits the uptake of dopamine (and, to a lesser extent, serotonin and norepinephrine; Paleacu, 2007). This drug was originally synthesized in the 1950s as an antischizophrenia drug; giving it to patients with HD alleviates their chorea, presumably because it depletes dopamine in neurons that project to the striatum and thus counteracts the effects of Huntington's disease on striatal function (Starr et al., 2008). Following up on this fortuitous observation, a clinical trial with 85 HD patients in 2006 revealed that tetrabenazine is relatively safe and effective (Huntington Study Group, 2006). These positive results led to US regulatory approval in 2008.

An interesting aspect of this process is that tetrabenazine had not been tested in HD models before the clinical trial because it had already been tested for safety in humans as part of the research on its potential to help schizophrenics. However, later work did show that tetrabenazine reduces symptoms in a line of HD model mice

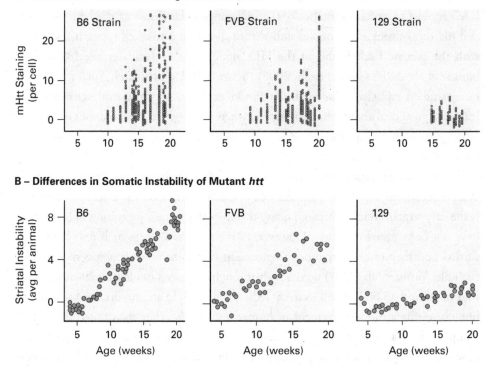

A – Differences in Mutant Huntingtin Accumulation

B6 Strain   FVB Strain   129 Strain

mHtt Staining (per cell)

B – Differences in Somatic Instability of Mutant *htt*

B6   FVB   129

Striatal Instability (avg per animal)

Age (weeks)   Age (weeks)   Age (weeks)

Figure 6.3

Strain differences in Huntington's disease (HD) model mice. (*A*) Differences in mutant *Huntingtin* protein (*mHtt*) accumulation. Ament et al. (2017) found that the intensity of antibody staining for mHtt in individual striatal cells increases with age in knock-in HD model mice (with 111 CAG repeats) but varies across strains. *mHtt* accumulation is fastest in the B6 (aka C57BL/6) mice and slowest in the 129 (aka 129/S2) mice. (*B*) Strain differences in *mHtt* accumulation are paralleled by differences in somatic instability, which is an increase over time in the length of the CAG repeat of the mutant *huntingtin* gene (*mHtt*). The graphs show how the degree of somatic instability for *mHtt* increases with age in the striatum of these mice (each data point represents an averaged instability index from a single mouse). It is unclear what causes the CAG repeats to expand in these postmitotic cells, but it probably involves malfunctions of the DNA repair machinery. Adapted from Ament et al. (2017).

(Wang et al., 2010), thus retroactively enhancing this model's predictive validity (see chapter 2). Later studies also showed that a related compound called deutetrabenazine works even better because it has a longer half-life and can, therefore, be taken less frequently; it was approved by the FDA in 2017 (Dean & Sung, 2018). No other drugs have been approved for the treatment of HD, but it is instructive to examine some trials that failed even though the preclinical research had been promising.

One failed HD trial examined coenzyme Q10, an antioxidant used by mitochondria to help them generate energy. Coenzyme Q10 had exhibited neuroprotective activity in several mouse HD models (Beal et al., 1994; Ferrante et al., 2002), and a small clinical trial showed it to be safe for human use (Feigin et al., 1996). Based on these findings, scientists in 2008 began a five-year clinical trial involving 600 HD patients. Unfortunately, the trial was halted early for futility, which means that preliminary analyses indicated a less than 5% chance of success (McGarry et al., 2017). In the meantime, a study on R6/2 mice also failed to demonstrate a positive effect for coenzyme Q10 (Menalled et al., 2010). Similar problems plagued research on creatine, which helps cells store metabolic energy. Creatine had yielded beneficial effects in three different mouse HD models, including two studies on R6/2 mice, but a clinical trial that enrolled 553 HD patients was halted for futility (Hersch et al., 2017).

A third important failed HD trial examined inhibition of sirtuin, which deacetylates various proteins. The effects of manipulating sirtuin expression had been studied in multiple HD models, including nematodes, flies, and mice, where it yielded somewhat contradictory results (Duan, 2013). In 2014, however, a sirtuin inhibitor called Selisistat was shown to be effective in both fly and mouse HD models (Smith et al., 2014). These data led to an exploratory clinical trial, which showed that Selisistat was safe (Süssmuth et al., 2015). Sadly, a larger subsequent trial revealed no statistically significant effects in key parameters, which caused the trial sponsor to halt the drug's development (Haider, 2018).

These trials may have failed because the test subjects were so far along in the disease that many neurons were irreversibly lost. It is also possible, however, that the HD model mice and flies are just too different from human HD patients to have strong predictive validity. To address this potential problem, it might be useful to examine monkeys expressing mutant *huntingtin*, as they might share with humans some features that are not shared by nonprimates. Indeed, researchers have used viruses to transfect mutant *huntingtin* fragments into the striatum of macaques (Palfi et al., 2007); a transgenic line of monkeys has also been created (Yang et al., 2008; Chan et al., 2014). These nonhuman primate models of HD have yielded some interesting results (Emborg, 2017), but most of the transgenic monkeys died long before they could reproduce.

An alternative approach is to focus more energy on the study of human cells, especially iPSCs derived from HD patients (HD iPSC Consortium, 2012; An et al., 2012; Csobonyeiova et al., 2020). Unfortunately, a major limitation of such studies is that they generally fail to replicate the in vivo interactions between different cell types (Creus-Muncunill & Ehrlich, 2019). This could be important because it is already known, for example, that mutant Huntingtin can affect the ability of striatal neurons to obtain important growth factors from cortical neurons and cholesterol from astrocytes (Zuccato et al., 2001; Valenza et al., 2015). In addition, the axonal connections between the cortex and the striatum are known to undergo significant changes as HD progresses (Raymond et al., 2011; Strong et al., 2012). Co-culture systems and organoids (see chapter 4) may be able to mitigate these limitations, but their development as HD models is only just beginning.

Meanwhile, recent efforts to develop HD therapies have focused mainly on suppressing *huntingtin* expression by means of synthetic RNAs and other oligonucleotides. Some of these approaches suppress both the mutant and the wild-type *huntingtin* allele (Carroll et al., 2011), which could be problematic because the wild-type protein performs a large variety of potentially important intracellular functions. Fortunately, studies in mice and nonhuman primates suggest that this will not be a significant problem (Drouet et al., 2009; Kordasiewicz et al., 2012; Stanek et al., 2014), and tests in a few humans have so far revealed no negative effects (Tabrizi et al., 2019); a large phase III trial is under way. A second concern is the possibility of off-target effects, which is to say the unintended suppression of other genes (Monteys et al., 2014). A third open question is how long the suppression will last, although using viruses to transfect the RNA constructs into the target cells should have quite durable effects (Wild & Tabrizi, 2017).

One way to avoid several of these problems is to use genome editing techniques to "correct" the *huntingtin* mutation in the most vulnerable cells (Monteys et al., 2017), but this genome editing approach is still in the preclinical stage. Overall, the history of HD therapy development offers both hope and lessons in humility.

6.2.2 Amyotrophic Lateral Sclerosis

Another intensively studied neurodegenerative disease is amyotrophic lateral sclerosis (ALS), which in 2014 affected about one in every 20,000 people in the United States (Mehta et al., 2018). It is also known as Lou Gehrig's disease, after a widely admired American baseball player who died from the disease in 1941. The principal symptom of ALS is progressive paralysis, caused by the degeneration of motor neurons, first in spinal cord but then also in the cerebral cortex. The disease typically manifests when people reach their 50s, although Lou Gehrig began to show symptoms at the age of 36.

Five percent to 10% of cases are classified as familial. Of those, 12% to 23% can be linked to mutations in a gene called *superoxide dismutase-1* (*SOD1*) (Andersen 2006; Tan et al., 2017; Gill et al., 2019). Another 16% of the familial cases have been linked to mutations in a gene called *C9orf72*, which also account for 6% to 8% of the sporadic ALS cases (presumably through mutations that were not inherited). In addition to *SOD1* and *C9orf72*, ALS has been linked to a handful of other genes, including *TDP43*, but mutations in these genes are even less common. Remarkably, many of the ASL-linked proteins colocalize with one another in the intraneuronal aggregates characteristic of ALS, even if those proteins are wild type (i.e., not mutated). Whether these aggregates are neuroprotective or toxic remains, as in the case of Huntington's disease, a matter of debate (Strong et al., 2005; Blokhuis et al., 2013).

To study the mechanisms underlying ALS and seek potential therapies, biologists have created a variety of transgenic mice, most of which express mutant forms of human *SOD1*. In these mice the transgene is usually inserted as multiple copies (as many as 25), which is quite different from the human condition. Moreover, transgene expression is abnormally high in most SOD1 mouse lines, and the degree of motor neuron degeneration and paralysis covaries with the level of mutant *SOD1* activity (Gurney et al., 1994; Gurney 1997; Dal Canto & Gurney, 1997). Specifically, the mice with the highest number of transgene copies and highest levels of *SOD1* activity die earliest and exhibit some pathologies not seen in human ALS. A related problem is that copy numbers and gene activity levels also vary between supposedly similar mouse lines, and most studies fail to specify those crucial aspects of their research animals (Zwiegers et al., 2014). A more recently developed mouse ALS model expresses mutant versions of human *TDP43* and exhibits progressive symptoms similar to those of human ALS (Wegorzewska et al., 2009). However, further study of these mice did not replicate some aspects of the earlier reports (Perrin, 2014) and revealed that these mice actually die from acute bowel obstruction, rather than skeletal muscle wasting (Esmaeili et al., 2013). Thus, the principal mouse models of ALS are not as similar to human ALS as one might hope. Some non-mammalian animal models of ALS have been created over the years, but their contributions to ALS research have thus far been minor (Patten et al., 2016).

In vitro models of ALS have historically involved a variety of immortal cell lines, but these tend to be quite different from the motor neurons that selectively die in ALS. Many researchers have, therefore, turned their attention to the study of human iPSCs that can be differentiated into motor neurons. Such studies have revealed, for example, that iPSC-derived motor neurons are more sensitive than some other types of neurons to *SOD1* aggregates (Benkler et al., 2018). Another study found that iPSC-derived motor neurons from a patient with sporadic ALS developed *TDP43*-positive aggregates that were quite similar to those observed in the patient at autopsy (Burkhardt et al., 2013). A key advantage of these iPSC models is that the cells can be subjected to high-throughput screens for drugs that modulate aggregate formation or cell death.

However, iPSC-based models also have limitations (Preza et al., 2016; Volpato & Webber, 2020) (see also section 4.2.5). As already noted in the section on Huntington's disease, patient-derived iPSCs vary genetically between patients. The methods used to reprogram adult cells into stem cells add to this heterogeneity because they alter the cells' genome and epigenome in ways that are not entirely predictable. The protocols used to differentiate iPSCs also vary between laboratories, and any given protocol is likely to produce cells that differ from the in vivo target cells, at least in maturity. Finally, when iPSCs are cultured for several generations, they tend to accumulate mutations that adapt them to the culture conditions.

The development of therapies for ALS has been a frustrating affair. More than 70 compounds have been reported to have beneficial effects in SOD1 model mice, but a meta-analysis has indicated that many of these studies were "methodologically poor" and subject to publication bias, meaning that negative results tended not to be published (Benatar, 2007). A nonprofit biotech company then tested dozens of promising drugs in thousands of SOD1 mice, but was unable to replicate the previously published results (Scott et al., 2008; Perrin, 2014) (see figure 1.1B). Even riluzole, which had long been the only drug approved for the treatment of ALS, does not appear to be effective in SOD1 mice, even though it has at least a modest beneficial effect in ALS patients. In retrospect, this is perhaps not surprising, given that the SOD1 mice generally express so much mutant *SOD1* that any treatments for them would have to be very robust (van der Worp et al., 2010). Lithium, which is often used to treat bipolar disorder (see section 6.3.2), did show promising effects in SOD1 mice and even in a pilot study of ALS patients (Fornai et al., 2008). These findings prompted three larger clinical trials on the effects of lithium on ALS, but all of them failed (Petrov et al., 2017).

In spite of all the frustration, ALS therapy development is not hopeless. Particularly promising is that the antioxidant edaravone was recently approved for ALS in both Japan and the United States. European regulators did have some concerns, however, and the drug's producer responded by withdrawing its application. Perhaps the costs of obtaining the requested evidence could not be justified, given the relative rarity of ALS. Another promising drug is masitinib, an anti-inflammatory kinase inhibitor that suppresses microglia. It can reduce some toxic interactions between microglia and motor neurons in vitro and is beneficial in SOD1 rats, reducing symptoms even when it is administered after the onset of paralysis (Trias et al., 2016). Masitinib (in combination with riluzole) passed a phase II/III trial in 2017, and large phase III trial is currently underway.

A third intriguing advance is a recent phase II trial of a drug cocktail that reduces cellular stress and seems to slow ALS progression, at least slightly (Paganoni et al., 2020). It is worth noting that this trial is the first to be funded through the popular Ice Bucket Challenge (2021) to support ALS research.

6.2.3 Alzheimer's Disease

Alzheimer's disease (AD) is by far the most common neurodegenerative disorder, with roughly 6 million patients currently in the United States alone and affecting roughly 10% of people older than 65 (Alzheimer's Association, 2019). Its hallmark symptom is progressive memory loss, but late stages of AD entail a general cognitive decline. Delineating AD from other age-related dementias is difficult, and a definitive diagnosis still requires postmortem examination (McKhann et al., 2011).

Within the brain, AD manifests as a progressive loss of neurons and synaptic connections, especially in the hippocampus and adjacent cortical areas. It also involves the degeneration of forebrain neurons that use acetylcholine as their neurotransmitter and project to the cortex. Another common feature of AD is the formation of two types of insoluble protein aggregates. Some aggregates consist mainly of beta-amyloid; these are called amyloid plaques. The other principal aggregates in AD brains are neurofibrillary tangles, which consist of twisted strands of a protein called tau.

Familial cases of AD have been linked to mutations in amyloid precursor protein (APP) and presenilin, which is involved in APP processing. These findings helped to spawn the amyloid cascade hypothesis, which posits that the primary cause of AD is the formation of beta-amyloid from APP, and that this beta-amyloid then accumulates in the form of amyloid plaques, damages neurons, and somehow triggers the formation of neurofibrillary tangles (Hardy & Higgins, 1992). Although most of the research on AD is based on this hypothesis (Herrup, 2015), some AD patients have few amyloid plaques, while others possess numerous plaques but relatively intact memory (Dickson et al., 1992; Giannakopoulos et al., 2003). Moreover, familial AD accounts for only about 1% of all AD cases; among these, only about 5% are linked to mutations in APP or presenilin (Drummond & Wisniewski, 2017; Mullane & Williams, 2019).

This leaves most cases of AD without a clear genetic basis. Numerous genetic and environmental risk factors for AD have now been identified (including old age, stress, genetic variation in apolipoprotein E, and a lack of mental and physical exercise), but synthesizing all this information into a coherent mechanistic picture has remained a serious challenge (Medina et al., 2017). In spite of this complexity, most of the preclinical AD research has focused on models that express mutant APP and presenilin. The hope has been that these genetic models can provide insights and yield potential therapies that will translate to AD resulting from other, more mysterious causes.

The vast majority of AD models are transgenic mice that express mutant forms of human APP (Jankowsky & Zheng, 2017; Mullane & Williams, 2019). Most of these mice display some AD-like symptoms, including amyloid deposition in the brain and reduced performance on some memory tasks. However, the mice generally express abnormally high levels of the transgene and fail to develop neurofibrillary tangles or extensive neuron loss. To overcome this limitation, scientists have created mice that combine as many as five different mutations in two or even three different AD-linked genes, including *presenilin* and *tau* (Oddo et al., 2003; Oakley et al., 2006). Triple transgenic mice do exhibit neurofibrillary tangles, but neuron loss is still quite limited.

Moreover, these mice are poor genetic models of the human disorder insofar as AD patients usually carry just a single mutation, in either APP *or* presenilin, and mutations in *tau* are generally associated with frontotemporal dementia rather than AD. Given these differences, it is not surprising that the changes in neuronal gene expression observed in AD model mice (compared with wild-type mice) tend to be quite different from those observed in humans with AD (Burns et al., 2015; Hargis & Blalock, 2017) (figure 6.4).

A newer, second-generation of mouse AD models modifies the endogenous AD-linked genes of mice rather than inserting additional human genes (Reaume et al., 1996; Baglietto-Vargas et al., 2021). These knock-in mice avoid the overexpression and random transgene insertion problems of the earlier models; but, as in the case of Huntington's disease (see section 6.2.1), their phenotypes tend to be much more subtle. To solve this problem, researchers have created knock-in mice that carry multiple mutations (Siman et al., 2000; Flood et al., 2002; Saito et al., 2014); unfortunately, such combinatorial models sacrifice fidelity at the genetic level for similarity at the neurobiological and behavioral levels.

The observation of milder phenotypes in knock-in AD model mice suggests that mice may be less vulnerable than humans to the effects of AD-linked mutations. This hypothesis is supported by some evidence (Rosen et al., 2016; Espuny-Camacho et al., 2017), but it has not been tested thoroughly. A potentially related observation is that beta-amyloid aggregates in mice tend to be more soluble than their human counterparts, have different binding properties, and generate less inflammation (McGeer, 2003; Jucker, 2010; Drummond & Wisniewski, 2017). Species differences in the *tau* gene may also play a role (Umeda et al., 2014). Alternatively, mice may simply not live long enough to develop AD unless the underlying mechanisms are artificially enhanced. The problem with this proposition is that life span generally scales with body size (Speakman 2005), and mice do age much faster than humans. The transcriptional signature of brain aging, for example, is quite similar between mice and humans despite the differences in chronological age (Burns et al., 2015).

The hypothesis that humans are exceptionally vulnerable to AD is consistent with the observation that few other species naturally develop this disease (Walker & Jucker, 2017). Several nonhuman primate species do accumulate amyloid deposits in their brains as they age, and very old macaques and chimpanzees exhibit neurofibrillary tangles (Edler et al., 2017; Paspalas et al., 2018), but whether these symptoms are associated with neuron loss remains unclear. Some very old mouse lemurs (i.e., strepsirrhine primates) exhibit extensive neurodegeneration, but their pathology is focused on the frontal lobes rather than the temporal lobe (where cell death is most prominent in human AD) (Pifferi et al., 2019).

A – Gene Expression Changes in AD vs. AD Models

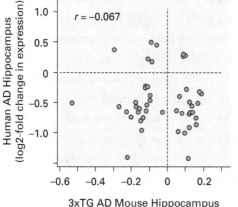

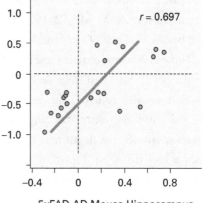

B – Correlations between Transcriptome Study Results

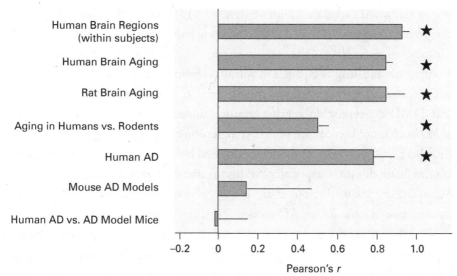

Figure 6.4

Transcriptomic comparisons between human Alzheimer's disease (AD) and mouse AD models. (A) Gene expression changes in AD versus AD models. Hargis and Blalock (2017) compared the transcriptional profiles obtained from the hippocampus of human AD patients to those obtained from the hippocampus of two transgenic mouse AD models (3xTG—triple transgenic mice; 5xFAD—mice carrying five mutations linked to familial AD). Each data point represents a single gene that is differentially expressed in both species. Whereas the transcriptional profile of the 5xFAD model correlates positively with that of human AD, the profile of the 3xTG model does not. (B) Correlations between transcriptome study results. Based on many such pairwise correlation analyses, Hargis and Blalock determined that the transcriptional profiles for young versus old rodents correlates pretty well with the profiles associated with human aging (stars indicate a statistically significant correlation), but that the mouse AD models correlate poorly with one another and, on average, are a poor reflection of human AD. Adapted from Hargis and Blalock (2017).

Outside of primates, AD-like phenotypes are rare. Some dogs do develop senile dementia and AD-like pathology, but they lack neurofibrillary tangles (Head 2013; Prpar Mihevc & Majdič, 2019). Structures similar to tangles have been reported in the brains of sheep, goats, and leopard cats, but whether these animals develop other AD-like symptoms remains unclear (Braak et al., 1994; Nelson et al., 1994; Chambers et al., 2012). Rats and mice do not normally develop signs of AD, but common degus (which are also rodents) do develop AD-like cognitive deficits, along with amyloid deposits and neurofibrillary tangles, when they get very old (Hurley et al., 2018). Of all these species, the degus may be the ethically most acceptable genetically natural AD model to work with. However, work with this model would be slow, because only a subset of individuals develop the disease and symptoms take years to emerge (much as in humans).

Given the challenges of working with natural AD models, it is not surprising that the development of AD therapies has relied mainly on the genetically modified mouse models. More than 200 different therapies have been reported as being effective in mouse AD models. Unfortunately, of 413 clinical trials examining 244 such compounds between 2002 and 2012, only one passed through to marketing approval (Cummings et al., 2014). The years since then have only produced additional failures, prompting some drug companies to withdraw from neurological research entirely. The one success during this period was memantine, an uncompetitive N-methyl-D-aspartic acid (NMDA) receptor blocker that improves some cognitive and behavioral symptoms of AD but has not been shown to delay progression of the underlying disease (Thomas & Grossberg, 2009). Intriguingly, memantine had been used since 1982 to treat a variety of other brain disorders, especially Parkinson's disease, but was found to be effective for AD in the late 1980s (Parsons et al., 1999). Subsequent studies then showed that it is also effective in some mouse AD models (Martinez-Coria et al., 2010).

The only other drugs that have received regular approval for AD are inhibitors of acetylcholinesterase (AChE), an enzyme that is critical for the synthesis of acetylcholine. These AChE inhibitors (e.g., donepezil) had long been used to boost cognition in humans, but their use for the treatment of AD accelerated after scientists discovered that cholinergic neurons (i.e., neurons that use acetylcholine as their transmitter) die relatively early in AD (Summers et al., 1986; Contestabile, 2011). Although these drugs modestly boost cognition in late stages of AD, they (like memantine) do not delay disease progression (Geerts, 2009; Cooper et al., 2013).

Of the clinical AD trials that failed, most involved drugs that were designed to reduce the synthesis of beta-amyloid or increase its clearance. Often this entailed either vaccination with beta-amyloid or the injection of synthetic antibodies against this peptide (Nicoll et al., 2003; McGeer, 2003). Although the tested drugs had been effective in AD model mice (Schenk et al., 1999; Oddo et al., 2004) and generally reduced levels of beta-amyloid in humans, they ultimately failed; several did so repeatedly (e.g., solanezumab). It is possible that the clinical trials failed because the AD patients were already too far advanced in their disease for the drugs to make much difference. However, several recent trials administered the putative therapies to patients with mild cognitive impairment, which can be a sign of incipient AD; sadly, they failed as well (Lowe, 2020; Bhandari 2020). A lot of hope is currently riding on aducanumab (aka Aduhelm™), another human antibody against beta-amyloid. Although aducanumab apparently did not perform entirely as hoped, it may have limited benefits. (Indeed, aducanumab received "accelerated approval" from the FDA in the summer of 2021, after this book had gone to press. However, this approval was an unusually controversial decision, going against the recommendation of the FDA's own advisory committee, and called for a post-approval, phase 4 confirmatory trial [Cavazzoni, 2021; Steinbrook, 2021].)

In light of these observations, alternatives to the amyloid cascade hypothesis should be considered carefully (Mullane & Williams, 2013). For example, it is interesting to note that mice expressing function-blocking antibodies against nerve growth factor (NGF) develop amyloid plaques, neurofibrillary tangles, and extensive cell loss as they get old (Capsoni et al., 2000). This process probably involves the cholinergic neurons that die in AD, because those neurons require NGF for survival. This unusual AD model is not entirely incompatible with the amyloid cascade hypothesis, but it suggests that modulating growth factors may be more effective in the treatment of AD than manipulating levels of beta-amyloid (Blurton-Jones et al., 2009). Of course, clinical trial failures may also stem from issues with the quality of the preclinical AD research, which does appear to suffer from systemic publication bias and poor, inconsistent methodology (Egan et al., 2016).

6.2.4 Parkinson's Disease

Parkinson's disease (PD) is the second most common neurodegenerative disease, affecting roughly 1% of the population in industrialized countries by age 65 and 4% to 5% by age 85 (Eriksen et al., 2003; de Lau & Breteler, 2006). Its principal symptoms are body rigidity, slow movements, a flexed body posture, tremor of the hands, and sleep disturbances. Within the brain, the principal correlate of PD is the progressive degeneration of dopaminergic neurons in the substantia nigra (SN), which project heavily to the striatum (Bové & Perier, 2012). The damaged SN neurons, as well as some less vulnerable neurons in other brain regions, tend to contain Lewy bodies, which are intracellular aggregates of various proteins, including alpha-synuclein (Spillantini et al., 1997).

Approximately 10% of all PD cases are classified as familial, and 30% of these can be linked to mutations in a specific gene. One of these is the gene that encodes alpha-synuclein (aSyn), which is a major component of the Lewy bodies mentioned in the previous paragraph. Subsequent research revealed that even overexpression of wild-type *aSyn* can trigger the formation of Lewy bodies and that aSyn accumulations are linked to increased cellular stress, but precisely how aSyn damages neurons remains debatable. A second gene linked to PD is *leucine-rich repeat kinase 2* (*LRRK2*; aka *PARK8*). Mutations in this gene are found in a greater percentage of PD patients than mutations in a*Syn* (Klein & Westenberger, 2012), but the problems caused by these mutations are likewise obscure (Tolosa et al., 2020). The mechanisms underlying sporadic PD apparently include various pesticides (e.g., paraquat and rotenone) and toxins that invade the nervous system via the gut. However, these nongenetic causes remain poorly understood and somewhat controversial (Ascherio & Schwarzschild, 2016; Rietdijk et al., 2017; Parashar & Udayabanu, 2017).

Researchers have created a variety of transgenic mice expressing mutant versions of human *aSyn* or overexpressing wild-type *aSyn* (Chesselet et al., 2012). They have also created mice that express mutant *LRRK2* (Jagmag et al., 2016). These genetically modified PD models do show some PD-like symptoms, including a reduction of striatal dopamine levels, but neuron loss in the SN is minor at best (Fleming et al., 2005; Potashkin et al., 2010).

Greater model fidelity has been achieved with a nongenetic model that was discovered by accident. A rogue graduate student in 1976 accidentally injected himself with a bad batch of a synthetic opioid and rapidly developed symptoms of advanced PD, including the loss of dopaminergic neurons (Davis et al., 1979). A similar incident with several heroin addicts in San Francisco produced analogous results and led to the identification of a putative chemical culprit called MPTP (1-methyl-4-phenyl-1,2,3,6-tetrahydropyridine; Langston et al., 1983). To the initial consternation of scientists, giving MPTP to rats did not cause PD-like symptoms, but subsequent experiments in macaques and several other nonhuman primates did kill the dopamine neurons and cause PD-like behavioral symptoms (Burns et al., 1983; Bové & Perier, 2012; Blesa et al., 2018). Further studies then clarified that rats and a few strains of mice are less susceptible than primates to MPTP poisoning, for reasons that still remain somewhat unclear (Giovanni et al., 1994; Hamre et al., 1999; Smeyne et al., 2005; Kasahara et al., 2013).

Biologists have also created non-mammalian and in vitro models of PD (Delenclos et al., 2019). They have, for example, expressed various PD-linked genes in fruit flies and observed that such flies exhibit protein aggregates resembling Lewy bodies, a progressive degeneration of dopaminergic neurons, and a loss of motor control (Feany & Bender, 2000). These and other non-mammalian PD models (Pienaar et al., 2010) were developed with an eye toward high-throughput drug screening. A yeast PD model, for instance, has been used in a large genetic screen to identify 55 genes that increase human *aSyn* toxicity and 29 that decrease it (Yeger-Lotem et al., 2009). Because many of the yeast genes have human homologs, these data may reveal useful information about the mechanisms that lead from *aSyn* to cellular toxicity. However, as mentioned previously, molecular homology need not imply conservation of molecular interactions. Moreover, yeast lack an endogenous *aSyn* homolog (Pienaar et al., 2010). One might argue that this actually makes yeast an ideal cellular "clean room" in which to study the functions of human *aSyn* without the confounds caused by an endogenous homolog (Hofer et al., 2018), but that clean room is surely quite different from *aSyn*'s cellular context in humans. We will come back to this in chapter 7.

The most common and oldest PD therapy is the drug L-dopa, which easily traverses the blood-brain barrier and is converted to dopamine within the brain. Once in the brain, L-dopa causes any remaining dopaminergic neurons to synthesize and release more dopamine, which ameliorates the symptoms of PD but does not slow the neurodegeneration. The animal research that revealed L-dopa's effectiveness in the late 1950s (Carlsson et al., 1957) relied on mice that had been given reserpine, which rapidly depletes the brain of dopamine and several other neurotransmitters (notably serotonin and norepinephrine). This reserpine model of PD fell out of favor as the transgenic and MPTP-based models emerged, but it predicted L-dopa's efficacy in humans quite well (Leal et al., 2016). Unfortunately, long-term use of L-dopa generally requires increases in dosage over time and can have negative side effects, including uncontrolled movements and impulsive behavior (Ahlskog & Muenter, 2001).

The principal treatment for L-dopa-resistant PD is deep brain stimulation (DBS). DBS involves implanting an electrode deep in the human brain and applying electrical stimuli to the targeted area. The method was originally developed to address diverse ailments (notably chronic pain), but two important refinements made it highly effective for PD (Gardner, 2013). First, scientists realized that high-frequency stimulation suppresses neuronal activity for some time afterward, rather than increasing it (as one might intuitively expect). Second, researchers focused their stimulation on the subthalamic nucleus, where DBS is most effective for relieving the motor symptoms of PD (Nashold & Slaughter, 1969; Benabid et al., 2009). Some observers have argued that these refinements could have been achieved without the benefit of prior animal research (Bailey, 2014), but the scientists most responsible for developing DBS into a PD therapy have argued forcefully that their approach was inspired by the earlier discovery that MPTP-treated monkeys exhibit a dramatic increase in subthalamic activity (Benabid et al., 2015a, 2015b; Blesa et al., 2018). If the loss of dopaminergic neurons increases activity in the subthalamic nucleus, they argued, then it stands to reason that lesioning or inhibiting this nucleus might reverse or alleviate some of the MPTP-associated symptoms. This was indeed the case (Bergman et al., 1990; Benazzouz et al.,

1993), and human trials soon followed. DBS was FDA approved as a general PD treatment in 2002 and, by 2013, it had been administered in more than 40,000 PD patients (Gardner, 2013). Closely related to DBS is the idea of using gene therapy to suppress subthalamic neurons, but this approach is still under development (Bartus et al., 2014).

Physicians have also tried to treat PD with cell implants. Beginning in the 1970s, scientists grafted cells from the ventral midbrain of very young embryos (where the substantia nigra usually develops) into the striatum of PD model rats, monkeys, and eventually PD patients (Perlow et al., 1979; Sladek et al., 1986; Hagell et al., 1999). They observed good survival and integration of the implanted cells, increased levels of dopamine, and long-term improvements in behavior (W. Li et al., 2016). However, two clinical trials in 2001 and 2003 failed to show the hoped-for clinical benefit, and some subjects actually developed motor problems that probably stemmed from excessive uncontrolled dopamine release (Olanow et al., 2003). These failures dampened enthusiasm for this form of PD therapy, but advances in stem cell biology have now rekindled hope. Specifically, researchers have begun to implant stem cell–derived dopaminergic neurons into the striatum of various PD models (Kriks et al., 2011; Kikuchi et al., 2017; Fan et al., 2020). The results have been promising, and new clinical trials are underway (Barker & TRANSEURO Consortium, 2019).

6.3 NEUROPSYCHIATRIC DISORDERS

Many neurological disorders are not accompanied by neuronal injury or degeneration. People tend to refer to them as behavioral, mental, or psychiatric disorders, but given that the brain is the principal organ of behavior control and thought, they can also be described as "brain disorders" (Insel et al., 2010). Delineating these disorders from one another is difficult because their symptoms tend to overlap and their genetic links remain largely obscure (Nestler & Hyman, 2010; Kas et al., 2007; Cuthbert & Insel, 2013). Definitions aside, neuropsychiatric disorders are real and cause enormous suffering. We here focus on just three of these disorders, namely major depressive disorder, bipolar disorder, and schizophrenia. They are all among the 10 most burdensome psychiatric disorders in terms of years living with disability (see figure 6.1).

6.3.1 Major Depressive Disorder

A key feature of major depression is that the depressive symptoms last for two or more weeks and seriously impair the patient's ability to work or fully function socially. Sadly, it afflicts roughly 14% of people in the United States at some point in their adult life (with women being more susceptible) (Kessler et al., 2012). Scientists used to think major depression is caused by a depletion of the brain's monoamine transmitters (which include dopamine, norepinephrine, and serotonin), but more recent work has called that relatively simple hypothesis into question without providing a new and widely accepted alternative. Genetic analysis linked 15 different genes to major depression in 2016, and a more recent, larger study identified 200 such genes (Hyde et al., 2016; Cai et al., 2020). However, all these genetic linkages are weak, meaning that mutations in each gene only slightly increase the risk of major depression. Because scientists know so little about the biological mechanisms underlying depression, finding or creating models for this debilitating disease is hampered by uncertainties.

Drugs that deplete one or more of the monoamine transmitters (e.g., reserpine) can induce an inability to express pleasure and other depression-like symptoms in animals (McKinney & Bunney, 1969). So can injections of lipopolysaccharides (LPS), which cause inflammation not just in the body (see chapter 5, section 5.1.4) but also in the brain (Yirmiya, 1996). Another way to trigger depression-like symptoms in rodents is to surgically remove their olfactory bulb, which deprives these animals of their main means for sensing external stimuli (Kelly et al., 1997). Although many of the symptoms emerging in these models can be alleviated by some known anti-depressant drugs (to be discussed shortly), the various drug- and surgery-induced models of depression do not provide a coherent picture of the mechanisms underlying major depression.

More promising are efforts to induce depression-like symptoms though manipulation of an animal's environment. Isolating individuals of highly social species (e.g., parrots or gorillas) can cause them to appear profoundly unhappy for long periods of time and to deteriorate physically (McKinney & Bunney, 1969). Similar effects can be achieved by exposing animals to other stressors, such as a more dominant "bully," repeated electric shocks, physical restraint, or wet bedding (Willner, 1984). These manipulations cause depression-like symptoms especially if they are unpredictable, inescapable, and combined with one another. Although such models of depression are euphemistically called "mild chronic stress" (Willner, 2005), they typically cause the animals to appear as if they had given up and lost interest in things they used to like. To quantify such changes in behavior, scientists have developed various tests that measure, for example, how long an animal will struggle when suspended by its tail or whether it prefers to drink sugar water over the usual fare; it is important, however, not to conflate these tests with the depression-inducing manipulations (Kalueff et al., 2007; Nestler & Hyman, 2010; Demin et al., 2019). The predictive validity of environmentally induced models of depression is demonstrated by the observation that the animals often "get better" when treated with drugs that are effective for depression in humans. Nonetheless, these models have thus far played only a minor role in the development of major new types of antidepressants.

The first major antidepressant drug was iproniazid (aka Marsilid). This compound was originally developed to treat tuberculosis and shown in vitro and in several mammalian species to be an inhibitor of monoamine oxidase, the enzyme that degrades monoamines (Zeller & Barsky, 1952). Observant clinicians in the 1950s discovered that, as a side effect, iproniazid often made patients euphoric and boosted their appetite. These observations prompted tests on depressed subjects in the late 1950s (Loomer et al., 1957), and the positive results of these studies led to the widespread use of iproniazid as an antidepressant (Sandler, 1990). Unfortunately, the drug causes dangerous spikes in blood pressure when it interacts with dietary tyramine, a compound found in cheese. On account of this cheese reaction, iproniazid was withdrawn from the market in 1961.

The second major antidepressant was imipramine, which is considered a tricyclic antidepressant because of the three rings in its chemical structure. Imipramine is structurally related to chlorpromazine, which is used to treat schizophrenia (which we will discuss shortly), and it was originally synthesized as part of a search for more effective antischizophrenia drugs. When tested in the 1950s on approximately 500 humans with a wide variety of mental disorders, imipramine was ineffective against schizophrenia but effective against major depression (though the improvement over placebo was only approximately 30%; Maxwell & Eckhardt, 2012). Imipramine was approved for treatment of major depression in 1959 and spawned a long series of studies looking for more effective tricyclic antidepressants. Many such drugs have come to market, but they all appear to be similarly effective (Arroll et al., 2005).

The third major class of antidepressant drugs to be discovered are the serotonin reuptake inhibitors (SRIs), which reduce the ability of neurons to recycle previously released serotonin and thus cause the released serotonin to hang around longer. The possibility that serotonin might have a special role in depression dates back to the 1960s, when scientists discovered that depressed suicide victims had low levels of serotonin in the brain. In addition, the earlier antidepressant imipramine had been suggested to function, at least in part, as an SRI (Wong et al., 2005). These observations led to an explicit search for additional serotonin reuptake inhibitors, using in vitro preparations of rat brain and in vivo assays. By 1974, fluoxetine (aka Prozac) emerged as one such drug (Wong et al., 1974). Studies with animal models and early human tests indicated good safety but poor effectiveness. After several additional trials, however, fluoxetine was FDA approved in 1987 for the treatment of major depression. Subsequent research produced a variety of additional SRIs, some of which were even more selective for serotonin (as opposed to other monoamines). However, nonselective SRIs actually tend to work better against depression than highly selective SRIs (Roth et al., 2004). Particularly puzzling is that these drugs, like the tricyclic antidepressants, take several weeks before the depression lifts (Machado-Vieira et al., 2010), even though the direct cellular effects can be observed almost immediately.

A very recent addition to our armamentarium against major depression is ketamine, which blocks NMDA-type glutamate receptors (Sleigh et al., 2014). Ketamine was synthesized in the 1960s as part of a search for new anesthetics. It was tested on diverse animals and first given to humans in 1964. Deemed relatively safe and effective, it was then used widely as an anesthetic in veterinary medicine. Even in early studies, however, doctors noticed that ketamine causes human subjects to become detached from reality, which explains why it is sometimes abused (e.g., under the name of Special K). Meanwhile, scientists in the 1990s began to suspect, on the basis of research on both humans and mice, that depression might be linked to changes in NMDA receptors. Prompted by this hypothesis, they gave ketamine to a few depressed patients and observed profoundly positive effects (Berman et al., 2000). A larger follow-up study confirmed these findings (Zarate et al., 2006), and ketamine is now recognized as a highly effective treatment for major depression, especially in cases where other treatments have failed. One of its significant advantages over other antidepressants is that its positive effects are felt within a few hours.

Finally, it is interesting that one of the first, and still one of the most effective, treatments for major depression is electroconvulsive therapy (ECT, aka electroshock therapy). It is still unclear exactly how ECT exerts its antidepressant effects, and it certainly has a host of risky side effects. However, it is widely recognized as being effective against major depression and, when administered in conjunction with anesthesia, is recommended as a measure of last resort (Roth et al., 2004; Li et al., 2020).

6.3.2 Bipolar Disorder

Bipolar disorder (BD) affects at least 1% to 2% of the world's population (Yutzy et al., 2012; Brown, 2019) and is characterized by dramatic mood swings that recur at variable intervals, ranging from weeks to years (McCormick et al., 2015). Periods of deep depression tend to alternate with periods of mania, which feature prolonged episodes of great energy and well-being as well as elevated feelings of self-worth, racing thoughts, and often great creativity. Indeed, the list of writers, poets, painters, and musicians who likely struggled with BD is long ("List of People with Bipolar Disorder," 2021). That said, BD is a severely debilitating disease that often ends in suicide. Unfortunately, its underlying mechanisms remain mysterious. Although monozygotic twins are far more likely than dizygotic twins to share a diagnosis of BD, only a few genetic mutations have been linked to this disease, and each of them only slightly elevates the disease risk (Barnett & Smoller, 2009; Orrù & Carta, 2018). It is not surprising, therefore, that animal models of BD are relatively scarce and often of dubious validity (Malkesman et al., 2009; Beyer & Freund, 2017). For example, some authors have tried to model BD by giving animals cocaine or amphetamines, but this only mimics some aspects of the mania and fails to capture the cyclic nature of BD. Others have manipulated some of the BD-linked genes in mice, and some of these animals respond to lithium, which is the main drug used to treat human BD. However, the extent to which these models can lead to novel therapies remains as yet unknown.

The story of how lithium came to be "psychiatry's most consistently effective medicine" (Draaisma, 2019, p. 584) is both strange and complex. Lithium salts have a long history as treatments for diseases such as gout, but the idea that they could be used to treat depressive disorders originated with the Danish psychiatrist Frederik Lange in 1894 (Shorter, 2009). The idea was then ignored for roughly 50 years, until a psychiatrist named John Cade injected the urine of mental patients (or an 8% solution of urea) into guinea pigs and found that treating those animals with lithium reduced the injection's toxicity (Cade, 1949; Brown, 2019). He further noticed that lithium injected into otherwise untreated animals rendered the animals lethargic, though "fully conscious" (Cade, 1949). Being interested in BD, Cade then made a huge conceptual leap and administered lithium to 10 patients suffering from mania. Although Cade reported that many of his patients became "practically normal" after a few weeks, most readers were skeptical. A few years later, however, Mogens Schou and his collaborators confirmed the essence of Cade's report (Schou et al., 1954). Still, skepticism persisted. It was only in 1970, after Schou and his collaborators showed in a large placebo-controlled clinical trial that lithium prevents the recurrence of manic/depressive episodes (Baastrup et al., 1970) that the medical community took note.

High doses of lithium do cause some serious side effects, but these can be controlled by monitoring drug levels in the blood. Although several alternatives to lithium have been developed over the years (e.g., valproate), lithium remains the drug of choice for many patients with this disorder (Baldessarini & Tondo, 2000).

6.3.3 Schizophrenia

Schizophrenia currently affects roughly 1% of the global population, and its symptoms typically begin in adolescence or early adulthood (with men being more susceptible) (R. Li et al., 2016). Definitions of schizophrenia have changed over the years (Carpenter & Koenig, 2008), but some of the disease's most common symptoms are psychotic episodes (e.g., delusions or hallucinations), reduced emotional expression, low motivation, and asocial behavior. Intriguingly, schizophrenia is among the most heritable psychiatric disorders. If one monozygotic twin is diagnosed with schizophrenia, the other twin's chance of sharing the disease is much higher than the rate for dizygotic twins (48% versus 4%) (Onstad et al., 1991; Sullivan et al., 2012). Nonetheless, the genetic underpinnings of schizophrenia are very complex, as indicated by the fact that more than 100 (perhaps as many as 1,000) genes have been linked to this disease (Pratt et al., 2012; Schizophrenia Working Group of the Psychiatric Genomics Consortium, 2014). Moreover, each schizophrenia-linked mutation accounts for only a relatively minor increase in disease risk, and diverse environmental factors (e.g., prenatal stress) are almost certainly involved as well (Sullivan, 2005). At the level of the brain, a few differences in brain structure and functional dynamics reportedly correlate with schizophrenia (Dietsche et al., 2017; Vanes et al., 2019), but linking this variation to specific genes or schizophrenia symptoms remains a tall order.

Creating animal models of schizophrenia likewise remains challenging, both because of the complex genetics and because we cannot really know when nonhumans are experiencing hallucinations, delusions, or disordered thought. Scientists have tried to model schizophrenia by injecting animals with amphetamines (yes, the same manipulation also used to model mania) or other drugs that can trigger psychotic episodes in humans (Pratt et al., 2012; Steeds et al., 2015). Others have stressed rats early in their development—for instance, by underfeeding the pregnant mothers, isolating the newborns, or lesioning the ventral hippocampus of juveniles (Powell, 2010). These efforts have allowed scientists to model select aspects of schizophrenia, but the core symptoms have remained intractable.

Hoping to surmount these problems, scientists have created mice with modified schizophrenia-linked genes (Kvajo et al., 2012). For example, they have produced transgenic mice that express function-blocking versions of a gene called *disrupted in schizophrenia-1* (*DISC1*) (Hikida et al., 2007). These mice exhibit some schizophrenia-like symptoms, but mice with humanlike mutations in the endogenous *DISC1* gene exhibit far more subtle symptoms (Koike et al., 2006). Moreover, *DISC1* is associated with multiple psychiatric disorders, mutated in relatively few schizophrenics, and mutated in many people who do not develop schizophrenia (Kvajo et al., 2012). Thus, despite the gene's name, it remains unclear to what extent *DISC1* mice are a good model for schizophrenia. Indeed, thus far none of the genetic or developmental models of schizophrenia have played a major role in the development of novel therapies.

Some of the earliest strategies for treating schizophrenia were seriously misguided. Prefrontal lobotomies, in particular, were widely touted as a way to make schizophrenics (and other agitated patients with mental disorders) more manageable, but the side effects were often severe (Acharya, 2004; Tan & Yip, 2014). This procedure, which surgically destroys the prefrontal cortex or disconnects it from the rest of the brain, was inspired by the results of frontal cortex lesions in chimpanzees (Jacobsen et al., 1935). However, this study was small and was never intended to serve as the basis for clinical translation. Nor were lobotomies tested in rigorous clinical trials (Tierney,

2000). They did help to reduce the population of schizophrenics in overcrowded mental institutions (Swayze, 1995), but in retrospect they were one of medicine's most irresponsible misadventures. Fortunately, lobotomies fell out favor in the late 1950s, in part because pharmacological treatments for schizophrenia became available.

Drug treatments for schizophrenia have focused primarily on controlling the psychotic aspects of the disease, and they generally involve the blocking of dopamine receptors (Carpenter & Koenig, 2008). The first major antipsychotic drug was called chlorpromazine (aka Thorazine). Historically it was preceded by promethazine, which had originally been developed as an antihistamine. The physician scientist Henri Laborit in 1949 recognized that this drug was a mild sedative and could be used as an adjunct to surgical anesthesia. Laborit then encouraged colleagues to develop additional antihistamines with even stronger drowsiness-inducing effects. This search was conducted with rats and soon led to chlorpromazine.

Laborit and his clinical colleagues then gave chlorpromazine to various patients and found that it consistently reduced the agitation of manic patients and quieted the delusions and hallucinations of schizophrenics. In the words of Laborit and colleagues, "in doses of 50–100 mg intravenously, it provokes not any loss in consciousness, not any change in the patient's mentality but a slight tendency to sleep and above all 'disinterest' for all that goes on around him" (1952; trans. Caldwell, 1970, p. 3). By the end of 1952, chlorpromazine had been given to roughly 2 million patients. It dramatically reduced the number of patients confined to mental institutions and remained popular for several years. Unfortunately, chlorpromazine and other, functionally similar "typical anti-psychotics" (Meyer & Simpson, 1997) had serious extrapyramidal side effects, including tremors, erratic movements, slurred speech, anxiety, and paranoia.

The discovery of "atypical antipsychotics" (aka second-generation antipsychotics) began with the development of clozapine. This drug was first synthesized in 1958, quickly tested in animals, and then shown in humans to be an effective antipsychotic that was less likely than chlorpromazine to cause extrapyramidal side effects (Shen, 1999; Crilly, 2007). Unfortunately, clozapine can lead to an alarming loss of white blood cells, a side effect that proved to be a major stumbling block in its development. Nevertheless, a large clinical trial in 1988 demonstrated that clozapine was more effective than chlorpromazine in patients who had not responded to prior treatment with haloperidol, a typical antipsychotic (Kane et al., 1989). Moreover, clozapine's negative side effects could be controlled by carefully monitoring white blood cell counts. Consequently, clozapine was granted US approval for the treatment of schizophrenia in 1990.

At a mechanistic level, clozapine differs from the typical antipsychotics mainly in that it blocks dopamine receptors more transiently and, in addition, blocks serotonin receptors (Meltzer, 1994). The latter property explains why clozapine and other atypical antipsychotics tend to block the psychotic effects of the hallucinogen LSD (lysergic acid diethylamide), which exerts its effects at least in part by activating serotonin receptors (Valeriani et al., 2015).

Considering the various typical and atypical antipsychotic drugs, patients and physicians now have at their disposal a variety of drugs to treat schizophrenia. Some work better in some patients than others, and often a combination of drugs is best. Overall, however, schizophrenia treatments remain inadequate. All the drugs have potentially serious side effects, and it remains unclear how their various mechanisms of action relate to schizophrenia's behavioral and cognitive symptoms. To illustrate some of the mystery that still surrounds this disease, schizophrenics tend to smoke more heavily than control subjects (Kumari & Postma, 2005; Manzella et al., 2015). They seem to be self-medicating with nicotine, but how and why this works remains unknown.

6.4 SUCCESSES, FAILURES, HOPES

As noted in this chapter's opening quote, frustration has been growing over the lack of progress in the development of treatments for neuropsychiatric disorders over the last half century. Progress on treatments for CNS injury and neurodegenerative diseases has likewise been meager. Although advances in the treatment of cancer and some other diseases have also been slower than one would like (see chapter 5), the neurological disorders have been especially refractory to therapy development. Why has progress been so slow?

One very likely explanation is that human brains exhibit virtually no adult neurogenesis (Bergmann et al., 2012; Sorrells et al., 2018), which means that lost neurons cannot be replaced. Therefore, neurological disorders involving neuron death should ideally be treated before a significant number of neurons has been lost. An analogous argument could be made for treating psychiatric disorders before they manifest, especially if they involve irreversible losses of neuronal connections (S. Li et al., 2019). Prophylactic interventions have worked well for the treatment of cancer and cardiovascular disease, and early intervention is quite common in animal research on neurodegenerative disorders (e.g., in mouse models of Parkinson's disease) (Zeiss et al., 2017). However, for neurological disorders in humans, this strategy would require clinical trials that include huge numbers of subjects because it is currently difficult to predict who will develop a neurological disorder later in life. Moreover, treating huge numbers of people for decades before they might develop symptoms would likely be extremely expensive.

In the following section, I discuss three additional potential explanations for the slow progress of neurological treatment development. The first is that drug development has historically involved a great deal of serendipity; perhaps we are currently just stuck in streak of bad luck. Second, some of the models may be actively misleading us, effectively trapping us in dead ends, while ignoring other research avenues. Third, most of the current models may be mimicking only parts of the targeted disease. All these explanations contain kernels of truth, but each on its own is incomplete.

6.4.1 The Role of Serendipity

Many of the drugs that have ended up being widely used to treat neurological disorders were originally developed for other purposes. Tetrabenazine and imipramine were designed to be antipsychotics but turned out to be at least somewhat effective against Huntington's disease and major depression, respectively. Iproniazid was developed to fight tuberculosis before it was discovered to have antidepressant activity; lithium had long been used a treatment for gout before it was repurposed to fight bipolar disorder; and chlorpromazine was created as part of a search for novel antihistamines. Ketamine, too, was originally developed as an anesthetic, not as a depression fighter. These were all serendipitous discoveries, but hardly dumb luck. They came as a result of insightful clinical observations and minds that were open to unexpected developments. As Pasteur famously opined in 1854, "in the fields of observation chance favors only the prepared mind" (see Pearce, 1912).

A related observation is that many neurological treatments were discovered before 1960, when it was still possible for clinicians to test drugs on themselves (e.g., Cade tested lithium on himself before he gave it to his patients) and on small groups of human volunteers without extensive prior animal testing (Janssen, 2009). The more extensive regulations of the modern era were imposed as justified responses to major medical disasters, but they make chance discoveries less likely. It is still possible to test already approved compounds for off-label purposes, hoping to find new uses for old drugs (Heemskerk et al., 2002), but the odds of making serendipitous discoveries have certainly gone down.

One might think that the problem is much less severe in preclinical research on non-mammalian or in vitro models because they often allow for high-throughput drug screening. However, the odds that those discoveries will translate to humans appear thus far to be quite low. Most likely, these models are just too different from the human condition to allow for consistent extrapolation.

6.4.2 Convenient Models That Mislead

Although it is a given that "all models are wrong" (recall the opening quote of chapter 2), my review of the preclinical research on neurological disorders suggests that many of our current model systems may be too wrong to be useful in the quest for therapies. The models may be convenient insofar as they develop symptoms quickly and reliably, thereby facilitating the rapid publication of results, but most of the models are more like caricatures of the human condition than facsimiles (Libby, 2015). The R6/2 mice, for example, develop symptoms far more rapidly than humans with Huntington's disease (and develop severe epilepsy as they age). Similarly, mice that combine multiple mutations linked to Alzheimer's disease are quite unnatural. Moreover, most of our models are based on genetic information gleaned from familial versions of the disease, but for most neurological disorders sporadic cases outnumber the familial ones. What if the sporadic diseases differ in ways that matter for therapy development? And, as we just discussed, what if the models that are most suitable for high-throughput drug screening are too divergent to produce more than promising hits?

In short, one may well wonder whether biomedical researchers have been behaving like the proverbial drunk who searches for his keys at night under a streetlight (Schnabel, 2008): when a policeman asks the drunk why he is looking there, the drunk responds that the keys were lost on the other side of the street but the light is better under the lamp. In the words of Hittner et al. (2019), "when it comes to scientific research, the 'streetlight effect' is no longer a joke" (p. 2).

Searching in darkness is difficult, of course. Nor should one underestimate the pressures on researchers to produce publications and tangible results. In the absence of those, grant funding quickly dries up, and ambitious students seek greener pastures. When that happens, it becomes essentially impossible to make any empirical discoveries. Work with model systems that more closely resemble the human condition often takes years before results emerge, especially if those systems involve nonhuman primates or other large animals. Such work requires great patience from administrators and granting agencies as well as the individual investigators. Working with nonhuman primates, dogs, or any other higher species also entails the risk of attacks by animal rights activists, as well as internal ethical conflicts. In vitro studies with human cells can yield results more rapidly and harbor fewer ethical quandaries, but constructing in vitro systems that even remotely approach the complexity of their in vivo target is a much slower, more treacherous path. Given these pressures, why not study whichever models are most easily studied and most likely to yield quick results?

The problem is that the most convenient models may sometimes lead researchers down what Shakespeare called the primrose path (Burian, 1993). One should certainly be wary of turning August Krogh's principle (see chapter 2 and the opening quotation of chapter 3) on its head: there may well be a "most convenient" model in which to study any biological problem (Krebs, 1975), but studying a very convenient model will not necessarily answer the questions that originally motivated the inquiry. Moreover, some of the convenient models may have unsuspected problems. For example, the FVB/N strain of mice is very convenient because the animals breed readily, but this strain (and a few others) also carries a gene causing blindness (Chen et al., 2013; Zeiss, 2015).

In general, it seems preferable to work with natural (aka spontaneous) disease models instead of artificially constructed models. Such models have been very useful, for example, in research on diabetes, hypertension, and atherosclerosis. Unfortunately, spontaneous models of neurological disorders are rare. This is especially frustrating because the artificial models of neurological disorders are, as we discussed, very likely to be misleading in at least some respects. As Felix d'Hérelle, a pioneer in the study of infectious diseases, once complained, "because all illnesses studied by significant authors were 'artificial' illnesses (neither the rabbit nor the guinea pig are affected by cholera or typhus in the natural environment) they have bearing only when talking about the artificial illness and not at all practical for application to real, natural illnesses which occur in humans" (Endersby 2007, p. 290). D'Hérelle turned out to be wrong about the utility of "artificial" models in vaccine development (see chapter 5), but his point may be applicable to more complex illnesses, especially those involving the nervous system.

6.4.3 Partial Models and Endophenotypes

Most biomedical researchers readily admit that their models are imperfect. More interesting is that some, especially those who study neurological disorders, tend to refer to their study systems as partial or incomplete models (Jucker, 2010; Zahs & Ashe, 2010; Drummond & Wisniewski, 2017). Mouse AD models, for example, are considered incomplete if they exhibit amyloid plaques but not neurofibrillary tangles or extensive cell death (Schwab et al., 2004). Similarly, animal models of schizophrenia are said to be partial insofar as they share with their human counterparts only a specific subset of symptoms (Roberts, 2007).

Although the idea of partial models seems innocuous, it does entail the debatable assumption that complex diseases are caused by a set of largely independent processes that can be modeled, and ultimately treated, individually. Moreover, it implies that the various models and treatments can be combined with additive effects (e.g., Oddo et al., 2003). This approach is potentially transformative, but its assumptions should be examined carefully. One might point out, for example, that a single change in a complex system may, over time, lead to a variety of interdependent effects and that a given effect may be produced by diverse causal processes.

A sophisticated and impactful version of the partial modeling approach is the Research Domain Criteria (RDoC) framework developed and promoted by the National Institute of Mental Health (Insel et al., 2010; Kozak & Cuthbert, 2016) (for an analogous European framework, see Schumann et al., 2014). This framework is based on the notion of endophenotypes, which I introduced in chapter 2 (section 2.4.1). The core idea is that between the genotype and the outwardly observable phenotype lies an intermediate level of organization, the endophenotype. Individual endophenotypes (i.e., traits at the endophenotype level) can be observed with special tools and behavioral tests.

They include, for example, the death of specific neurons, changes in neuronal connectivity or function, and deficits in attention. Some authors have argued that endophenotypes must be heritable and have adaptive functions (Gottesman & Gould, 2003; Cuthbert & Insel, 2013); however, in practice these two criteria are rarely treated rigorously. The framework's most important aspect, in my view, is that it formally justifies the modeling of traits that contribute to a disorder without having to claim that one has modeled the entire disorder; that is, it justifies the creation of partial models, including so-called pathway models (Schnabel, 2008). The endophenotypes can then, in theory, be linked to underlying genes and, going in the other direction, to specific mental or behavioral symptoms. Therapy development is envisioned as targeting specific endophenotypes, which implies that the ideal treatment for complex disorders will require multiple simultaneous treatments and that any one treatment may be of use in multiple disorders.

The RDoC framework originated, and is presently employed, mainly in biological psychiatry. This is understandable because psychiatric disorders tend to have a large number of overlapping diagnostic criteria (Lilienfeld, 2014) and, as we have reviewed, remain difficult to treat. The framework may well be applicable to other types of complex diseases, but this remains to be seen. Some authors have hailed the RDoC framework as a new paradigm in which complex diseases like schizophrenia are deconstructed into multiple domains without a "unifying pathophysiology" that ties them together (Carpenter & Koenig, 2008). This is an exciting idea, but the RDoC framework has been criticized for being difficult to falsify, in the sense of providing no clear criteria for determining whether or not it is "mapping well onto the state of nature" (Lilienfeld & Treadway, 2016).

Ultimately, the framework's utility is an empirical question: will it lead to new treatments that the traditional paradigm would not have revealed? This question remains open. What is already clear, however, is that the RDoC framework has allowed research on fundamental aspects of neurobiology and behavior to be considered translational, even if it does not explicitly target a specific psychiatric disorder (Rubio et al., 2010; Anderzhanova et al., 2017). Thus, the RDoC framework facilitates the integration of basic and translational research (McArthur & Borsini 2008) (see chapter 1, section 1.1). Such a rapprochement could well be useful far beyond psychiatry.

7 DIAGNOSIS AND RECOMMENDATIONS

The high Phase III negative outcome rate problem [of clinical trials] is endemic, affecting sepsis, stroke, cancer, cardiology and orthopaedics research, to name just a few. These persistent failures have had a chilling effect on pharmaceutical industry investments in new drug development and the costs to pharma and government are staggering. It is no wonder that funding agencies and policy-makers in both sectors are deeply concerned and realize that continuing to do the same things in the same way cannot go on. Despite much recent discussion, little has been done to change the situation for the better. This paralysis could be due in part to the lack of any broad consensus about where the most basic problems lie.

—STEIN (2015), P. 1259

Although biomedical research has made enormous progress over the last 200 years, alleviating a great deal of human suffering, it does have a problem. As stated in the chapter's opening quotation, far too many therapies show promise in animal and in vitro experiments but fail in clinical trials. Findings in toxicology also do not translate from animal and in vitro systems to humans as reliably as one would hope. The human, animal, and financial costs of these failures are high, but charting a more successful path is difficult. Different people are likely to offer different remedies, ranging from calls for increased rigor in preclinical research to championing experiments on human volunteers, rather than animals. I, too, feel torn between the various options. The reduction of animal suffering is certainly a worthy goal, especially if animals are dying in vain; yet human suffering is also a significant concern. If we do accept the need for animal research, should that research focus on species that are most similar to us, given that those animals are also most likely to feel and think the way we do? If we opt for studying more distant relatives, or even in vitro systems, are we inviting more failures of translation, thus wasting time, resources, and lives?

I struggled mightily with these questions as I was working on this book but cannot, in the end, offer a simple, clear remedy. Instead, I have opted for giving you, the reader, a trove of information and ideas that seem relevant to the problem of translational failure and success. Overall, I hope the book will help you think this problem through and make well-informed, deliberate decisions about your own personal way forward. If you are in a position of influence, this book will hopefully help you appreciate the problem's nuances and make wise decisions on policy. With those overarching goals in mind, I use this last chapter to present four very different perspectives on the use of material models in biology, followed by my own attempt at synthesis and some specific recommendations.

7.1 FOUR PERSPECTIVES ON MODELS IN BIOLOGY

How should researchers use animal and cellular models to develop effective treatments for human disease? Given that progress in this area has been slower than we want, do we need a radical course correction, or are we simply going through a spell of mediocre luck? As the opening quotation asks, where do the most basic problems lie, and how can they be fixed? When we try to answer these questions, each of us is likely to come at the problem from a somewhat different perspective, each addressing only part of the problem. We are like the proverbial blind men examining an elephant, each coming to quite different conclusions. In the following sections, I sketch four of these disparate perspectives and briefly critique them. Although my descriptions are certainly strawmen cartoons, delineating these viewpoints should, I hope, help us appreciate the problem's full complexity.

7.1.1 The Animal Welfare Perspective

In this book, we discussed many instances in which data from animal research failed to predict a compound's safety or efficacy in humans. A paradigm example is the trial of TGN1412, which had passed safety tests in rodents and monkeys but caused life-threatening harm in human volunteers (see chapter 1). One response to such translational failures is to argue that animal experiments in general cannot be trusted and should, therefore, be stopped (Shanks et al., 2009; Pound & Bracken, 2014). A milder, more widely accepted version of this anti-vivisectionist stance is to argue that researchers should reduce the number of animals they use and minimize their suffering. In particular, they should experiment on presumably less sentient animals—that is, animals thought to be not very capable of having feelings (Harnad, 2016)—or insentient materials whenever feasible. I call this the animal welfarist's perspective.

Complete replacement of sentient animals with insentient materials in biological research was the original recommendation made by Russell and Burch (1959) in their influential book on the 3R approach (refine, reduce, replace). However, the recommendation to use *relatively* insentient animals that are "low on the phylogenetic scale" has become more widely accepted by the scientific community (Tannenbaum & Bennet, 2015; Franco et al., 2018). Indeed, the scale-based approach to replacement has been codified in the *Guide for the Care and Use of Laboratory Animals* (National Research Council, 2011b), which is used to regulate animal research in the United States; many other countries have similar guidelines. Among the general public, too, the vast majority (80% to 90%) accept the use of animals in medical research so long as animal welfare is optimized and alternatives have been explored (Festing & Wilkinson, 2007). Importantly, there appears to be a broad consensus that research on "lower animals"—notably pests and animals we kill for food—is more acceptable than research on nonhuman primates and animals we like to keep as pets (see chapter 3). This view is what one might call a moderated form of the welfarist's perspective.

This more moderate, scale-based approach to animal replacement can be criticized for its reliance on the notion of a phylogenetic scale, which is inconsistent with how evolution actually works (see chapter 2, section 2.5.1) and is shaped by rather subjective preferences of some species over others (a form of speciesism, although this term is usually restricted to a bias in favor of humans alone) (Singer, 1990; Oberg, 2016). However, even if we reject the concept of a single, linear phylogenetic scale, it does seem reasonable to suppose that the degree of sentience (if we accept that sentience comes in degrees) correlates with nervous system complexity (Bullock, 2002; Mather, 2008) and that, therefore, monkeys or chimpanzees will likely suffer more than, for example, a nematode, a fly, or a zebrafish larva when undergoing analogous experimental procedures. Similarly, most people would agree that cultured cells—even in the form of organoids or organs-on-a-chip (see chapter 4)—are less sentient than vertebrates. If that is true, then why not replace animals that society considers to be highly sentient with less sentient animals or insentient cells? Even if our judgments about degrees of sentience are somewhat subjective (Striedter, 2016), is this not a prudent, righteous strategy?

The problem is that shifting research to less sentient or insentient systems implies a shift toward systems that are more different from humans, at least on average. When it comes to biomedical research, the overall similarity between an animal or in vitro model and the human condition—what Russell and Burch (1959) called the model's "fidelity"—tends to decrease with its position on the phylogenetic scale or, to be more objective, its phylogenetic and genetic distance from us (see section 7.1.2). Similarly, even the most elaborate in vitro systems are a far cry from their in vivo counterparts in terms of overall complexity. This decrease in model fidelity is a problem if we assume that our ability to translate findings from a model system to humans correlates with model fidelity.

That said, successful models do not need to be similar to their target in all respects; it is important only that they mimic the target in the features one is trying to model.

For example, viruses, bacteria, and yeast were extremely useful models in the early days of molecular biology because the phenomena of scientific interest at the time turned out to be broadly conserved. For example, the ability of X-rays to damage DNA could be studied in a wide variety of simple animals and even cultured cells because the underlying mechanisms are very general. In fact, microbes and cultured cells were far more convenient for this kind of research than more complicated animals, and thus were better models in practical terms. As we reviewed in chapter 3, the situation changed only as researchers turned their attention to more complex biological problems that were less broadly shared and, therefore, required research on more complex animals, such as fruit flies. One might even argue that biomedical progress over the last 50 years entailed a steady increase in the complexity of the problems being investigated and that this increase required a shift to ever more complex model systems, notably mammals. At least biologists are now tackling a broader array of problems, which requires a broader array of species for research.

Given all these considerations, using mice as model animals seems like a reasonable compromise between having a relatively high-fidelity model and relatively low ethical barriers. After all, mice are mammals like us, yet widely reviled as pests (Little, 1935). Maybe we can exclude them from our concern for animal welfare (as US legislation does; see chapter 2). Moreover, their rapid rate of reproduction and relative tolerance of inbreeding make mice well suited to strain standardization and genetic analyses (see chapter 3). It makes sense, therefore, that mice have played such a prominent role in biomedical research over the last few decades (Libby, 2015). Another, very different compromise between high model fidelity and low ethical concerns can be achieved by performing experiments on cultured human cells. They are genetically human yet presumably insentient (so long as cultured mini-brains remain quite different from human brains; see chapter 4).

Presented with these two main compromise options, one may debate their relative merits. In practice, however, it is more sensible to embrace a tiered approach in which one begins research with relatively simple, low-fidelity models and then proceeds to progressively "higher" models. The idea is that the simple systems allow researchers to explore and develop hypotheses that can then be tested for generality. This tiered, sequential research strategy is common in toxicology and is often pursued by companies involved in drug development. Moreover, it is no accident that regulatory agencies often require safety and efficacy data from two different species, one of which is not a rodent. Obtaining the nonrodent (often nonhuman primate) data is usually the last step before proceeding to clinical trials.

Many advocates for animal welfare are not in favor of the tiered research strategy, however, because it does involve some higher animals, at least at late stages of translational research. They would prefer that scientists shift farther away from animal research and instead focus more of their energy (and research funds) on clinical research. Indeed, careful clinical observations have played a major role in drug development—for example, by revealing unexpected beneficial drug effects (see chapter 6). Epidemiological studies have also played critical roles in many different areas of medical research, including cancer biology, the discovery of risk factors for cardiovascular disease, and toxicology (see chapter 5). In addition, genetic analyses of human subjects can lead to the identification of some disease mechanisms. The cause of sickle cell anemia, for example, was greatly clarified when research showed that the disease in humans is associated with abnormally shaped red blood cells and a mutation in the hemoglobin gene (Frenette & Atweh, 2007).

Unfortunately, purely observational studies on humans are limited in what they can tell us about the underlying disease mechanisms. As Lord Moulton—a famous mathematician and great advocate for medical research—supposedly once said, "when we are reduced to observation, Science crawls" (Edsall, 1969, p. 467). Moulton was right, but experimental studies on humans must be carefully monitored and tightly controlled (e.g., Kalm & Semba, 2005). History is replete with incidents of human experimentation that were later regarded as unethical ("Unethical Human Experimentation," 2020). For example, the psychiatrist Robert Heath in the 1950s and 1960s conducted poorly controlled deep brain stimulation experiments in humans without clear medical benefits. In his most notorious experiment, Heath attempted to "cure" a man of homosexuality by stimulating his reward circuitry while the subject engaged in sex with a female prostitute (Oliveria, 2018). Ethical aberrations such as these led the World Health Association in 1964 to draft the Declaration of Helsinki, which was amended in 1975 to state explicitly that "concern for the interests of the subject must always prevail over the interests of science and society" (Carlson et al., 2004, p. 709). Although this declaration continues to be debated and revised (World Medical Association, 2013), it is widely accepted and codified in regulations around the globe.

Even now, however, unethical human experiments remain a matter of concern. For instance, the physician scientist Paolo Macchiarini between 2008 and 2014 transplanted stem cell–covered plastic tracheas into multiple patients without being transparent about his results, which were quite often lethal (Rasko & Power, 2017). As an indicator of how serious this problem was, Macchiarini has now been indicted in Sweden for aggravated assault (Schneider, 2020). Even when conducted with the best of intentions, human experimentation often creates complex ethical dilemmas (Edsall, 1969). In short, shifting research from nonhumans to humans does not cause ethical concerns to disappear entirely.

7.1.2 Animal Nepotists: Favoring Our Relatives

Rosenblueth and Wiener wrote in 1945 that "the best material model for a cat is another, or preferably the same cat" (p. 320). Similarly, one may well argue that the best material model for a human is another human and that researchers should therefore emphasize clinical studies over research on nonhumans. This argument is often made with an eye toward promoting animal welfare (see section 7.1.1). However, as we discussed in chapter 2, the principal motivation behind the use of models in science is that they are more accessible to experimental investigation than the target system; and as we just discussed, human experimentation is perforce limited. So if we cannot experiment with humans, why not use the next best thing: nonhuman primates? We may not want to perform invasive procedures on our closest relatives, the chimpanzees, or even other apes, but why not macaques or marmosets (see chapter 3)? Averaging over all traits, they are bound to be more similar to humans than any nonprimate, simply because we are more closely related to other primates than to nonprimates.

There is some logic to this argument, as similarity does decrease on average with the square root of phylogenetic distance (Letten & Cornwell, 2014; Striedter, 2019). It certainly seems reasonable to say that "in general, species that have diverged most recently have the closest resemblances in DNA sequences and functions of protein and RNA derived from these sequences" (National Research Council, 1985, p. 16). Within primates, the Old World primates (including both macaques and us) are more similar to one another, on average, than they are to the more distantly related New World monkeys (e.g., marmosets). Specifically, the genomewide identity of protein coding sequences is 94% for humans versus macaques and 91.7% for marmosets versus humans (Preuss, 2019). Of course, these observations hold only on average, and some traits do not follow the similarity-distance rule. For example, marmosets, like humans but unlike macaques, form strong male-female bonds and exhibit extensive, prolonged paternal care (Fernandez-Duque et al., 2009).

A more general critique of the nepotist's view would focus on the fact that many of the traits that are shared by all primates are also shared by nonprimates and can, therefore, be modeled in them. It becomes important know, therefore, which traits are phylogenetically conservative (i.e., widely shared) and which evolved more recently, either with the origin of primates or within primates. This is no easy task, because we cannot know for certain which trait is present in which species until we look, a paradox sometimes called the "extrapolator's circle" (Steel, 2007; Bolker, 2009). However, we do know by now that some organ systems evolved more rapidly than others within the primate lineage. Especially the immune and central nervous systems have diverged far more than, say, the cardiovascular and skeletal systems. Moreover, it seems reasonable to assume that traits associated with the more conservative systems have a greater likelihood of being modeled successfully in nonprimates. At least one might adopt a tiered approach in which nonprimates are studied first, and primates are called upon only if those initial efforts prove unsuccessful. For example, the observation that the human immunodeficiency virus (HIV) and hepatitis C viruses infect only humans and chimpanzees has been used to justify research on the latter species for those two specific diseases (Institute of Medicine & National Research Council, 2011).

We can conclude that, as a general rule, the odds of successful translation from a model to its target increase with phylogenetic relatedness, but it is never guaranteed. Even when we compare humans and chimpanzees, one or the other species may well be divergent in any specific trait.

Another very important caveat to the nepotist's perspective is that rates of evolutionary change tend to vary across lineages because of differences in generation time, DNA repair mechanisms, population size, and various other factors (Thomas et al., 2006, 2010). This variation in evolutionary rates can confound the general relationship between overall similarity and phylogenetic relationship. For example, it is well established that rodents and primates are more closely related to one another than to pigs, but the genomes of humans and pigs are more similar to one another than they are to that of mice (figure 7.1), mainly because mice have a much shorter generation time (and hence a faster rate of evolutionary change) than either humans or pigs. Similarly, it is not the case that humans are more closely related to dogs than to mice, even though the human and dog genomes and proteins are more similar to one another (Thomas et al., 2003; Asher et al., 2009).

Given these considerations, it is tempting to replace the outdated notion of a phylogenetic scale with the notion of a genetic divergence scale, in which mice and rats rank lower than cats, dogs, pigs, cows, and all the nonhuman primates (see figure 2.2). However, to the nepotist's distress, neither phylogenetic divergence nor overall genetic similarity are perfect predictors of how well a given trait will translate from a model species to humans.

7.1.3 The Perspective of a Pragmatic Optimist

I suspect that many biologists would listen to the arguments presented in the previous two sections and conclude that, yes, working with model systems is complicated and does not always translate effectively, but with the flood of recent technical advances success is probably just around the corner. The failures of translation are mainly due to technical limitations, poorly designed preclinical research, and faulty clinical trials. The heavy emphasis on mice and a few other model organisms is not to blame because model selection always depends on the question that is being asked, and biologists usually choose whichever model is most convenient for answering the question at

A – DNA Distances

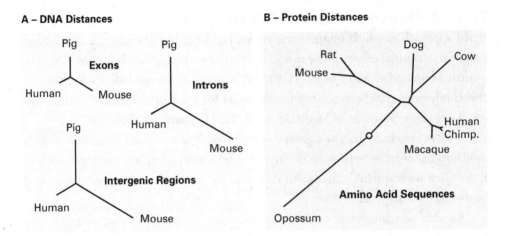

Exons

Pig · Human · Mouse

Introns

Pig · Human · Mouse

Intergenic Regions

Pig · Human · Mouse

B – Protein Distances

Rat · Mouse · Dog · Cow · Human · Chimp. · Macaque · Opossum

Amino Acid Sequences

C – Phylogenetic Relationships

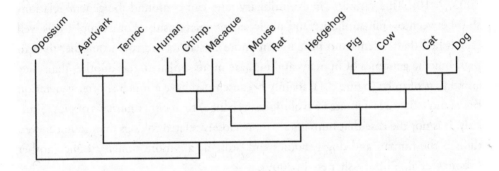

Opossum · Aardvark · Tenrec · Human · Chimp. · Macaque · Mouse · Rat · Hedgehog · Pig · Cow · Cat · Dog

Figure 7.1

Genetic distance versus phylogenetic distance. (*A*) DNA distances. Wernersson et al. (2005) compared the DNA sequences of humans and mice with a large fraction of the pig genome (it had not all been sequenced at the time). The branch lengths in these diagrams represent the degree of estimated sequence divergence and demonstrate that the genetic distance between humans and pigs is less than that between humans and mice. (*B*) Protein distances. An analysis of more species by Cannarozzi et al. (2007) compared protein (rather than DNA) sequences, but again found that rodents are genetically quite divergent from primates. (*C*) Phylogenetic distances. Despite the findings shown in A and B, most biologists agree that humans are more closely related to rodents than to other nonprimates. The reason for the discrepancy between genetic distance and phylogenetic distance (see also figure 2.2) is mainly that rodents reproduce more rapidly than larger mammals and therefore diverge more rapidly. Adapted from (A) Wernersson et al. (2005); (B) Cannarozzi et al. (2007); (C) Asher et al. (2009).

hand. Such optimists are likely to acknowledge that model selection is also influenced by various other concerns, such as researcher training and experience, animal welfare, funding, and the need to produce publications at a reasonable rate. However, they do not think it is imperative to consider species, sex, or other differences between model systems, at least so long as the research is addressing "fundamental" questions about molecular interactions and cellular phenomena underlying disease.

It certainly is true that recent advances in biological techniques have been astonishing. In fact, progress is so rapid that few can keep up, which is one reason why collaborative research has become more important than ever (Wu et al., 2019). Moreover, the new techniques do overcome many of the limitations of the older technology. It is now much easier, for example, to modify specific DNA sequences in a variety of species, generate and differentiate stem cells, and culture three-dimensional tissues. Even some species differences between humans and the models can be overcome by humanizing specific genes or part of the model's immune system (see chapter 3). In light of these advances, some optimism is clearly warranted.

Likewise, it is true that clinical trials can be improved. Clinical trials have already come a long way in terms of being carefully designed, monitored, and analyzed (Junod, 2008; Chow & Liu, 2013). There are, however, ongoing discussions about how heterogeneous the subject pools should be, how far along in the disease process subjects ought to be, how long they should be monitored, and whether tests for efficacy should be performed earlier in the trial pipeline (rather than in the phase III trials). Unfortunately, a major obstacle to many of the proposed changes is increased cost, which is already very high for any late-phase clinical trial. Nonetheless, better enrollment strategies and analytical methods are being developed (Lin & Lee, 2020).

Discussions about improving the rigor and reliability of preclinical research are even more prominent. The topic is too large for us to discuss in depth, but the upshot is that many critics suggest preclinical studies should be more like clinical trials. That is, the sample sizes should be large enough to observe the expected effects (i.e., the studies should be adequately powered), the subjects should be randomly assigned to different treatment groups, proper control groups ought to be included, the data should be analyzed blindly (i.e., without knowing a subject's group assignment), and outcome measures should be specified ahead of time. Ideally, the entire experiment should be registered in advance, and the results should be published regardless of how they turn out. Naturally, such "preclinical trials" (Mogil & Macleod, 2017) would be labor intensive and expensive. They might also discourage exploratory research. Therefore, such rigorous experimental designs should be implemented only for a "final, crucial, confirmatory experiment" (Mogil & Macleod, 2017) that would, if successful, lead to a clinical trial. Although this strategy would be costly and reduce publication rates for the participating investigators, it would cut down on the extremely frustrating experience of realizing after a failed clinical trial that the preclinical data had been inadequate (Perrin, 2014; O'Collins et al., 2017).

Despite these valid points, blaming insufficient experimental rigor for the profusion of trial failures is problematic because it neglects the fact that successful translation requires replication in a *different* species or system. As Charles Leathers put it in 1990, "the most brilliant design, the most elegant procedures, the purest reagents, along with investigator talent, public money, and animal life are all wasted if the choice of animal is incorrect" (p. 68). It is relatively easy for an investigator to write a sentence at the end of their research paper about how the findings suggest a promising new therapy or treatment approach for a human disease; it is more difficult to know whether the reported results are truly likely to translate or are simply another red herring (Drucker, 2016). To quote Franz van der Staay (2017), "It has hardly ever been questioned in this distending stream of critical reviews addressing lack of translatability, whether the appropriate model animal species have been used" (p. 73). As we discussed in chapter 6, preclinical research may at times suffer from the streetlight effect. That is, the investigators may sometimes be looking for answers in very convenient models that, inconveniently, differ from humans in ways that impede successful translation.

Another potential criticism of the optimist's perspective is that the advantages of novel techniques and therapies are often touted before their limitations become apparent. A classic example is that heroin was originally marketed as a cough suppressant (Sneader, 1998). More recently (and less dramatically) stem cell transplantation showed tremendous initial promise, but using this technique to replace lost cells has proven far more difficult than expected, especially in human brains (Stoker et al., 2017). Similarly, it is distressing to discover that many anticancer drugs currently in clinical trials kill tumor cells in mice even when their putative target molecules are deleted genetically, presumably because the RNA knockdown techniques originally used to identify those putative targets are not as selective as researchers had believed (Lin et al., 2019).

Although this process of diminishing excitement about novel techniques and therapies is natural (part of the "hype cycle"), it becomes a problem when science advances so rapidly that the limitations become a mere afterthought and researchers eager to try the latest techniques move rapidly from one set of inflated expectations to the next. Under such conditions, it becomes difficult for the field at large to know (or even to ask) why some of the promises were never fulfilled.

7.1.4 The View from Comparative Biology

Some biologists bemoan the fact that biomedical research has become heavily focused on a small number of species generally referred to as "model organisms." These dozen or so species (e.g., *Mus musculus*, *Drosophila melanogaster*, *Caenorhabditis elegans*, *Saccharomyces cerevisiae*, *Escherichia coli*, and *Arabidopsis thaliana*) achieved their special status mainly because they reproduce rapidly, are exceptionally amenable to genetic analyses, and managed to attract a research community that was willing to develop shared resources to facilitate research on that species (Davis, 2004; Leonelli & Ankeny, 2013). Although these model organisms are distributed across multiple plant and animal phyla (Müller & Grossniklaus, 2010), they represent a minute fraction of the total species diversity.

This kind of taxonomic bias is common in many areas of biology (Pawar, 2003; Rosenthal et al., 2017), but it has become especially severe in biomedical research, where mice have nearly displaced most other research animals. Comparative biologists tend to decry this "murine 'model' monotheism" (Libby, 2015) and advocate for more species diversity in biological research, including biomedical research (Bolker, 1995; Preuss, 2000; Manger et al., 2008; Brenowitz & Zakon, 2015; Yartsev, 2017). This goal of studying a more diverse set of species has been made more attainable by the ever-increasing number of sequenced genomes (National Center for Biotechnology Information, 2020) and the recent advances in genome editing, which can be applied to many different species (Hsu et al., 2014).

An obvious risk to increasing species diversity in biological research, however, is that we might sacrifice depth for breadth. Research on a small set of model organism produces detailed knowledge about these systems at multiple levels of analysis (National Research Council, 1985; Ankeny & Leonelli 2020). Distributing the same amount of research effort across a large number of species would necessarily create more shallow knowledge (Schaffner, 1998). To mitigate this problem, comparative biologists tend to accept a compromise that allows for in-depth studies of a few species that then serve as "reference species" for comparative studies (Striedter et al., 2014). The idea is not simply to compare the heavily studied model organisms to one another, but to examine additional species that are near the reference species phylogenetically. One might, for example, study insects other than *Drosophila melanogaster*, ray-finned fishes other than the zebrafish, and rodents other than *Mus musculus*. The benefit of starting with a reference species is that our deep knowledge of that species can guide the comparative inquiry. Knowing what to look for (and how to look for it) makes the comparative research much easier, even if it ultimately reveals some species differences. In general, the comparative research reveals patterns of species similarities and differences that suggest phylogenetic scenarios and provide important clues about which findings are likely to be conserved across which species (Miller et al., 2019). This knowledge, in turn, should facilitate translational success.

Nonetheless, most comparative biologists are less interested in translation itself than in the discovery of general principles (National Research Council, 1985). Although comparative biology is sometimes viewed dismissively as being analogous to stamp collecting (Lewin, 1982), it has a long, distinguished history of seeking and discovering general biological principles. As long as one knows only about a single species, it is impossible to know how general one's findings are (Beach, 1950). Even when species differences exist, comparative biologists can discover how they represent different manifestations of conserved principles. As Claude Bernard put it in 1865, "The problem of science will consist precisely in this, to seek the unitary character of physiological and pathological phenomena in the midst of the infinite variety of their particular manifestations" (National Research Council, 1985).

August Krogh, for example, was interested in how the general physicochemical principles of gas exchange were implemented in organisms that lived in very different environments (Krogh, 1941). Later observers tend to emphasize that Krogh liked to study "extreme" organism with special adaptations that made them particularly amenable to experimentation (Krebs, 1975; Pollak, 2014; Green et al., 2018). This is true, but his ulterior aim was the discovery of general principles, not idiosyncrasies. There is nothing wrong with scientific analyses of phenomena that are found only in a few unusual species—they may be inherently fascinating or lead to biologically inspired (e.g., biomimetic) technology (Benyus, 2002)—but it would be wrong to think of comparative biology as being concerned primarily with the study of oddities. A good illustration of this point is Peter Getting's (1988) review of the neural mechanisms that animals use to generate rhythmic activity (e.g., walking, flying, breathing). Although his comparative analysis revealed a variety of mechanisms rather than just one, Getting was relieved to find a set of conserved "building blocks" that were combined in diverse ways.

A major challenge for comparatists who do want to extrapolate their findings across species is that such efforts often depend critically on the details of how the general principles are implemented. The devil, as they say, lies in the details. Even seemingly minor changes in the parameters that govern the interactions between conserved molecules, cells, or organ systems can influence whether a specific compound is toxic or safe, effective or inactive, even if the general principles are conserved. Again, the trial of TGN1412 is a good example: the drug had tested safe in monkeys but was harmful to humans because of a species difference in the expression level of a specific protein (CD28) on a specific type of immune cell (Eastwood et al., 2010). Similarly, chocolate is far more toxic to dogs than to humans, not because dogs lack the enzymes to break down theobromine (the potentially toxic ingredient), but because this compound is degraded and excreted more slowly in dogs (Finlay & Guiton, 2005). Returning to matters of human health, two of the metabolic enzymes that collectively metabolize more than 70% of all marketed drugs differ in complexity, expression levels, and catalytic activity between humans and mice (Gonzalez & Yu, 2006). Such seemingly minor, but translationally relevant, species differences are likely to be even more pronounced as we try to extrapolate across more distant relatives (Cross et al., 2011).

One productive way in which comparative biologists have dealt with this variation is to discover "principles of variation" that allow them to predict species differences (rather than similarities) and thus facilitate cross-species extrapolation. For example, they have used comparative neurodevelopmental data to derive a statistical model that can be used to "translate time" across a broad range of mammalian species (Workman et al., 2013). This model helps investigators identify equivalent developmental stages in different species, independently of how rapidly they develop in terms of absolute time (e.g., marsupials develop much more slowly than placental mammals but go through similar stages of development). Such considerations clarify, for example, why thalidomide toxicity tests in diverse species had initially provided confusing results: the drug's effects depend on the developmental stage of the fetus, not on their absolute age (Monamy, 2000). The same general approach reveals the folly of arguing that Alzheimer's model mice do not exhibit the full spectrum of AD symptoms observed in humans because they do not live as long (Foidl & Humpel, 2020): just as dogs age more quickly than humans (T. Wang et al., 2020), so do mice—except more so. As a general rule, life span scales with body size (Speakman, 2005)!

Principles that facilitate cross-species extrapolation have also been discovered in the area of drug dose scaling. To illustrate the problem, consider the fate of Tusko, the adult male elephant who was given 297 mg of LSD in a misguided experiment that killed the elephant within 2 hours of the drug injection (West et al., 1962). The problem was that the investigators calculated the elephant's dose on a milligram per kilogram basis, scaling up linearly from doses known to be safe in humans and cats. This was an inexcusable mistake, for physiologists had long advised that drug doses scale more tightly with metabolic rate than body weight, and that metabolic rate scales against body weight with an exponent of approximately 0.75 (i.e., with negative allometry, rather than linearly). According to the metabolic scaling rule, Tusko should have received only about 1% to 2% of the delivered dose (Harwood, 1963; Boxenbaum & DiLea, 1995).

Later studies showed that factors other than metabolic rate—especially the rate at which a drug is cleared from the body—scale with a variety of different exponents. Therefore, many researchers nowadays obtain data on a drug's clearance rate and several other parameters (notably drug absorption, distribution, and excretion) from multiple species, graph them against body weight, and then use best-fit lines to calculate the predicted parameters for their species and drug of interest (figure 7.2). The predicted parameters are then used to calculate the best-guess dose for administration. In practice, different authors use somewhat different procedures to make their predictions, adjusting them for different classes of animals (Caldwell, 1981; Mahmood, 2007; Sinha et al., 2008). Even

A – Allometric Scaling of Drug Clearance Rates

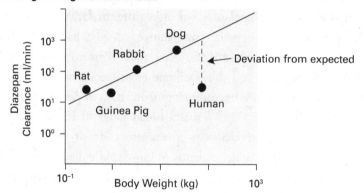

B – Drugs that Deviate from Allometric Predictions

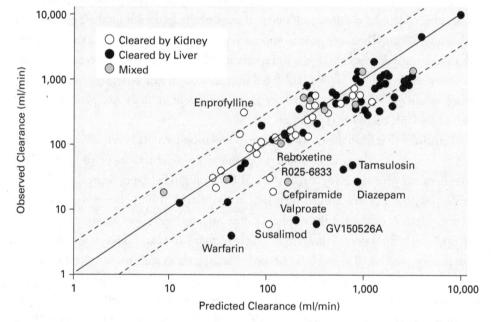

Figure 7.2

Predicting drug clearance with allometry. (*A*) Allometric scaling of drug clearance rates. Extrapolating drug doses across species frequently involves allometric analyses, in which relevant parameters (such as drug clearance rates) are determined in several species. Those data are then graphed against body size and subjected to a linear regression analysis (usually in log-log plots). The resultant best-fit line, in turn, can be used to predict the examined parameter in other species, based on their body weight. This technique works well for many drugs and species, but, as shown here, the clearance rate of diazepam (aka valium) is unusually low in humans. (*B*) Drugs that deviate from allometric predictions. The bottom graph shows the allometric prediction errors for 102 drugs whose clearance rates have been examined in humans and at least three other species (most often dogs and rats). Adapted from (A) Boxenbaum & DiLea (1995); (B) Huang & Riviere (2014), based on data in Tang & Mayersohn (2006).

so, exceptions abound (Aitken, 1983; Calabrese, 1991). For example, the high sensitivity of cats to aspirin defies the predictive procedures, because cats lack an enzyme that is critical to aspirin degradation (Sharma & McNeill, 2009). Still, it is undeniable that the more sophisticated scaling procedures predict safe doses more effectively than the linear extrapolation method that doomed Tusko (Hunter, 2010). Moreover, knowing the scaling rules makes it much easier to identify the "true outliers" (figure 7.2). In aggregate, these comparative studies indicate that data from macaques predict human responses to drugs more faithfully than data from dogs, rabbits, or guinea pigs (Caldwell, 1981), an observation that might please the animal nepotists described in section 7.1.2.

Another medically relevant principle of variation involves cancer susceptibility. A comparative analysis of diverse human tissues has shown that the risk of cells becoming cancerous increases with the number of times their progenitors divided during development (Tomasetti & Vogelstein, 2015). This explains why epithelial tumors, for example, are more common than neuronal tumors (because human neurons tend not to divide after birth). Given this consideration, one would expect large animals with long life spans to develop cancer at a much higher rate than smaller, short-lived species (Seluanov et al., 2018). However, cancer rates do not scale this way. For example, humans undergo about 10^5 more cell divisions in a lifetime than mice, yet the cancer rates of humans and mice are similar. In the words of Rangarajan and Weinberg (2003, p. 952), "about 30% of laboratory rodents have cancer by the end of their 2–3 year lifespan and about 30% of people have cancer by the end of their 70 to 80 year lifespan."

The answer to this puzzle, which is called Peto's paradox (Peto, 1977, 2016), is that humans and other large, long-lived species have evolved various mechanisms that reduce cancer risk (Holliday, 1996; Caulin et al., 2015; Nunney et al., 2015). For example, the number of mutations required to make cells cancerous is larger in humans than in mice (Hahn & Weinberg, 2002). In addition, humans have shorter telomeres and reduced telomerase activity (Hornsby, 2007; Vanhooren & Libert, 2013). Thus, recognition of Peto's paradox predicted and now explains the existence of some species differences in cellular and molecular mechanisms.

7.1.5 A Dialectical Approach: Big Tent Biology

Although the four perspectives I just presented are very different from one another, they are not mutually exclusive. I, for one, oscillate among them. Ultimately, each person—at least each biomedical researcher—must somehow balance the competing interests and attitudes that these perspectives embody. As we pursue this balance, it is important to have access to a broad range of accurate information and to keep in mind that societal attitudes toward the use of models in biology have shifted repeatedly over the course of history—and may shift again.

For me personally, the biggest challenge is to balance my concerns about animal welfare with the desire to alleviate the suffering of sick humans (as well as that of animals in veterinary care). Although an animal or cellular model can be useful even if it does not mimic its target system perfectly, it seems fair to say that successful translation is more likely when the model resembles its targets in as many respects as feasible. It is unfortunate, therefore, that the animals most similar to us (on average) also warrant the greatest ethical concerns. The dilemma was nicely summarized by Karen Rader (2004): "What animals are enough like us to make laboratory results obtained from them generalizable to humans, but not so much like us that we ethically prohibit their being the subjects of experiments?" (p. 22).

There is not one correct answer to this question, as it depends on the research problem under investigation and the stage of the research program. Therefore, I advocate for the use of multiple species in biomedical research, as well as the use of in vitro systems. In effect, I argue for a big-tent biology in which concentrated research on a few reference species is complemented by comparative studies (Striedter et al., 2014), and in which translational research coexists harmoniously with studies that, at first blush, lack direct relevance to human health (i.e., basic biology; Zoghbi, 2013). This inclusive perspective may seem designed to please as many stakeholders as possible (except presumably the anti-vivisectionists), but that is not my intention. Instead, my aim is to encourage every reader—especially the junior scientists (Yartsev, 2017) and shapers of research policy—to think deeply about the relevant issues and then decide for themselves which questions they want to address, which models they consider most promising, and what other kinds of research should be supported.

As we contemplate these questions, we drift toward philosophy. Although biologists generally express little interest in formal philosophy, "scientists in search of answers to real-world issues . . . make decisions constantly that are based on a 'philosophical' stance as to how to do science. . . . For better or for worse, some kind of philosophy is an integral part of the doing of biology" (Orzack, 2012, p. 170). Ideally, those philosophical stances should be developed consciously and, if possible, stated explicitly. To that end, I offer a few recommendations (for an analogous list, see Bolker, 2017).

7.2 RECOMMENDATIONS

The selection of a model system is one of the most influential decisions biologists make in their research, and there are many options. The first step should be to identify a large question of interest. The challenge then becomes finding one or more models that are optimal for addressing this research question. The task requires a careful assessment of one's own and societal attitudes toward animal research as well as a variety of practical considerations, such as the availability of appropriate animal care or cell culture facilities, the feasibility of the required experiments (given the tools available), and the amount of time and resources one can devote to the research. In addition, it is important to learn as much as possible about the strengths and weaknesses of the available model systems and to imagine which new systems might be superior.

Swamped by all these considerations, the overarching question sometimes gets lost. The long-term goals tend to recede as the research homes in on specific puzzles posed by earlier research. This narrowing of focus is understandable and often highly productive, but must questions posed by previous research in a particular model system be answered in the same system? Might a different system offer a better chance of yielding an answer? What would be the costs and benefits of switching to a different model, or adding one? As one contemplates such questions, some points are good to keep in mind.

7.2.1 Know Your Animals, Know Your Cells!

Biologists nowadays tend to order their research animals or cells from a vendor, much like laboratory supplies. In fact, animals and cells are often listed under "materials and supplies" in grant applications. This attitude is ethically problematic, at least when dealing with presumably sentient animals. Moreover, even when the ethical concerns are minimal, the view of animals and cells as research supplies reduces the likelihood that the researchers will inform themselves thoroughly about the peculiarities of their research subjects. Those peculiarities, in turn, can have significant effects on the results of an experiment, especially if they remain unknown and uncontrolled.

In short, biologists can benefit from learning as much as possible about their research animals and cells. It is important to know, for example, that mice are not small rats. These two taxa evolved separately for about 20 million years and diverged considerably in both genome and behavior (Gibbs et al., 2004; Ellenbroek & Youn, 2016). For example, male mice of many strains (including wild mice) are more likely than male rats to fight with other males, which has implications for how to house them in the laboratory (Crawley, 2007; Kondrakiewicz et al., 2019). Another important difference is that rats swim frequently in nature, whereas mice tend to avoid the water; this probably explains why mice are not as good as rats on spatial memory tasks that require swimming (Whishaw & Tomie, 1996). Many other examples of potentially important species differences are mentioned in the preceding pages.

Within a species, strain differences can also be significant. Particularly obvious is that laboratory strains are very different from their wild relatives. For one thing, the laboratory animals tend to be larger and less aggressive. For another, they exhibit numerous differences in physiology, behavior, and gene expression in the brain (Chalfin et al., 2014). The many different strains of laboratory mice also differ from one another (Pugh et al., 2004; Crawley, 2007). Some become blind or deaf as they mature, which researchers should know before they subject these animals to tests requiring those sensory modalities (Brown, 2007). Others vary in their responses to stress (Van Bogaert et al., 2006; Mozhui et al., 2010). Not surprisingly, strain differences can also be dramatic at the genetic level. In the words of Cutler et al. (2007),

> over a hundred regions of a strain's genome may be amplified or deleted in relation to the C57BL/6J reference [mouse strain]. This intra-species variability results in many genes being completely or partially deleted, while many other genes are present at increased copy number in a given mouse strain. . . . This surprising variability in the gene content of inbred mouse strains provides both challenges and opportunities in the use of these strains as models for understanding disease and gene function. (pp. 1743–1744)

The genetic variation across strains becomes especially problematic when mutant lines are maintained on mixed genetic backgrounds, because then genetic tests may be needed to determine which animals are likely to develop which traits (Jankowsky & Zheng, 2017). This may not seem like a significant problem because researchers are typically not focused on strain differences, but the genetic background effects can be quite dramatic. For example, one line of mice carrying a mutant gene linked to Alzheimer's disease was developed on a mixed background, but the animals died prematurely when bred onto the pure background of Black-6 mice (Carlson et al., 1997). Even substrains of the popular Black-6 mice differ in a number of potentially important physiological, biochemical, and behavioral respects (Mekada et al., 2009; Simon et al., 2013). Yet another complication is that mutant model mice may over time diverge genetically from the initial stock to the point where in some cases the animals no longer exhibit the phenotype that originally justified their use as a model (Lutz & Osborne, 2014; the Jackson Laboratory, 2021).

As if all this genetic variation were not troublesome enough, biologists must also worry about the many environmental factors that can significantly modify their research animals. For example, laboratory mice are typically raised in rather barren cages that offer little room for exercise but plenty of monotonous food. Under those conditions, the animals tend to become obese, physiologically compromised, and cognitively impoverished, relative to animals that grow up under more natural, enriched conditions. These differences, in turn, can affect how the animals perform in various tests and respond to treatments such as exercise (Würbel, 2007; Martin et al., 2010; Schellinck et al., 2010). As mentioned in chapter 2 (section 2.2.2), researchers also tend to keep their mice in conditions that are colder than the mice would like (i.e., subthermoneutral conditions), which can influence the outcome of experiments (see figure 2.2; Kokolus et al., 2013). In addition, rearing conditions can affect an animal's microbiome and modify its immune system (Tao & Reese, 2017; Masopust et al., 2017). Finally, researchers should know that many animals are influenced by their cagemates (Baud et al., 2017) and by the humans who handle them. It is startling to realize, for example, that the performance of mice in some behavioral tasks is influenced by the experimenter's sex (Sorge et al., 2014).

Recognizing that experimental outcomes often depend on a large variety of environmental and genetic factors, good researchers tend to control their research animals and experimental conditions rather tightly, which typically means keeping them highly standardized (Nelson, 2018). This strategy makes it easier to obtain consistent results, even if many of the variation-inducing factors are incompletely understood. Unfortunately, results that were obtained only under very specific conditions are unlikely to generalize to other species—that is, they are unlikely to translate well (see section 7.2.2).

Analogous considerations apply to in vitro research. It is rather alarming, for example, how frequently researchers fail to realize that their cell cultures are overrun by HeLa cells or other rapidly dividing cells, even though reports of such cross-contamination have been published repeatedly since 1968 (Gartler, 1968). One study estimates that as many as 20% of all cell lines are misidentified, and another study lists 360 such lines (Hughes et al., 2007; Capes-Davis et al., 2010). It is also common for researchers to use cells that have multiplied in culture for so long that, unbeknownst to the experimenter, they are likely to have changed substantially from their original state (Kaur & Dufour, 2012; Singer et al., 2017). Most cancer cell lines, for example, have diverged somewhat from the original tumors, and some are "hypermutated"—yet they are commonly used (Domcke et al., 2013). Of course, important discoveries can be made with cells that have evolved divergently under in vitro conditions. HeLa cells, for instance, are highly derived (see chapter 4) but have made innumerable contributions to biology. Still, it is clearly a good idea for biologists to learn as much as they can about the cells that they are working with and then to make a conscious decision about their suitability for answering the questions at hand (Krishna et al., 2014).

As part of that decision process, researchers should contemplate the sex of their cells. An astonishing 75% of in vitro studies and 80% of the rodent cell lines distributed by major cell line repositories fail to specify the sex of the cultured cells (Shah et al., 2014; Park et al., 2015). Yet male cells differ from female cells in how they respond to toxins, for example (Nunes et al., 2014). As long as sex remains so commonly unspecified, discovering such differences is difficult. In addition, researchers should consider that homologous cells from different species are likely to differ in diverse ways (Yu & Thomson, 2008; Yang et al., 2011). For example, as mentioned previously, "there are several important differences in the hardwiring of the growth-controlling circuitry of human and mouse cells. . . . Although humans and mice share a common set of protein components, the regulation of their function is distinct enough to generate quite different rules governing their transformation [into tumor cells]" (Hahn & Weinberg, 2002, p. 337).

Induced pluripotent stem cells (iPSCs) derived from human subjects avoid the species difference problem, but they are nonetheless variable and hence difficult to "know." Even when derived from the same individual, induced stem cell lines may vary because of differences in the genome or epigenome of the original cells, because of random changes introduced during the pluripotency induction process, because of mutations incurred while the cells were dividing in vitro, because of differences in the protocol used to differentiate the cells, and because of how far along the cells are in the differentiation process (Carey et al., 2011; Liang & Zhang, 2013; Merkle et al., 2017). To appreciate the importance of this variability, ask yourself: Which cells should be used as controls in studies with patient-derived iPSCs? How can researchers separate the disease-linked features of their cells from other, potentially confounding variables? In the case of Huntington's disease, where the disease-causing mutation is readily identified, control iPSCs can be created by "editing out" the mutation (An et al., 2012). Unfortunately, this is not possible for most other diseases, where the genetic basis is less well defined.

7.2.2 Standardize, but Not Too Much!

Just as the animals and cells used in research vary in diverse ways, so do the methods used to study them. Collectively, all this variation makes it difficult to replicate results, both within a laboratory and across laboratories (see chapter 1). As mentioned in the previous section, scientists often try to combat this problem by calling for stricter standardization of the research subjects and testing methodology (Global Biological Standards Institute, 2013; Freedman & Inglese, 2014). Historically, the extensive inbreeding of laboratory rats, mice, and flies was clearly motivated by a desire to standardize the animals (see chapter 3); similarly, a major motivation behind the creation of immortal stem cell lines was increased standardization of the cells. Researchers also strive to standardize the environmental conditions in which their animals are raised or their cells are maintained, and they try to standardize experimental assays and tests. Yet, despite strong efforts and extensive collaboration between laboratories, variability persists (Crabbe et al., 1999; Wahlsten et al., 2003; Egan et al., 2016; Kafkafi et al., 2018) (figure 7.3). Not everyone agrees that this is bad.

Although increased standardization can certainly be profitable, it does have some downsides. For one thing, premature standardization can leave investigators stuck with inferior systems and methods (Kalueff et al., 2007). For another, extensive standardization

A – Variation in Equipment Across Studies

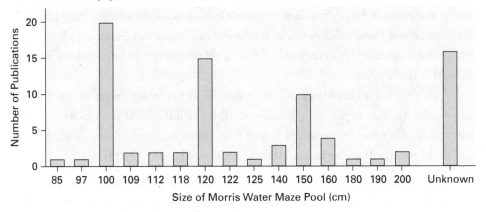

B – Variation Across Strains and Laboratories

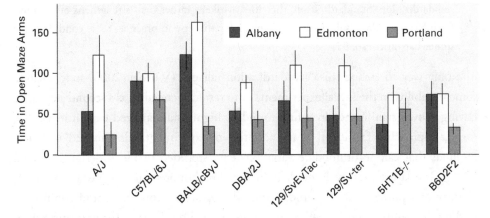

Figure 7.3

Variation across laboratories and mouse strains. (*A*) Variation in equipment across studies. Egan et al. (2016) found that the diameter of the pools used in Morris Water Maze memory experiments performed with Alzheimer's disease model mice varied significantly across 83 published studies (or was not reported). (*B*) Variation across strains and laboratories. Crabbe et al. (1999) compared how mice of different strains and in different laboratories (located in Edmonton, Portland, and Albany) performed on an elevated plus maze with two open and two closed arms. The illustrated measure is the amount of time spent in the open arms, which is a measure of "anxiety" as it is commonly measured in mice; the error bars represent standard errors. Evidently, the variation is substantial, both across strains and across laboratories, despite extensive efforts to standardize experimental procedures. Adapted from (A) Egan et al. (2016); (B) Crabbe et al. (1999).

can lead to the reporting of very small effects that are unlikely to be of use in the real world. For example, highly standardized tumor transplantation studies in mice (see chapter 5) can reveal minute differences in tumor size or growth, but in human clinical trials tumor shrinkage must be on the order of 30% for the treatment to be considered a success (Gould et al., 2015).

A more insidious drawback of standardization is that removing as much variation as possible from an experiment leaves open the possibility that the results may hold only under those specific conditions, which would reduce replicability (Richter et al., 2010). In the words of the famous statistician Ronald Fisher (1953),

> The exact standardization of experimental conditions, which is often thoughtlessly advocated as a panacea, always carries with it the disadvantage that a highly standardized experiment supplies direct information only in respect of the narrow range achieved by standardization. Standardization, therefore, weakens rather than strengthens our ground for inferring a like result, when, as is invariably the case in practice, these conditions are somewhat varied. (p. 99)

The only way to escape this "standardization fallacy" (Würbel, 2000) is to include some variability in the initial experiments or to test for generality as a second step, after having shown significance in a first round of highly standardized experiments. The former approach is facilitated by the adoption of experimental designs and analytical methods that can accommodate variation (e.g., stratified or block designs; Garner, 2014). The latter strategy is implicit in the time-honored (if not always followed) strategy of replicating studies under somewhat different conditions and confirming findings in multiple species before transitioning to clinical trials. The anticancer drug trastuzumab, for example, was shown to be effective in multiple mouse models by several different laboratories before it was tested in humans (Pegram & Ngo, 2006).

One area in which I personally would like to see more standardization is the statistical analysis of similarities and differences across species, strains, or cellular systems. For example, the extent of overlap between lists of differentially expressed genes in two or more systems (e.g., in multiple models of a specific disease) can be evaluated in several different ways (Kuhn et al., 2007; Plaisier et al., 2010; Handler & Haynes, 2019), but all of the approaches entail different sets of assumptions (Lu et al., 2009; Djordjevic et al., 2016; Hargis & Blalock, 2017).

Given this profusion of methodologies, the best approach is probably to compare different models and the target to one another using a single methodology, and then evaluate their relative degrees of similarity. Using such an approach, Burns et al. (2015) determined that some AD model mice are much better than others in terms of modeling human Alzheimer's disease. It is unclear, however, whether different methods of analysis would yield equivalent results. Similarly, Kuhn et al. (2007) compared the gene expression changes observed in humans with Huntington's disease and several mouse HD models. Although the data clearly indicate some differences between the models (especially in the time course of symptom development), the authors focused on their similarities. As in the controversy surrounding mouse models of human sepsis (see chapter 5, section 5.1.4), the lack of generally accepted standards for comparative transcriptome analyses increases the likelihood that some biologists will see significant similarities where others see great differences.

7.2.4 Learn from Clinical Trial Failures!

Scientific models, both abstract and material (see chapter 2), can promote discovery by revealing phenomena that are evident in the model but not yet known for the target. Thus, a useful model generates predictions that can be tested on the target system. This benefit is widely recognized, but models can also be useful when their predictions are not borne out, when the extrapolation fails. In that case, researchers must reevaluate the model's assumptions and go back to the drawing board.

In biomedical research, the process of learning from failed extrapolations is a form of reverse translation, although this term is often used more generally to describe situations where patient data inspire the creation of disease models (Malkesman et al., 2009; Nadeau & Auwerx, 2019). Despite the general interest in reverse translation, the most common response to failed clinical trials is to start additional trials (with the same or other drugs) or to withdraw from such research entirely. Cook et al. (2014), for instance, report that a major pharmaceutical company used to accept that a certain percentage of its clinical trials would fail and attempted to address the problem simply by conducting a greater number of trials, hoping for a few blockbuster drugs. When this approach did not succeed, the company embarked on a more systematic evaluation of its successes and failures.

Although post hoc analyses of clinical trial failures are rarely published, a few failures did receive detailed follow-up attention. The drug thalidomide, for example, caused a large number of miscarriages and birth defects because it inhibits blood vessel formation in the developing limbs of embryos. Once this mechanism was understood, researchers hypothesized that thalidomide might be effective against cancer by preventing growing tumors from becoming vascularized (Rehman et al., 2011). Indeed, thalidomide is now used in the treatment of multiple myeloma (Palumbo et al., 2008). Another good example is presented by the anticancer drug gefitinib (Ledford, 2008). A clinical trial that had averaged its results across all enrolled patients showed this drug to be ineffective against lung cancer. However, subsequent studies showed that the drug is effective in a subset of patients who have intact, wild-type versions of the *EGFR* (epidermal growth factor receptor) gene (Paez et al., 2004; T. J. Lynch et al., 2004). Apparently, patients with mutations in *EGFR* had been common enough in the initial trial to confound the results. Thanks to this insight, gefitinib is now approved for the treatment of lung cancer in patients with wild-type *EGFR*.

The analysis of clinical trial failures is likely to reveal mainly information that is specific to the tested drug. However, closer scrutiny of many different failed trials might reveal more general patterns. Many researchers have concluded, for example, that the clinical trials for neurological disorders may have enrolled patients too late in the progression of their disease for the putative treatments to have the hoped-for effect (see chapter 6). They are now testing this hypothesis by enrolling patients earlier, but identifying who is likely to develop a neurological disorder is not a trivial task. Families or isolated populations who have a high prevalence of a specific disease, such as Huntington's disease, may prove useful in this regard (Castilhos et al., 2016).

Along a very different line, I suspect that the high failure rate for drugs targeting neurodegenerative diseases (see chapter 6) stems, at least in part, from species differences in how human and mouse neurons respond to cellular stressors such as the presence of misfolded proteins. Human neurons might be intrinsically more sensitive, or the human brain's immune system might be less effective at preventing cell death. Unfortunately, the literature so far contains only a few hints consistent with this hypothesis (Smith & Dragunow, 2014; Burns et al., 2015; Espuny-Camacho et al., 2017).

7.2.5 Embrace Diversity!

Historically, many biologists—especially molecular biologists—have treated variation between species as a nuisance to be reduced or ignored (Davis, 2003). While this "diversity denial" (Murray et al., 2016) persists in some corners of biology, the attitude toward variation has begun to shift with the rise of comparative genomics and, more generally, the maturation of molecular biology. An important factor in this shift is that genome editing and other technological advances now make it much easier to manipulate the genomes of many different species and cells (Juntti, 2019; Matthews & Vosshall 2020).

Biologists have also started paying more attention to sex, strain, and cell type differences. Such differences have long been reported, but were then neglected. As Hynds et al. (2018) put it in their review of cancer cell lines, the discovery that cancer cells can change dramatically in cell culture should "signal an end to the era in which researchers informally acknowledge that different strains of cancer cell lines are heterogeneous and unstable over long-term passage, but in practice treat them as though they were clonal entities" (p. 4). Of course, similarities should not be neglected either. As Francis

Bacon, the founding father of the scientific method, observed in 1620, some scientists tend to focus on differences, others on similarities; the trick is not to "fall into excess" (see chapter 2, section 2.6).

In this book, I certainly have emphasized differences over similarities, but I have done so with the intent of countering biology's lingering bias in favor of similarities. In general, I think it is best to view similarities and differences as reciprocally illuminating one another (Lehrman, 1971). When confronted with a trove of unfamiliar information, our natural inclination is to seek patterns amid the chaos. However, once those regularities have become apparent, the deviations from the expected come into focus, setting off another round of searching for patterns amid the exceptions. Thus, it is the tension between similarities and differences, between order and disorder, that causes science to advance (Bohm, 1957). The aim, therefore, should be to seek unity within diversity, but not at the expense of diversity.

In particular, I focus in this book on differences that are likely to be medically relevant. Some biologists have recognized such differences for a long time. For example, the biochemist Efraim Racker wrote in 1954 that

> the steadily growing recognition of the existence of alternate pathways, of qualitative and quantitative differences in enzymatic patterns, of differences in submicroscopic cell structure, permeability and rate of cell division have been quoted in favor of a "disunity in biochemistry." The assessment of those features that are not common to various cells might serve to provide us with a better understanding of the disease process as well as its control. (Friedmann, 2004, p. 57)

Extending this perspective, it becomes apparent that there is a considerable amount of disunity not just in biochemistry but in all of biology. For example, many genes have been lost, duplicated, or created de novo in various lineages. Even the broadly conserved "disease genes" have sometimes diverged substantially (see chapter 2). So have their regulatory regions and many of the other genes and proteins with which those disease genes interact. Consistent with this divergence, 27 out of 120 genes that are essential for survival or reproduction in humans are nonessential in mice (Liao & Zhang, 2008). One such gene is *SOD1* (Dickinson et al., 2016), which probably helps to explain why findings from SOD1 model mice have not translated well to human ALS (see chapter 6). Similarly, humans who possess three copies of the wild-type *APP* (amyloid beta precursor protein) gene develop early onset Alzheimer's disease (associated with Down's syndrome), but this is not the case in mice expressing three copies of their own wild-type APP (Wiseman et al., 2015). Recognizing such variation is interesting, but more important is to learn from it. In the words of Fisher and Bannerman (2019), we need to recognize variation and use it as a source of insight. Variation has hitherto been seen as a problem and something that should be diminished and reduced at all costs. . . . We suggest that variation could be an opportunity that may allow us to understand and identify disease mechanisms and risk factors and, at the same time, to elucidate treatment strategies on an individual by individual basis. Embracing and understanding variation may be of great benefit for translation. (p. 11)

One way to gain translationally useful insights from variation is to study species, strains, or individuals that are either highly susceptible or resistant to specific diseases (Green et al., 2018). Naked mole rats, for example, are unusually resistant to cancer, which helps explain why they live up to 30 years, far longer than other small rodents. Detailed studies have revealed a variety of mechanisms that contribute to this cancer resistance, including an increased tendency for cells to stop dividing when they contact other cells (Seluanov et al., 2009; Tian et al., 2013). Some of those anticancer defenses in naked mole rats may ultimately lead to cancer treatments in humans. Similarly, researchers have begun to explore why rats and some strains of mice are much less sensitive than monkeys to MPTP, the neurotoxin that induces Parkinsonian symptoms in primates (see chapter 6) (Giovanni et al., 1994; Smeyne et al., 2005). Another interesting line of medically relevant comparative research is the study of individual differences in the propensity of rats to become addicted to drugs (Deroche-Gamonet et al., 2004). Paralleling humans, some rats are far more likely than others to become compulsive drug users, and researchers have started to decipher which factors might account for those individual differences (Belin et al., 2011).

A very different but also promising line of comparative research is the use of recombinant inbred strains (Churchill et al., 2004; Threadgill et al., 2011; Collaborative Cross Consortium, 2012). For this work biologists have assembled "panels" of many different inbred lines (e.g., of mice or fruit flies) that were created by randomly intercrossing two or more different parental strains. Because each line is inbred, researchers reap all the benefits of working with "standardized" animals. However, the lines in the panel differ from one another in both genotype and phenotype. Crucially, the lines have already been well characterized so that phenotypic variation between the strains can readily be correlated with genotypic variation (i.e., mapped). This procedure is especially useful for the genetic dissection of complex traits that involve the interactions of multiple genes, which are difficult to study with traditional genetics (Flint & Mackay, 2009). It also facilitates the study of gene-environment interactions. From a biomedical perspective, working with recombinant inbred lines is an excellent (albeit labor-intensive) way of determining the molecular mechanisms underlying disease resistance or susceptibility (Souza et al., 2020).

Despite the benefits of studying a wide array of species, it is important to reiterate that not everything one finds in our distant relatives will be broadly conserved. Many people seem to believe that evolution creates complex organisms by merely adding novel features to simpler ancestors, which implies that all the features we find in simple organisms should be conserved in more complex creatures. However, evolution often modifies old features, even if they are "fundamental" (i.e., involved in an essential cellular or physiological process). The molecular interactions that control the cell cycle, for example, have diverged substantially between plants, yeast, and multicellular animals (see Striedter, 2019). Similarly, the genome of *Drosophila* features only four chromosomes and 14,000 protein coding genes, but it is more complex than originally thought (e.g., as a result of extensive alternative splicing) and has diverged considerably between the various *Drosophila* species (Holloway et al., 2008; Hales et al., 2015).

Even the two major types of yeast used in research (*Schizosaccharomyces cerevisiae* and *S. pombe*) have diverged to the point where roughly 20% of the genes in either species have no clear homologs in the other (Wood et al., 2002). This rampant divergence between distantly related species (note that the two yeasts have evolved separately for 330–420 million years) does not contradict the notion that similarity tends, on average, to correlate with phylogenetic relatedness. However, it implies that the tiered approach to model systems research is fraught: many of the features we discover in simple, distant relatives may differ substantially from our own. Or they may not. Again, we cannot be sure until we do the comparative analysis.

7.2.6 Reckon with Complexity!

Since the early days of fly genetics (see chapter 3) and the discovery of DNA in 1953 (Watson & Crick, 1953), scientists have tended to embrace a gene-centered view of biology. Although this general approach has undeniably yielded significant progress, it is ripe for reexamination. In particular, we might want to question whether it is appropriate to

> think of each gene as a word in a language, the common language of biology. A word has a specific meaning but it can appear in different contexts. So the word *house* has the same meaning in a very simple sentence that it would have in a complex sentence. Think of genes as the vocabulary of biology. Once this vocabulary is understood, it can be the starting point to understand how genes are assembled to create organisms with a range of different complexities. To learn this vocabulary, we are best off understanding it by reading the simplest text possible. For that reason, we start with something simple like the worm. (Bargmann 2000, p. 520)

This statement is not, strictly speaking, wrong, especially if we emphasize the "starting point" phrase. However, the philosophy of science embedded in this analogy is problematic for two reasons. First, it disregards the many derived (i.e., not broadly conserved) features of the so-called simple organisms (as we just discussed). Second, it implies that the meanings of words and the functions of genes are largely independent of the context in which they are deployed. This is certainly not true for many words (Gennari et al., 2007). The word "house," for instance, can be a noun or a verb, depending on how it is deployed in a sentence. Genes, too, often have different functions (i.e., meanings) in different molecular and cellular contexts.

Although most genes and proteins are thought to interact with a conserved core set of other molecules, the overall "wiring diagram" of genetic and protein-protein interactions can vary across cell types, cell states, developmental stages, species, and other dimensions (Bandyopadhyay et al., 2010; Ideker & Krogan, 2012; Diss et al., 2013). For example, the transcription factor *Myc* can stimulate epidermal cells either to proliferate or to stop dividing and differentiate, depending on the timing and level of *Myc* expression (Watt et al., 2008).

Gene functions can also vary with the availability of other, interacting genes and proteins. The transcription factor REST (repressor element-1 silencing transcription factor), for instance, targets very different sets of genes in embryonic stem cells versus neural progenitors, most likely because of major differences in chromatin organization between the two cell types (Johnson et al., 2008). Then there are changes in gene sequence that can directly modify a protein's interactions with other molecules. Mutations in the *huntingtin* gene, for example, expand the corresponding protein's interactome (Basu et al., 2013). Major changes in gene function also occur across species, as illustrated by the evolution of the Huntingtin-associated protein Hap1. This protein underwent significant sequence changes in its functional domains that were associated with major changes in the protein's interactome and physiological functions (Lumsden et al., 2016). Another interesting example of evolutionary change in gene function is that deletion of the *retinoblastoma* (*Rb*) gene causes retinal degeneration in humans but has the same effect in mice only if it is deleted in conjunction with an additional tumor repressor gene (Robanus-Maandag et al., 1998; Hahn & Weinberg, 2002). This finding implies some significant genetic rewiring between humans and mice.

One could list many additional examples of genes that have somewhat different functions in different cell types, at different developmental stages, in different species, and so forth, but no matter how long the list, the examples would likely be dismissed (by some) as being exceptions to the rule. Indeed, there is surprisingly little hard evidence to provide empirical estimates of genetic conservation versus change. However, some relevant findings have emerged from studies that use abstract, mathematical models to understand networks of interacting genes and proteins. Specifically, scientists have modeled several well-defined molecular systems, such as insulin or growth factor signaling, with a set of differential equations (Sedaghat et al., 2002; Brown et al., 2004). Such models typically contain many "free" parameters that are not known but could affect the system's behavior.

A clever way around this problem is to systematically explore the entire parameter space (i.e., test many randomly selected sets of parameters) and ask how differences in specific parameters affect the system's behavior. Such studies have revealed that most of these biological systems are subject to "sloppy control," which means that a relatively small number of parameters are disproportionately influential (Brown et al., 2004; Gutenkunst et al., 2007). The flip side is that many parameters exert little influence over the system's behavior and, therefore, do not need to be empirically determined if one wants to model the system mathematically.

Even more interesting, for present purposes, is that the importance of changes in any one parameter tends to depend on the constellation of the other parameters (figure 7.4). That is, a parameter that is influential within one randomly selected set of parameters may well be negligible in the context of a different parameter set. Moreover, when such parameter sets are subjected to simulated (i.e., in silico) evolution, parameters that exerted great control over the system's behavior in one generation may become unimportant in a later generation, or vice versa. This phenomenon, called "causal drift" (Wagner, 2015), is consistent with the finding that essential genes in one species may be nonessential in another (see section 7.2.5) and that gene knockout effects often depend on the mutation's genetic background (Gerlai, 1996).

More generally, the computational modeling suggests that context-dependent gene and protein functions might be the rule rather than the exception, at least for biological networks of modest size (which is all we can currently model). In the context of translational research, this means that "any one disease determinant that is crucial in one [genetic] background will be modestly important in another and virtually irrelevant in yet another background. In an evolving population that explores the parameter space of such a circuit through DNA mutations, genetic determinants of disease can vary randomly over time" (Wagner, 2015, pp. 2–3).

In other words, we should not be surprised that different species, sexes, and strains sometimes respond differently to different drug treatments, disease triggers, or other manipulations, even when the underlying principles (i.e., the applicable differential equations) remain conserved. It is a consequence of biological complexity!

How should this insight influence the course and conduct of translational research? I do not know, but I submit that biologists are already on the right track insofar as they are paying more attention to both biological complexity and variability (especially sex differences; see chapter 1). In terms of drug development, I agree with Roth et al. (2004) and others (Scannell et al., 2012; Reddy & Zhang, 2013) that we might benefit

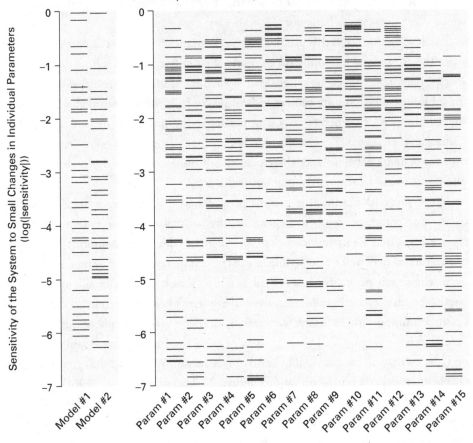

Figure 7.4

Sloppy control and context-dependence in complex systems. (*A*) Two models. Gutenkunst et al. (2007) performed a sensitivity analysis for two biological systems that have been modeled quantitatively—namely, rat growth factor signaling and *Drosophila* segment polarity development. Each horizontal bar indicates how sensitive the model is to small changes in a specific parameter (the closer to zero the plotted data point, the greater the sensitivity). For both models, some parameters have a lot of influence over the system, whereas others exert very little control; this is called "sloppy control" (Brown et al., 2004). (*B*) Fifty versions of the same model. Wagner (2015) performed a similar analysis on a model of cellular insulin signaling, but he did this for multiple, randomly selected, viable parameter sets. Each column shows the sensitivity distribution of a specific model parameter across the 50 randomly generated iterations of the model. Across iterations, the individual parameters varied widely in how much control they exert over the system. The general conclusion is that the importance of any parameter within a complex system depends heavily on the system's other parameters. Adapted from (A) Gutenkunst et al. (2007); (B) Wagner (2015).

from shifting our search image from the highly selective magic bullets promoted by Paul Ehrlich and others (Strebhardt & Ullrich, 2008) to magic shotguns that affect multiple processes, at least for complex diseases. Such multitarget drugs may be more difficult to discover in the laboratory, but history suggests that some fungi, plants, or animals may have already "discovered" compounds with the requisite properties naturally, though natural selection (e.g., aspirin from willow tree bark). I also believe that we have much to gain from further efforts to fight disease by strengthening the body's natural defenses, as in the case of cancer immune checkpoint therapy (see chapter 5). Improving our ability to manipulate the brain's immune system (mainly microglia) would be especially helpful in the fight against neurological disorders.

As biologists proceed with this hard work, I hope that they will increasingly adopt an organismal and comparative perspective, rather than the gene-centered view that has dominated biomedical research for the last 70 years. In essence, I would love to see a merger of the comparative physiological perspective that August Krogh (1929) and other early physiologists embraced (see chapter 2, section 2.6) with the insights and techniques of modern molecular biology. Whether the emerging field of systems biology will expand to incorporate this broader perspective remains to be seen (Joyner, 2011).

7.3 CONCLUSION

My overarching aim in this book has been to increase awareness of the issues that surround the selection and use of model systems in biomedical research. As announced in chapter 1, my focus has been on biomedical research, rather than biology in general, because the high rate of translational failures in biomedicine represents an urgent problem that warrants more debate and analysis. As stated in this chapter's opening quotation, "continuing to do the same things in the same way cannot go on" (Stein, 2015, p. 1259). The question of how to change research conduct or policy is difficult, of course, and I do not have definitive answers. Most of my recommendations are offered not with a sense of certainty, but with the conviction that, after such extensive study of the topic, I really ought to offer at least a few suggestions. Even if my efforts simply prompt some readers to give the topic more explicit thought, I would be pleased.

Libby (2015) and others have pointed out that many of our disease models are not really models as much as they are tools biologists can use to gain pathophysiological insights. I agree with this assessment, but describing animals and cells as "tools" downplays the fact that they have evolutionary and developmental histories that make them different from one another. We can use genetic techniques and environmental manipulations to alter our models, but we do not build them from scratch; they have their own complexities and idiosyncrasies, many of which remain unknown. Therefore, when we use these material models to gain insights into the human condition, we will sometimes stumble over differences that make a difference. As they say in veterinary medicine, it is "all one medicine until it is different" (Hunter, 2010). Of course, biomedical researchers should try to extrapolate from their models to humans, but they should do so humbly and endeavor to learn from their mistakes. In particular, they should seek principles that can account for the observed variation. They should also be prepared to change their models and, when feasible, use multiple models.

In a stinging critique of biomedical research, David Horrobin (2003) accused the scientists engaged in animal and in vitro research of being unconcerned with real-world clinical problems, but this is unfair. They do care, and most are deeply frustrated when their best efforts at biomedical translation fail. It is simply that there is no easy fix. Still, as in so many aspects of life, becoming more aware of where the problems lie tends to reveal a more productive road forward.

APPENDIX

Winners of the Nobel Prize in Physiology or Medicine, their celebrated discoveries, and their principal model systems. The listed information is based mainly on compilations by the National Research Council (1985) and the Nobel Foundation (https://www.nobelprize.org/prizes/lists/all-nobel-laureates-in-physiology-or-medicine/). Please note that most of the winners worked with multiple model systems, not all of which are listed here.

Year	Winner(s)	Discovery or Research Area	Experimental System(s)
1901	von Behring	Serum therapy	Guinea pig
1902	Ross	Mechanisms of malaria	Mosquito, human, diverse birds
1903	Finsen	Light radiation therapy	Bacteria (various)
1904	Pavlov	Physiology of digestion	Dog
1905	Koch	Mechanisms of tuberculosis	Cattle, sheep
1906	Golgi, Cajal	Fine structure of the nervous system	Human, other mammals, birds, reptiles
1907	Laveran	Protozoa as causes of disease	Human, protozoa, mammal, bird, fish
1908	Metchnikov, Ehrlich	Immune mechanisms	Guinea pig
1909	Kocher	Thyroid gland function and disease	Human
1910	Kossel	Proteins and nucleic acids	Bird
1911	Gullstrand	Dioptrics of the eye	Human
1912	Carrel	Blood vessel and organ transplantation	Dog
1913	Richet	Anaphylaxis	Dog, rabbit
1914	Barany	Vestibular physiology and pathology	Human
1919	Bordet	Immunity	Guinea pig, horse, rabbit
1920	Krogh	Regulation of capillary flow	Frog
1922	Hill, Meyerhof	Muscle physiology	Frog

(continued)

Year	Winner(s)	Discovery or Research Area	Experimental System(s)
1923	Banting, Macleod	Insulin	Dog, rabbit, fish
1924	Einthoven	Electrocardiogram	Human, dog, frog
1926	Fibiger	*Spiroptera carcinoma* (a parasitic roundworm)	Rat, cockroach
1927	Wagner-Jauregg	Malaria inoculation to treat dementia paralytica	Human
1928	Nicolle	Typhus	Monkey, chimpanzee, pig, rat, mouse
1929	Eijkman, Hopkins	Antineuritic and growth-stimulating vitamins	Bird
1930	Landsteiner	Human blood groups	Human
1931	Warburg	Respiratory enzymes	Yeast
1932	Sherrington, Adrian	Neuron function	Frog, dog, cat
1933	Morgan	Role of chromosomes in heredity	Fruit fly
1934	Whipple, Minot, Murphy	Liver therapy in cases of anemia	Human, dog
1935	Spemann	Organizer effects in embryonic development	Amphibian
1936	Dale, Loewi	Chemical transmission by neurons	Cat, mammal, frog, reptile, bird
1937	Szent-Gyorgyi Nagyrápolt	Cellular combustion and vitamin C	Potato, pepper, lemon
1938	Heyman	Neural regulation of breathing	Dog
1939	Domagk	Antibacterial effects of prontosil	Mouse, rabbit
1943	Dam, Doisy	Vitamin K	Rat, mouse, dog, chick, alfalfa, fish meal
1944	Erlanger, Gasser	Properties of single nerve fibers	Frog, cat
1945	Fleming, Chain, Florey	Penicillin	*Penicillium* bacteria
1946	Muller	Mutations produced by X-ray irradiation	Fruit fly
1947	Cori, Cori, Houssay	Catalytic conversion of glycogen	Frog, toad, dog
1948	Müller	DDT as a an insecticide	Beetle, fly, aphid, gnat
1949	Hess, Moniz	Hypothalamus; prefrontal lobotomies	Cat, human
1950	Kendall, Reichstein, Hench	Adrenal cortical hormones	Cattle, dog, human
1951	Theiler	Yellow fever	Monkey, mouse
1952	Waksman	Streptomycin against tuberculosis	Guinea pig
1953	Krebs, Lipmann	Citric acid cycle; coenzyme A	Pigeon
1954	Enders, Weller, Robbins	Cultivation of the polio virus	Monkey, human, mouse, polio virus
1955	Theorell	Oxidation enzymes	Horse
1956	Cournand, Forssmann, Richards	Cardiovascular pathology and heart catheterization	Human
1957	Bovet	Antihistamines	Dog, rabbit
1958	Beadle, Tatum, Lederberg	Genetic analysis of metabolism; bacterial genetics	Fruit fly, snapdragon, neurospora, *Escherichia coli*

Year	Winner(s)	Discovery or Research Area	Experimental System(s)
1959	Ochoa, Kornberg	Biological synthesis of nucleic acids	*Azotobacter, E. coli*, snake, mycobacterium, bacteriophage T2, calf
1960	Burnet, Medawar	Acquired immune tolerance	Rabbit, cattle, mouse, sheep, chick
1961	von Bekesy	Mechanisms of cochlear function	Guinea pig
1962	Wilkins, Watson, Crick	Structure and function of nucleic acids	Turnip yellow mosaic virus, tobacco mosaic virus, *E. coli*
1963	Eccles, Hodgkin, Huxley	Ionic basis of neuronal signaling; neuronal inhibition	Cat, frog, squid, crab
1964	Bloch, Lynen	Cholesterol and fatty acid metabolism	Rat, yeast
1965	Jacob, Lwoff, Monod	Genetic control of enzyme and virus synthesis	Bacteriophage, *E. coli*
1966	Rous, Huggins	Tumor-inducing viruses; hormonal treatment of prostate cancer	Rat, chicken, rabbit, virus, dog, rat
1967	Granit, Hartline, Wald	Visual processing in eyes	Frog, rabbit, snake, eel, horseshoe crab, fish, squid, chicken
1968	Holley, Khorana, Nirenberg	Interpretation of the genetic code; protein synthesis	Rat, *E. coli*, yeast, virus, frog, guinea pig
1969	Delbrück, Hershey, Luria	Viral genetics	Bacteria, bacteriophage
1970	Katz, von Euler, Axelrod	Neurotransmitter storage, release, and inactivation	Squid, Frog, Cat, Rat
1971	Sutherland	Hormone action	Mammalian liver
1972	Edelman, Porter	Chemical structure of antibodies	Guinea pig, human, rabbit
1973	von Frisch, Lorenz, Tinbergen	Organization and elicitation of behavior patterns	Bee, wasp, bird, fish
1974	Claude, de Duve, Palade	Cellular organization	Chicken, guinea pig, rat
1975	Baltimore, Dulbecco, Temin	Tumor viruses	Poliovirus, monkey, virus, mouse, chicken
1976	Blumberg, Gajdusek	Origin and transmission of infectious diseases	Human, chimpanzee
1977	Guillemin, Schally, Yalow	Peptide hormones in the brain; peptide radioimmunoassays	Sheep, pig, human
1978	Arber, Smith, Nathans	Restriction enzymes	Bacteria
1979	Cormack, Hounsfield	Computer-assisted tomography	Human, pig
1980	Benacerraf, Dausset, Snell	Immune reactions to cell surface complexes	Human, guinea pig, mouse
1981	Sperry, Hubel, Wiesel	Functional lateralization and visual processing of the cerebral cortex	Human, cat, monkey
1982	Bergstrom, Samuelson	Prostaglandins	Soybean, sheep, rabbit
1983	McClintock	Mobile genetic elements	Corn
1984	Milstein, Kohler, Jerne	Monoclonal antibody production	Mouse, plasma cells
1985	Brown, Goldstein	Regulation of cholesterol metabolism	Human cells

(continued)

Year	Winner(s)	Discovery or Research Area	Experimental System(s)
1986	Cohen, Levi-Montalcini	Growth factors	Bird, mouse, in vitro
1987	Tonegawa	Generating antibody diversity	Mouse, in vitro
1988	Black, Elion, Hitchings	Drug treatment principles	Human, guinea pig, mouse, rat, bacteria, in vitro
1989	Bishop, Varmus	Retroviral oncogenes	Virus, chicken
1990	Murray, Thomas	Organ and cell transplantation	Human
1991	Neher, Sakman	Functions of single ion channels	Mammalian cells
1992	Fischer, Krebs	Reversible protein phosphorylation	In vitro
1993	Roberts, Sharp	Split genes	Adenovirus
1994	Gilman, Rodbell	G-proteins and signal transduction	In vitro
1995	Lewis, Nüsslein-Vollhard, Wieschaus	Genetic control of embryonic development	Fruit fly
1996	Doherty, Zinkernagel	Cell-mediated immune defense	Virus, mouse, in vitro
1997	Prusiner	Prions	Human, hamster, mouse
1998	Furchgott, Ignarro, Murad	Nitric oxide signaling in the cardio-vascular system	Rabbit, cattle, rat, in vitro
1999	Blobel	Intracellular transport signals	In vitro
2000	Carlsson, Greengard, Kandel	Neuronal signal transduction	Human, mouse, rat, *Aplysia*
2001	Hartwlell, Hunt, Nurse	Cell cycle regulation	Yeast, sea urchin
2002	Brenner, Horvitz, Sulston	Genetic regulation of development; programmed cell death	*Caenorhabditis elegans*
2003	Lauterbur, Mansfield	Magnetic resonance imaging	Magnets and analytical techniques
2004	Axel, Buck	Odorant receptors	Rat, mouse
2005	Marshall, Warren	*Heliobacter pylori*, gastritis, and peptic ulcers	*Heliobacter*, human
2006	Fire, Mello	Gene silencing by double-stranded RNA	*C. elegans*
2007	Capecchi, Evans, Smithies	Targeted modification of genes in embryonic stem cells	Mouse, mammalian cells
2008	zur Hausen, Barré-Sinoussi, Montagnier	Link between human papilloma virus and cancer; human immuno-deficiency virus	Viruses, human
2009	Blackburn, Greider, Szostak	Telomere function and regulation	Protozoa (*Tetrahymena*), yeast, human
2010	Edwards	In vitro fertilization	Human
2011	Beutler, Hoffmann, Steinman	Activation of innate immunity	Fruit fly, mouse
2012	Gurdon, Yamanaka	Reprogramming of cells to pluripotency	Frog, mouse
2013	Rothman, Schekman, Südhof	Regulation of intracellular vesicle trafficking	Yeast, rat, cow, mammalian cells

Year	Winner(s)	Discovery or Research Area	Experimental System(s)
2014	O'Keefe, Moser, Moser	Hippocampal place cells	Rat
2015	Campbell, Omura, Youyou	Roundworm infection therapy	Bacteria, human, mammals, plant (*Artemisia*)
2016	Ohsumi	Mechanisms of autophagy	Yeast
2017	Hall, Rosbach, Young	Molecular mechanisms of circadian rhythms	Fruit fly
2018	Allison, Honjo	Cancer immune checkpoint therapy	Mouse, human
2019	Kaelin, Ratcliffe, Semenza	Oxygen sensing by cells	Mouse, mammalian cells
2020	Alter, Houghton, Rice	Hepatitis C virus	Virus, human, chimpanzee

REFERENCES

Abbott, A. (2014). The changing face of primate research. *Nature, 506*, 24–26.

Abel, J. J., Rowntree, L. G., & Turner, B. B. (1914). On the removal of diffusible substances from the circulating blood of living animals by dialysis. *Journal of Pharmacology and Experimental Therapeutics, 5*, 275–316.

Abolins, S., King, E. C., Lazarou, L., Weldon, L., Hughes, L., Drescher, P., Raynes, J. G., Hafalla, J. C. R., Viney, M. E., & Riley, E. M. (2017). The comparative immunology of wild and laboratory mice, *Mus musculus domesticus. Nature Communications, 8*, Article 14811.

Abraham, J. J. (1948). Some account of the history of the treatment of syphilis. *British Journal of Venereal Diseases, 24*, 153–161.

Acharya, H. J. (2004). The rise and fall of the frontal lobotomy. In W. A. Whitelaw (Ed.), *The Proceedings of the 13th Annual History of Medicine Days* (pp. 32–41). Faculty of Medicine, University of Calgary.

Adami, H.-O., Berry, S. C. L., Breckenridge, C. B., Smith, L. L., Swenberg, J. A., Trichopoulos, D., Weiss, N. S., & Pastoor, T. P. (2011). Toxicology and epidemiology: Improving the science with a framework for combining toxicological and epidemiological evidence to establish causal inference. *Toxicological Sciences, 122*, 223–234.

Adriaens, E., Verstraelen, S., Alépée, N., Kandarova, H., Drzewiecka, A., Gruszka, K., Guest, R., Willoughby, J. A., & Van Rompay, A. R. (2018). CON4EI: Development of testing strategies for hazard identification and labelling for serious eye damage and eye irritation of chemicals. *Toxicology in Vitro, 49*, 99–115.

Agosto, L. M., Uchil, P. D., & Mothes, W. (2015). HIV cell-to-cell transmission: Effects on pathogenesis and antiretroviral therapy. *Trends in Microbiology, 23*, 289–295.

Agoston, D. V. (2017). How to translate time? The temporal aspect of human and rodent biology. *Frontiers in Neurology, 8*, Article 92.

Ahlskog, J. E., & Muenter, M. D. (2001). Frequency of levodopa-related dyskinesias and motor fluctuations as estimated from the cumulative literature. *Movement Disorders, 16*, 448–458.

Ahuja, C. S., Mothe, A., Khazaei, M., Badhiwala, J. H., Gilbert, E. A., Kooy, D. van der, Morshead, C. M., Tator, C., & Fehlings, M. G. (2020). The leading edge: Emerging neuroprotective and neuroregenerative cell-based therapies for spinal cord injury. *Stem Cells Translational Medicine, 9*, 1509–1530.

Aitken, M. M. (1983). Species differences in pharmacodynamics: Some examples. *Veterinary Research Communications, 7*, 313–324.

Alberts, A. W., Chen, J., Kuron, G., Hunt, V., Huff, J., Hoffman, C., Rothrock, J., Lopez, M., Joshua, H., Harris, E., et al. (1980). Mevinolin: A highly potent competitive inhibitor of hydroxymethylglutaryl-coenzyme A reductase and a cholesterol-lowering agent. *Proceedings of the National Academy of Sciences of the United States of America, 77,* 3957–3961.

Alberts, A. W., MacDonald, J. S., Till, A. E., & Tobert, J. A. (1989). Lovastatin. *Cardiovascular Drug Reviews, 7,* 89–109.

Altrock, P. M., Liu, L. L., & Michor, F. (2015). The mathematics of cancer: Integrating quantitative models. *Nature Reviews: Cancer, 15,* 730–745.

Alzheimer's Association. (2019). 2019 Alzheimer's disease facts and figures. *Alzheimer's and Dementia, 15,* 321–387. https://www.alz.org/media/documents/alzheimers-facts-and-figures-2019-r.pdf.

Amar, D., Shamir, R., Yekutieli, D. (2017). Extracting replicable associations across multiple studies: Empirical Bayes algorithms for controlling the false discovery rate. *PLoS Computational Biology, 13,* Article e1005700.

Ament, S. A., Pearl, J. R., Cantle, J. P., Bragg, R. M., Skene, P. J., Coffey, S. R., Bergey, D. E., Wheeler, V. C., MacDonald, M. E., Baliga, N. S., et al. (2018). Transcriptional regulatory networks underlying gene expression changes in Huntington's disease. *Molecular Systems Biology, 14,* Article e7435.

Ament, S. A., Pearl, J. R., Grindeland, A., St Claire, J., Earls, J. C., Kovalenko, M., Gillis, T., Mysore, J., Gusella, J. F., Lee, J.-M., et al. (2017). High resolution time-course mapping of early transcriptomic, molecular and cellular phenotypes in Huntington's disease CAG knock-in mice across multiple genetic backgrounds. *Human Molecular Genetics, 26,* 913–922.

An, M. C., Zhang, N., Scott, G., Montoro, D., Wittkop, T., Mooney, S., Melov, S., & Ellerby, L. M. (2012). Genetic correction of Huntington's disease phenotypes in induced pluripotent stem cells. *Cell Stem Cell, 11,* 253–263.

Andersen, P. M. (2006). Amyotrophic lateral sclerosis associated with mutations in the CuZn superoxide dismutase gene. *Current Neurology and Neuroscience Reports, 6,* 37–46.

Anderson, A. J., & Cummings, B. J. (2016). Achieving informed consent for cellular therapies: A preclinical translational research perspective on regulations versus a dose of reality. *Journal of Law, Medicine, and Ethics, 44,* 394–401.

Anderson, A. J., Piltti, K. M., Hooshmand, M. J., Nishi, R. A., & Cummings, B. J. (2017). Preclinical efficacy failure of human CNS-derived stem cells for use in the pathway study of cervical spinal cord injury. *Stem Cell Reports, 8,* 249–263.

Anderson, C. (1993). Pasteur notebooks reveal deception. *Science, 259,* 1117–1117.

Anderzhanova, E., Kirmeier, T., & Wotjak, C. T. (2017). Animal models in psychiatric research: The RDoC system as a new framework for endophenotype-oriented translational neuroscience. *Neurobiology of Stress, 7,* 47–56.

André, W., Sandt, C., Nondier, I., Djian, P., & Hoffner, G. (2017). Inclusions of R6/2 mice are not amyloid and differ structurally from those of Huntington disease brain. *Analytical Chemistry, 89,* 5201–5209.

Anisimov, V. N., Ukraintseva, S. V., & Yashin, A. I. (2005). Cancer in rodents: Does it tell us about cancer in humans? *Nature Reviews: Cancer, 5,* 807–819.

Ankeny, R. A. (2007). Wormy logic: Model organisms as case-based reasoning. In A. N. H. Creager, E. Lunbeck, & M. N. Wise (Eds.), *Science without Laws* (pp. 46–58). Duke University Press.

Ankeny, R. A., & Leonelli, S. (2011). What's so special about model organisms? *Studies in History and Philosophy of Science, 42,* 313–323.

Ankeny, R. A., & Leonelli, S. (2020). *Model Organisms*. Cambridge University Press.

APHIS Annual Report Animal Usage by Year (2019). https://www.aphis.usda.gov/animal_welfare/annual
-reports/2019/fy19-summary-report-column-F.pdf

Aquinas, T. (1947). *Summa Theologica* (Fathers of the English Dominican Province, Trans.). Benzinger Bros.

Arikha, N. (2007). *Passions and Tempers: A History of the Humours*. HarperCollins.

Armstrong, C. (1939). Successful transfer of the Lansing strain of poliomyelitis virus from the Cotton Rat to the White Mouse. *Public Health Report, 1896–1970, 54*, 2302–2305.

Arnegard, M. E., Whitten, L. A., Hunter, C., & Clayton, J. A. (2020). Sex as a biological variable: A 5-year progress report and call to action. *Journal of Women's Health, 29*, 858–864.

Arrasate, M., Mitra, S., Schweitzer, E. S., Segal, M. R., & Finkbeiner, S. (2004). Inclusion body formation reduces levels of mutant huntingtin and the risk of neuronal death. *Nature, 431*, 805–810.

Arrieta, M.-C., Walter, J., & Finlay, B. B. (2016). Human microbiota-associated mice: A model with challenges. *Cell Host and Microbe, 19*, 575–578.

Arroll, B., Macgillivray, S., Ogston, S., Reid, I., Sullivan, F., Williams, B., & Crombie, I. (2005). Efficacy and tolerability of tricyclic antidepressants and SSRIs compared with placebo for treatment of depression in primary care: A meta-analysis. *Annals of Family Medicine, 3*, 449–456.

Ascherio, A., & Schwarzschild, M. A. (2016). The epidemiology of Parkinson's disease: Risk factors and prevention. *Lancet Neurology, 15*, 1257–1272.

Ashburner, M., Ball, C. A., Blake, J. A., Botstein, D., Butler, H., Cherry, J. M., Davis, A. P., Dolinski, K., Dwight, S. S., Eppig, J. T., et al. (2000). Gene Ontology: Tool for the unification of biology. *Nature Genetics, 25*, 25–29.

Asher, R. J., Bennett, N., & Lehmann, T. (2009). The new framework for understanding placental mammal evolution. *BioEssays, 31*, 853–864.

Atlasi, Y., & Stunnenberg, H. G. (2017). The interplay of epigenetic marks during stem cell differentiation and development. *Nature Reviews: Genetics, 18*, 643–658.

Auman, J. T. (2010). Cancer pharmacogenomics: Do cancer cell lines have the right stuff? *Pharmacogenomics, 11*, 1035–1037.

Avramides, A. (2019). Other minds. In E. N. Zalta (Ed.), *The Stanford Encyclopedia of Philosophy*. Metaphysics Research Lab, Stanford University. https://plato.stanford.edu/entries/other-minds/.

Baastrup, P. C., Poulsen, J. C., Schou, M., Thomsen, K., & Amdisen, A. (1970). Prophylactic lithium: Double blind discontinuation in manic-depressive and recurrent-depressive disorders. *The Lancet, 2*, 326–330.

Baglietto-Vargas, D., Forner, S., Cai, L., Martini, A. C., Trujillo-Estrada, L., Swarup, V., Nguyen, M. M. T., Do Huynh, K., Javonillo, D. I., Tran, K. M., et al. (2021). Generation of a humanized Aβ expressing mouse demonstrating aspects of Alzheimer's disease-like pathology. *Nature Communications, 12*, Article 2421.

Bahar, F. G., Ohura, K., Ogihara, T., & Imai, T. (2012). Species difference of esterase expression and hydrolase activity in plasma. *Journal of Pharmaceutical Sciences, 101*, 3979–3988.

Bahr, J. M. (2008). The chicken as a model organism. In P. M. Conn (Ed.), *Sourcebook of Models for Biomedical Research* (pp. 161–167). Humana Press.

Bailey, D. (2010, October). The making of a miracle drug [prontosil]. *Smells Like Science*. https://web.archive
.org/web/20201112032954/https://www.smellslikescience.com/the-making-of-a-miracle-drug/.

Bailey, H. (2020, May 21) Being a pig farmer was already hard. Then came coronavirus. *The Washington Post*.

Bailey, J. (2014). Monkey-based research on human disease: The implications of genetic differences. *Alternatives to Laboratory Animals, 42*, 287–317.

Bailey, J., Knight, A., & Balcombe, J. (2005). The future of teratology research is in vitro. *Biogenic Amines, 19*, 97–145.

Bailey, M., Christoforidou, Z., & Lewis, M. C. (2013). The evolutionary basis for differences between the immune systems of man, mouse, pig and ruminants. *Veterinary Immunology and Immunopathology, 152*, 13–19.

Bakhle, Y. S. (1968). Conversion of angiotensin I to angiotensin II by cell-free extracts of dog lung. *Nature, 220*, 919–921.

Baldessarini, R. J., & Tondo, L. (2000). Does lithium treatment still work? Evidence of stable responses over three decades. *Archives of General Psychiatry, 57*, 187–190.

Balijepalli, A., & Sivaramakrishan, V. (2017). Organs-on-chips: Research and commercial perspectives. *Drug Discovery Today, 22*, 397–403.

Ballentine, C. (1981, June). Sulfanilamide disaster. *FDA Consumer Magazine*. https://www.fda.gov/about-fda/histories-product-regulation/sulfanilamide-disaster.

Bandyopadhyay, S., Mehta, M., Kuo, D., Sung, M.-K., Chuang, R., Jaehnig, E. J., Bodenmiller, B., Licon, K., Copeland, W., Shales, M., et al. (2010). Rewiring of genetic networks in response to DNA damage. *Science, 330*, 1385–1389.

Banker, G., & Goslin, K. (1998). Types of nerve cell cultures: Their advantages and limitations. In G. Banker & K. Goslin (Eds.), *Culturing Nerve Cells* (pp. 11–35). MIT Press.

Bar, S., & Benvenisty, N. (2019). Epigenetic aberrations in human pluripotent stem cells. *EMBO Journal, 38*, Article e101033.

Baraban, S. C., Dinday, M. T., & Hortopan, G. A. (2013). Drug screening in Scn1a zebrafish mutant identifies clemizole as a potential Dravet syndrome treatment. *Nature Communications, 4*, Article 2410.

Barbaro, B. A., Lukacsovich, T., Agrawal, N., Burke, J., Bornemann, D. J., Purcell, J. M., Worthge, S. A., Caricasole, A., Weiss, A., Song, W., et al. (2015). Comparative study of naturally occurring huntingtin fragments in *Drosophila* points to exon 1 as the most pathogenic species in Huntington's disease. *Human Molecular Genetics, 24*, 913–925.

Bard, J., Wall, M. D., Lazari, O., Arjomand, J., & Munoz-Sanjuan, I. (2014). Advances in Huntington disease drug discovery: Novel approaches to model disease phenotypes. *Journal of Biomolecular Screening, 19*, 191–204.

Bargmann, C. (2000). Simple organisms. *Neurobiology of Disease, 7*, 520–522.

Barker, R. A., & TRANSEURO Consortium. (2019). Designing stem-cell-based dopamine cell replacement trials for Parkinson's disease. *Nature Medicine, 25*, 1045–1053.

Barnes, G. T., Duyao, M. P., Ambrose, C. M., McNeil, S., Persichetti, F., Srinidhi, J., Gusella, J. F., & MacDonald, M. E. (1994). Mouse Huntington's disease gene homolog (Hdh). *Somatic Cell and Molecular Genetics, 20*, 87–97.

Barnett, J. A. (2007). A history of research on yeasts 10: Foundations of yeast genetics. *Yeast, 24*, 799–845.

Barnett, J. H., & Smoller, J. W. (2009). The genetics of bipolar disorder. *Neuroscience, 164*, 331–343.

Bar-On, Y. M., Phillips, R., & Milo, R. (2018). The biomass distribution on Earth. *Proceedings of the National Academy of Sciences of the United States of America, 115*, 6506–6511.

Bartus, R. T., Weinberg, M. S., & Samulski, R. J. (2014). Parkinson's disease gene therapy: Success by design meets failure by efficacy. *Molecular Therapy, 22*, 487–497.

Bashyam, H. (2007). Hormones and breast cancer: Controlling the danger within. *Journal of Experimental Medicine, 204*, 699.

Basu, M., Bhattacharyya, N. P., & Mohanty, P. K. (2013). Comparison of modules of wild type and mutant huntingtin and TP53 protein interaction networks: Implications in biological processes and functions. *PLoS One, 8*, Article e64838.

Baud, A., Mulligan, M. K., Casale, F. P., Ingels, J. F., Bohl, C. J., Callebert, J., Launay, J.-M., Krohn, J., Legarra, A., Williams, R. W., & Stegle, O. (2017). Genetic variation in the social environment contributes to health and disease. *PLoS Genetics, 13*, Article e1006498.

Beach, F. A. (1950). The snark was a boojum. *American Psychologist, 5*, 115–124.

Beal, M. F., Henshaw, D. R., Jenkins, B. G., Rosen, B. R., & Schulz, J. B. (1994). Coenzyme Q10 and nicotin-amide block striatal lesions produced by the mitochondrial toxin malonate. *Annals of Neurology, 36*, 882–888.

Beebe, D. J., Mensing, G. A., & Walker, G. M. (2002). Physics and applications of microfluidics in biology. *Annual Review of Biomedical Engineering, 4*, 261–286.

Beery, A. K. (2018). Inclusion of females does not increase variability in rodent research studies. *Current Opinion in Behavioral Sciences, 23*, 143–149.

Begley, C. G., & Ellis, L. M. (2012). Raise standards for preclinical cancer research. *Nature, 483*, 531–533.

Behar, S. M., & Sassetti, C. (2020). Tuberculosis vaccine finds an improved route. *Nature, 577*, 31–32.

Beigel, J. H., Tomashek, K. M., Dodd, L. E., Mehta, A. K., Zingman, B. S., Kalil, A. C., Hohmann, E., Chu, H. Y., Luetkemeyer, A., Kline, S., et al. (2020). Remdesivir for the treatment of COVID-19—preliminary report. *New England Journal of Medicine, 383*, 1813–1826.

Belin, D., Berson, N., Balado, E., Piazza, P. V., & Deroche-Gamonet, V. (2011). High-novelty-preference rats are predisposed to compulsive cocaine self-administration. *Neuropsychopharmacology, 36*, 569–579.

Bell, C. C., Hendriks, D. F. G., Moro, S. M. L., Ellis, E., Walsh, J., Renblom, A., Fredriksson Puigvert, L., Dankers, A. C. A., Jacobs, F., Snoeys, J., et al. (2016). Characterization of primary human hepatocyte spheroids as a model system for drug-induced liver injury, liver function and disease. *Scientific Reports, 6*, Article 25187.

Belzung, C., & Lemoine, M. (2011). Criteria of validity for animal models of psychiatric disorders: Focus on anxiety disorders and depression. *Biology of Mood and Anxiety Disorders, 1*, Article 9.

Benabid, A. L., Chabardes, S., Mitrofanis, J., & Pollak, P. (2009). Deep brain stimulation of the subthalamic nucleus for the treatment of Parkinson's disease. *Lancet Neurology, 8*, 67–81.

Benabid, A. L., DeLong, M., & Hariz, M. (2015a). Letter to the editor. *Alternatives to Laboratory Animals, 43*, 205–206.

Benabid, A. L., DeLong, M., & Hariz, M. (2015b). Letter to the editor. *Alternatives to Laboratory Animals, 43*, 427–431.

Benatar, M. (2007). Lost in translation: Treatment trials in the SOD1 mouse and in human ALS. *Neurobiology of Disease, 26*, 1–13.

Benazzouz, A., Gross, C., Féger, J., Boraud, T., & Bioulac, B. (1993). Reversal of rigidity and improvement in motor performance by subthalamic high-frequency stimulation in MPTP-treated monkeys. *European Journal of Neuroscience, 5*, 382–389.

Ben-David, U., Ha, G., Tseng, Y.-Y., Greenwald, N. F., Oh, C., Shih, J., McFarland, J. M., Wong, B., Boehm, J. S., Beroukhim, R., & Golub, T. R. (2017). Patient-derived xenografts undergo mouse-specific tumor evolution. *Nature Genetics, 49,* 1567–1575.

Ben-David, U., Siranosian, B., Ha, G., Tang, H., Oren, Y., Hinohara, K., Strathdee, C. A., Dempster, J., Lyons, N. J., Burns, R., et al. (2018). Genetic and transcriptional evolution alters cancer cell line drug response. *Nature, 560,* 325–330.

Benjamini, Y., & Hochberg, Y. (1995). Controlling the false discovery rate: A practical and powerful approach to multiple testing. *Journal of the Royal Statistical Society: Series B, Statistical Methodology, 57,* 289–300.

Benkler, C., O'Neil, A. L., Slepian, S., Qian, F., Weinreb, P. H., & Rubin, L. L. (2018). Aggregated SOD1 causes selective death of cultured human motor neurons. *Scientific Reports, 8,* Article 16393.

Bentham, J. (1789). *The Principles of Morals and Legislation.* T. Payne and Son.

Bentley, R. (2009). Different roads to discovery; Prontosil (hence sulfa drugs) and penicillin (hence β-lactams). *Journal of Industrial Microbiology and Biotechnology, 36,* 775–786.

Benyus, J. M. (2002). *Biomimicry: Innovation Inspired by Nature.* HarperCollins.

Benzer, S. (1959). On the topology of the genetic fine structure. *Proceedings of the National Academy of Sciences of the United States of America, 45,* 1607–1620.

Bergman, H., Wichmann, T., & DeLong, M. R. (1990). Reversal of experimental parkinsonism by lesions of the subthalamic nucleus. *Science, 249,* 1436–1438.

Bergmann, O., Liebl, J., Bernard, S., Alkass, K., Yeung, M. S. Y., Steier, P., Kutschera, W., Johnson, L., Landén, M., Druid, H., et al. (2012). The age of olfactory bulb neurons in humans. *Neuron, 74,* 634–639.

Berman, R. M., Cappiello, A., Anand, A., Oren, D. A., Heninger, G. R., Charney, D. S., & Krystal, J. H. (2000). Antidepressant effects of ketamine in depressed patients. *Biological Psychiatry, 47,* 351–354.

Bernard, C. (1957). *An Introduction to the Study of Experimental Medicine* (H. C. Greene, Trans.). Dover. (Original work published 1927)

Berry, A., Vitale, A., Carere, C., & Alleva, E. (2015). EU guidelines for the care and welfare of an "exceptional invertebrate class" in scientific research. *Annali dell'Istituto Superiore di Sanità, 51,* 267–269.

Beyer, D. K. E., & Freund, N. (2017). Animal models for bipolar disorder: From bedside to the cage. *International Journal of Bipolar Disorders, 5,* 35.

Bhaduri, A., Andrews, M. G., Mancia Leon, W., Jung, D., Shin, D., Allen, D., Jung, D., Schmunk, G., Haeussler, M., Salma, J., et al. (2020). Cell stress in cortical organoids impairs molecular subtype specification. *Nature, 578,* 142–148.

Bhandari, T. (2020, February 10). Investigational drugs didn't slow memory loss, cognitive decline in rare, inherited Alzheimer's, initial analysis indicates. *Washington University School of Medicine: News Hub.* https://medicine.wustl.edu/news/alzheimers-diantu-trial-initial-results/.

Bhatia, S., Naved, T., & Sardana, S. (2019). Introduction to animal tissue culture science. In S. Bhatia, T. Naved, & S. Sardana (Eds.), *Introduction to Pharmaceutical Biotechnology* (Vol. 3, pp. 1–30). IOP Publishing.

Bhatnagar, P., Wickramasinghe, K., Williams, J., Rayner, M., & Townsend, N. (2015). The epidemiology of cardiovascular disease in the UK 2014. *Heart, 101,* 1182–1189.

Bhullar, K. S., Lagarón, N. O., McGowan, E. M., Parmar, I., Jha, A., Hubbard, B. P., & Rupasinghe, H. P. V. (2018). Kinase-targeted cancer therapies: Progress, challenges and future directions. *Molecular Cancer, 17,* Article 48.

Bister, K. (2015). Discovery of oncogenes: The advent of molecular cancer research. *Proceedings of the National Academy of Sciences of the United States of America, 112*, 15259–15260.

Bitar, M., Kuiper, S., O'Brien, E. A., & Barry, G. (2019). Genes with human-specific features are primarily involved with brain, immune and metabolic evolution. *BMC Bioinformatics, 20*, Article S406.

Black, J. (1989). Drugs from emasculated hormones: The principles of syntopic antagonism. *Science, 245*, 486–493.

Black, J. W., Crowther, A. F., Shanks, R. G., Smith, L. H., & Dornhorst, A. C. (1964). A new adrenergic beta-receptor antagonist. *The Lancet, 283*, 1080–1081.

Blair, S. S. (2008). Segmentation in animals. *Current Biology, 18*, R991–R995.

Blattner, F. R., Plunkett, G., Bloch, C. A., Perna, N. T., Burland, V., Riley, M., Collado-Vides, J., Glasner, J. D., Rode, C. K., Mayhew, G. F., et al. (1997). The complete genome sequence of *Escherichia coli* K-12. *Science, 277*, 1453–1462.

Blesa, J., Trigo-Damas, I., del Rey, N. L.-G., & Obeso, J. A. (2018). The use of nonhuman primate models to understand processes in Parkinson's disease. *Journal of Neural Transmission, 125*, 325–335.

Blits, K. C. (1999). Aristotle: Form, function, and comparative anatomy. *Anatomical Record: Part B. New Anatomist, 257*, 58–63.

Blokhuis, A. M., Groen, E. J. N., Koppers, M., van den Berg, L. H., & Pasterkamp, R. J. (2013). Protein aggregation in amyotrophic lateral sclerosis. *Acta Neuropathologica, 125*, 777–794.

Blount, Z. D. (2015). The unexhausted potential of *E. coli*. *eLife, 4*, Article e05826.

Blurton-Jones, M., Kitazawa, M., Martinez-Coria, H., Castello, N. A., Müller, F.-J., Loring, J. F., Yamasaki, T. R., Poon, W. W., Green, K. N., & LaFerla, F. M. (2009). Neural stem cells improve cognition via BDNF in a transgenic model of Alzheimer disease. *Proceedings of the National Academy of Sciences of the United States of America, 106*, 13594–13599.

Bode, G., Clausing, P., Gervais, F., Loegsted, J., Luft, J., Nogues, V., & Sims, J. (2010). The utility of the minipig as an animal model in regulatory toxicology. *Journal of Pharmacological and Toxicological Methods, 62*, 196–220.

Bohm, D. (1957). *Causality and Chance in Modern Physics*. University of Pennsylvania Press.

Bolker, J. A. (1995). Model systems in developmental biology. *BioEssays, 17*, 451–455.

Bolker, J. A. (2009). Exemplary and surrogate models: Two modes of representation in biology. *Perspectives in Biology and Medicine, 52*, 485–499.

Bolker, J. A. (2017). Animal models in translational research: Rosetta Stone or stumbling block? *BioEssays, 328*, Article 1700089.

Bolker, J. A. (2019). Selection of models: Evolution and the choice of species for translational research. *Brain, Behavior and Evolution, 93*, 82–91.

Boodman, E. (2020, March 11). Coronavirus vaccine clinical trial starting without usual animal data. *STAT.* https://www.statnews.com/2020/03/11/researchers-rush-to-start-moderna-coronavirus-vaccine-trial-without-usual-animal-testing.

Bortoff, G. A., & Strick, P. L. (1993). Corticospinal terminations in two New-world primates: Further evidence that corticomotoneuronal connections provide part of the neural substrate for manual dexterity. *Journal of Neuroscience, 13*, 5105–5118.

Bosch, F., & Rosich, L. (2008). The contributions of Paul Ehrlich to pharmacology: A tribute on the occasion of the centenary of his Nobel Prize. *Pharmacology, 82*, 171–179.

Botstein, D., Chervitz, S. A., & Cherry, J. M. (1997). Yeast as a model organism. *Science, 277,* 1259–1260.

Botstein, D., & Fink, G. R. (1988). Yeast: An experimental organism for modern biology. *Science, 240,* 1439–1443.

Botstein, D., & Fink, G. R. (2011). Yeast: An experimental organism for 21st century biology. *Genetics, 189,* 695–704.

Botting, J. (2015). *Animals and Medicine.* Open Book.

Boutin, M. E., Hampton, C., Quinn, R., Ferrer, M., & Song, M. J. (2019). 3D engineering of ocular tissues for disease modeling and drug testing. In K. Bharti & K. Bharti (Eds.), *Pluripotent Stem Cells in Eye Disease Therapy* (pp. 171–193). Springer Nature.

Bové, J., & Perier, C. (2012). Neurotoxin-based models of Parkinson's disease. *Neuroscience, 211,* 51–76.

Bowater, L. (2016, November 8). The great pox. *Microbiology Today.* https://microbiologysociety.org/publi cation/past-issues/the-mobile-microbe/article/the-great-pox-mobile-microbe.html.

Box, G., Luceño, A., & Paniagua-Quinones, M. del C. (2009). *Statistical Control by Monitoring and Adjustment* (2nd ed.). Wiley.

Boxenbaum, H., & DiLea, C. (1995). First-time-in-human dose selection: Allometric thoughts and perspectives. *Journal of Clinical Pharmacology, 35,* 957–966.

Boyd, W. A., Smith, Marjolein V., Co, Caroll A., Pirone, Jason R., Rice, Julie R., Shockley, Keith R., & Freedman, Jonathan H. (2016). Developmental effects of the ToxCast™ Phase I and Phase II chemicals in *Caenorhabditis elegans* and corresponding responses in zebrafish, rats, and rabbits. *Environmental Health Perspectives, 124,* 586–593.

Braak, H., Braak, E., & Strothjohann, M. (1994). Abnormally phosphorylated tau protein related to the formation of neurofibrillary tangles and neuropil threads in the cerebral cortex of sheep and goat. *Neuroscience Letters, 171,* 1–4.

Braasch, I., Gehrke, A. R., Smith, J. J., Kawasaki, K., Manousaki, T., Pasquier, J., Amores, A., Desvignes, T., Batzel, P., Catchen, J., et al. (2016). The spotted gar genome illuminates vertebrate evolution and facilitates human-teleost comparisons. *Nature Genetics, 48,* 427–437.

Bracken, M. B. (1992). Pharmacological treatment of acute spinal cord injury: Current status and future prospects. *Paraplegia, 30,* 102–107.

Bradford, Y. M., Toro, S., Ramachandran, S., Ruzicka, L., Howe, D. G., Eagle, A., Kalita, P., Martin, R., Taylor Moxon, S. A., Schaper, K., & Westerfield, M. (2017). Zebrafish models of human disease: Gaining insight into human disease at ZFIN. *ILAR Journal, 58,* 4–16.

Bradley, A., Mennie, N., Bibby, P. A., & Cassaday, H. J. (2020). Some animals are more equal than others: Validation of a new scale to measure how attitudes to animals depend on species and human purpose of use. *PLoS One, 15,* Article e0227948.

Bragge, P., Synnot, A., Maas, A. I., Menon, D. K., Cooper, D. J., Rosenfeld, J. V., & Gruen, R. L. (2016). A state-of-the-science overview of randomized controlled trials evaluating acute management of moderate-to-severe traumatic brain injury. *Journal of Neurotrauma, 33,* 1461–1478.

Braithwaite, V. (2010). *Do Fish Feel Pain?* Oxford University Press.

Brandt, A. M. (2012). Inventing conflicts of interest: A history of tobacco industry tactics. *American Journal of Public Health, 102,* 63–71.

Bray, D. (1973). Branching patterns of individual sympathetic neurons in culture. *Journal of Cell Biology, 56,* 702–712.

Bregman, B. S., Kunkel-Bagden, E., Schnell, L., Dai, H. N., Gao, D., & Schwab, M. E. (1995). Recovery from spinal cord injury mediated by antibodies to neurite growth inhibitors. *Nature, 378*, 498–501.

Brenner, S. (2002). Nature's gift to science. *Nobel Lecture.* https://www.nobelprize.org/prizes/medicine/2002/brenner/lecture/.

Brenowitz, E. A., & Zakon, H. H. (2015). Emerging from the bottleneck: Benefits of the comparative approach to modern neuroscience. *Trends in Neurosciences, 38*, 273–278.

Brock, T. D. (1990). *The Emergence of Bacterial Genetics.* Cold Spring Harbor Laboratory Press.

Brodbelt, D. C., Blissitt, K. J., Hammond, R. A., Neath, P. J., Young, L. E., Pfeiffer, D. U., & Wood, J. L. N. (2008). The risk of death: The confidential enquiry into perioperative small animal fatalities. *Veterinary Anaesthesia and Analgesia, 35*, 365–373.

Brown, K. S., Hill, C. C., Calero, G. A., Myers, C. R., Lee, K. H., Sethna, J. P., & Cerione, R. A. (2004). The statistical mechanics of complex signaling networks: Nerve growth factor signaling. *Physical Biology, 1*, 184–195.

Brown, L. D., Stricker-Kongrad, A., & Bouchard, G. F. (2013). Minipigs: Applications in toxicology. *Current Protocols in Toxicology, 56*, 1.11.1–1.11.19.

Brown, R. E. (2007). Behavioural phenotyping of transgenic mice. *Canadian Journal of Experimental Psychology, 61*, 328–344.

Brown, W. A. (2019). *Lithium: A Doctor, a Drug, and a Breakthrough.* Liveright.

Brunham, L. R., & Hayden, M. R. (2013). Hunting human disease genes: Lessons from the past, challenges for the future. *Human Genetics, 132*, 603–617.

Brunner, D., Balcı, F., & Ludvig, E. A. (2012). Comparative psychology and the grand challenge of drug discovery in psychiatry and neurodegeneration. *Behavioural Processes, 89*, 187–195.

Bryan, J. (2009). From snake venom to ACE inhibitor—the discovery and rise of captopril. *Pharmaceutical Journal, 282*, Article 455.

Bućan, M., & Abel, T. (2002). The mouse: Genetics meets behaviour. *Nature Reviews: Genetics, 3*, 114–123.

Bullock, T. H. (2002). Grades in neural complexity: How large is the span? *Integrative and Comparative Biology, 42*, 757–761.

Burian, R. M. (1993). How the choice of experimental organism matters: Epistemological reflections on an aspect of biological practice. *Journal of the History of Biology, 26*, 351–367.

Burkart, J. M., & Finkenwirth, C. (2015). Marmosets as model species in neuroscience and evolutionary anthropology. *Neuroscience Research, 93*, 8–19.

Burkhardt, M. F., Martinez, F. J., Wright, S., Ramos, C., Volfson, D., Mason, M., Garnes, J., Dang, V., Lievers, J., Shoukat-Mumtaz, U., et al. (2013). A cellular model for sporadic ALS using patient-derived induced pluripotent stem cells. *Molecular and Cellular Neurosciences, 56*, 355–364.

Burns, R. S., Chiueh, C. C., Markey, S. P., Ebert, M. H., Jacobowitz, D. M., & Kopin, I. J. (1983). A primate model of parkinsonism: Selective destruction of dopaminergic neurons in the pars compacta of the substantia nigra by *N*-methyl-4-phenyl-1,2,3,6-tetrahydropyridine. *Proceedings of the National Academy of Sciences of the United States of America, 80*, 4546–4550.

Burns, T. C., Li, M. D., Mehta, S., Awad, A. J., & Morgan, A. A. (2015). Mouse models rarely mimic the transcriptome of human neurodegenerative diseases: A systematic bioinformatics-based critique of preclinical models. *European Journal of Pharmacology, 759*, 101–117.

Burns, T. C., & Verfaillie, C. M. (2015). From mice to mind: Strategies and progress in translating neuroregeneration. *European Journal of Pharmacology, 759*, 90–100.

Butterworth, K. T. (2019). Evolution of the supermodel: Progress in modelling radiotherapy response in mice. *Clinical Oncology, 31*, 272–282.

Button, K. S., Ioannidis, J. P. A., Mokrysz, C., Nosek, B. A., Flint, J., Robinson, E. S. J., & Munafò, M. R. (2013a). Power failure: Why small sample size undermines the reliability of neuroscience. *Nature Reviews: Neuroscience, 14*, 365–376.

Button, K. S., Ioannidis, J. P. A., Mokrysz, C., Nosek, B. A., Flint, J., Robinson, E. S. J., & Munafò, M. R. (2013b). Confidence and precision increase with high statistical power. *Nature Reviews: Neuroscience, 14*, 585–586.

Cade, J. F. (1949). Lithium salts in the treatment of psychotic excitement. *Medical Journal of Australia, 2*, 349–352.

Cahill, L. (2014). Equal ≠ the same: Sex differences in the human brain. *Cerebrum 2014*, Article 5.

Cahill, L., & Hall, E. D. (2017). Is it time to resurrect "lazaroids"? *Journal of Neuroscience Research, 95*, 17–20.

Cai, N., Choi, K. W., & Fried, E. I. (2020). Reviewing the genetics of heterogeneity in depression: Operationalizations, manifestations and etiologies. *Human Molecular Genetics, 29*, R10–R18.

Calabrese, E. J. (1988). Comparative biology of test species. *Environmental Health Perspectives, 77*, 55–62.

Calabrese, E. J. (1991). *Principles of Animal Extrapolation*. CRC Press.

Caldwell, A. E. (1970). *Origins of Psychopharmacology from CPZ to LSD*. Thomas.

Caldwell, J. (1981). The current status of attempts to predict species differences in drug metabolism. *Drug Metabolism Reviews, 12*, 221–237.

Calhoun, J. B. (1963). *The Ecology and Sociology of the Norway Rat*. Public Health Service, US Department of Health, Education, and Welfare.

Callaghan, J. C., & Bigelow, W. G. (1951). An electrical artificial pacemaker for standstill of the heart. *Annals of Surgery, 134*, 8–17.

Cambau, E., & Drancourt, M. (2014). Steps towards the discovery of *Mycobacterium tuberculosis* by Robert Koch, 1882. *Clinical Microbiology and Infection, 20*, 196–201.

Camerer, C. F., Dreber, A., Holzmeister, F., Ho, T.-H., Huber, J., Johannesson, M., Kirchler, M., Nave, G., Nosek, B. A., Pfeiffer, T., et al. (2018). Evaluating the replicability of social science experiments in *Nature* and *Science* between 2010 and 2015. *Nature Human Behavior, 2*, 637–644.

Camhi, J. M. (1984). *Neuroethology: Nerve Cells and the Natural Behavior of Animals*. Sinauer Associates.

Campbell, A. M. (1986). Bacteriophage lambda as a model system. *BioEssays, 5*, 277–280.

Campenot, R. B. (1977). Local control of neurite development by nerve growth factor. *Proceedings of the National Academy of Sciences of the United States of America, 74*, 4516–4519.

Cannarozzi, G., Schneider, A., & Gonnet, G. (2007). A phylogenomic study of human, dog, and mouse. *PLoS Computational Biology, 3*, Article e2.

Capecchi, M. R. (2005). Gene targeting in mice: Functional analysis of the mammalian genome for the twenty-first century. *Nature Reviews: Genetics, 6*, 507–512.

Capes-Davis, A., Theodosopoulos, G., Atkin, I., Drexler, H. G., Kohara, A., MacLeod, R. A. F., Masters, J. R., Nakamura, Y., Reid, Y. A., Reddel, R. R., & Freshney, R. I. (2010). Check your cultures! A list of cross-contaminated or misidentified cell lines. *International Journal of Cancer, 127*, 1–8.

Capsoni, S., Ugolini, G., Comparini, A., Ruberti, F., Berardi, N., & Cattaneo, A. (2000). Alzheimer-like neurodegeneration in aged antinerve growth factor transgenic mice. *Proceedings of the National Academy of Sciences of the United States of America, 97*, 6826–6831.

Carey, B. W., Markoulaki, S., Hanna, J. H., Faddah, D. A., Buganim, Y., Kim, J., Ganz, K., Steine, E. J., Cassady, J. P., Creyghton, M. P., et al. (2011). Reprogramming factor stoichiometry influences the epigenetic state and biological properties of induced pluripotent stem cells. *Cell Stem Cell, 9*, 588–598.

Carlson, G. A., Borchelt, D. R., Dake, A., Turner, S., Danielson, V., Coffin, J. D., Eckman, C., Meiners, J., Nilsen, S. P., Younkin, S. G., & Hsiao, K. K. (1997). Genetic modification of the phenotypes produced by amyloid precursor protein overexpression in transgenic mice. *Human Molecular Genetics, 6*, 1951–1959.

Carlson, R. V., Boyd, K. M., & Webb, D. J. (2004). The revision of the Declaration of Helsinki: Past, present and future. *British Journal of Clinical Pharmacology, 57*, 695–713.

Carlsson, A., Lindqvist, M., & Magnusson, T. (1957). 3,4-Dihydroxyphenylalanine and 5-Hydroxytryptophan as reserpine antagonists. *Nature, 180*, 1200–1200.

Carpenter, W. T., & Koenig, J. I. (2008). The evolution of drug development in schizophrenia: Past issues and future opportunities. *Neuropsychopharmacology, 33*, 2061–2079.

Carr, C. E., & Konishi, M. (1988). Axonal delay lines for time measurement in the owl's brainstem. *Proceedings of the National Academy of Sciences of the United States of America, 85*, 8311–8315.

Carrel, A., & Burrows, M. T. (1911). Cultivation of tissues in vitro and its technique. *Journal of Experimental Medicine, 13*, 387–396.

Carroll, J. B., Warby, S. C., Southwell, A. L., Doty, C. N., Greenlee, S., Skotte, N., Hung, G., Bennett, C. F., Freier, S. M., & Hayden, M. R. (2011). Potent and selective antisense oligonucleotides targeting single-nucleotide polymorphisms in the Huntington disease gene/allele-specific silencing of mutant huntingtin. *Molecular Therapy, 19*, 2178–2185.

Carty, N., Berson, N., Tillack, K., Thiede, C., Scholz, D., Kottig, K., Sedaghat, Y., Gabrysiak, C., Yohrling, G., von der Kammer, H., et al. (2015). Characterization of HTT inclusion size, location, and timing in the zQ175 mouse model of Huntington's disease: An in vivo high-content imaging study. *PLoS One, 10*, Article e0123527.

Caruana SJ, L. (2020). Different religions, different animal ethics? *Animal Frontiers, 10*, 8–14.

Casjens, S. R., & Hendrix, R. W. (2015). Bacteriophage lambda: Early pioneer and still relevant. *Virology, 479–480*, 310–330.

Cassar, S., Adatto, I., Freeman, J. L., Gamse, J. T., Iturria, I., Lawrence, C., Muriana, A., Peterson, R. T., Van Cruchten, S., & Zon, L. I. (2020). Use of zebrafish in drug discovery toxicology. *Chemical Research in Toxicology, 33*, 95–118.

Cassim, S., Raymond, V.-A., Lapierre, P., & Bilodeau, M. (2017). From in vivo to in vitro: Major metabolic alterations take place in hepatocytes during and following isolation. *PLoS One, 12*, Article e0190366.

Castelli, W. P., Anderson, K., Wilson, P. W. F., & Levy, D. (1992). Lipids and risk of coronary heart disease: The Framingham study. *Annals of Epidemiology, 2*, 23–28.

Castilhos, R. M., Augustin, M. C., Santos, J. A., Perandones, C., Saraiva-Pereira, M. L., Jardim, L. B., & Rede Neurogenética. (2016). Genetic aspects of Huntington's disease in Latin America: A systematic review. *Clinical Genetics, 89*, 295–303.

Catania, K. (2020). *Great Adaptations: Star-Nosed Moles, Electric Eels, and Other Tales of Evolution's Mysteries Solved.* Princeton University Press.

Caulin, A. F., Graham, T. A., Wang, L.-S., & Maley, C. C. (2015). Solutions to Peto's paradox revealed by mathematical modelling and cross-species cancer gene analysis. *Philosophical Transactions of the Royal Society of London: Series B, Biological Sciences, 370,* Article 20140222.

Cavaliere, F., & Matute, C. (2011). Utility of organotypic slices in Parkinson's disease research. In D. Finkelstein (Ed.), *Towards New Therapies for Parkinson's Disease* (pp. 101–112). InTech Open.

Cavazzoni, P. (2021, June 7). FDA's decision to approve new treatment for Alzheimer's disease. US Food and Drug Administration. https://www.fda.gov/drugs/news-events-human-drugs/fdas-decision-approve-new -treatment-alzheimers-disease.

Cepko, C. L. (1989). Immortalization of neural cells via retrovirus-mediated oncogene transduction. *Annual Review of Neuroscience, 12,* 47–65.

Chain, E., Florey, H. W., Gardner, A. D., Heatley, N. G., Jennings, M. A., Orr-Ewing, J., & Sanders, A. G. (1940). Penicillin as a chemotherapeutic agent. *The Lancet, 236,* 226–228.

Chalfin, L., Dayan, M., Levy, D. R., Austad, S. N., Miller, R. A., Iraqi, F. A., Dulac, C., & Kimchi, T. (2014). Mapping ecologically relevant social behaviours by gene knockout in wild mice. *Nature Communications, 5,* Article 4569.

Chambers, J. K., Uchida, K., Harada, T., Tsuboi, M., Sato, M., Kubo, M., Kawaguchi, H., Miyoshi, N., Tsuji-moto, H., & Nakayama, H. (2012). Neurofibrillary tangles and the deposition of a beta amyloid peptide with a novel N-terminal epitope in the brains of wild Tsushima leopard cats. *PLoS One, 7,* Article e46452.

Chan, A. W., Xu, Y., Jiang, J., Rahim, T., Zhao, D., Kocerha, J., Chi, T., Moran, S., Engelhardt, H., Larkin, K., et al. (2014). A two years longitudinal study of a transgenic Huntington disease monkey. *BMC Neuroscience, 15,* Article 36.

Chan, J. C. Y., Piper, D. E., Cao, Q., Liu, D., King, C., Wang, W., Tang, J., Liu, Q., Higbee, J., Xia, Z., et al. (2009). A proprotein convertase subtilisin/kexin type 9 neutralizing antibody reduces serum cholesterol in mice and nonhuman primates. *Proceedings of the National Academy of Sciences of the United States of America, 106,* 9820–9825.

Chapman, K., Pullen, N., Graham, M., & Ragan, I. (2007). Preclinical safety testing of monoclonal antibodies: The significance of species relevance. *Nature Reviews: Drug Discovery, 6,* 120–126.

Chargaff, E. (1976). Triviality in science: A brief meditation on fashions. *Perspectives in Biology and Medicine, 19,* 324–333.

Chemelli, R. M., Willie, J. T., Sinton, C. M., Elmquist, J. K., Scammell, T., Lee, C., Richardson, J. A., Williams, S. C., Xiong, Y., Kisanuki, Y., et al. (1999). Narcolepsy in orexin knockout mice: Molecular genetics of sleep regulation. *Cell, 98,* 437–451.

Chen, M., Kretzschmar, D., Verdile, G., & Lardelli, M. (2013). Models of Alzheimer's disease. In P. M. Conn (Ed.), *Animal Models for the Study of Human Disease* (pp. 595–632). Elsevier.

Chesselet, M.-F., Richter, F., Zhu, C., Magen, I., Watson, M. B., & Subramaniam, S. R. (2012). A progressive mouse model of Parkinson's disease: The Thy1-aSyn ("Line 61") mice. *Neurotherapeutics, 9,* 297–314.

Chien, A., Edgar, D. B., & Trela, J. M. (1976). Deoxyribonucleic acid polymerase from the extreme thermo-phile *Thermus aquaticus. Journal of Bacteriology, 127,* 1550–1557.

Chimpanzee Sequencing and Analysis Consortium. (2005). Initial sequence of the chimpanzee genome and comparison with the human genome. *Nature, 437,* 69–87.

Choi, S. H., Kim, Y. H., Hebisch, M., Sliwinski, C., Lee, S., D'Avanzo, C., Chen, H., Hooli, B., Asselin, C., Muffat, J., et al. (2014). A three-dimensional human neural cell culture model of Alzheimer's disease. *Nature, 515,* 274–278.

Chongtham, A., Bornemann, D. J., Barbaro, B. A., Lukacsovich, T., Agrawal, N., Syed, A., Worthge, S., Purcell, J., Burke, J., Chin, T. M., & Marsh, J. L. (2020). Effects of flanking sequences and cellular context on subcellular behavior and pathology of mutant HTT. *Human Molecular Genetics, 29*, 674–688.

Chow, S.-C., & Liu, J.-P. (2013). *Design and Analysis of Clinical Trials: Concepts and Methodologies.* John Wiley & Sons.

Christensen, D. (2004). Scientists divided on relevance of race in medical research. *Nature Medicine, 10*, Article 1266.

Churchill, G. A., Airey, D. C., Allayee, H., Angel, J. M., Attie, A. D., Beatty, J., Beavis, W. D., Belknap, J. K., Bennett, B., Berrettini, W., et al. (2004). The Collaborative Cross, a community resource for the genetic analysis of complex traits. *Nature Genetics, 36*, 1133–1137.

Clause, B. T. (1993). The Wistar Rat as a right choice: Establishing mammalian standards and the ideal of a standardized mammal. *Journal of the History of Biology, 26*, 329–349.

Clayton, J. A., & Collins, F. S. (2014). NIH to balance sex in cell and animal studies. *Nature, 509*, 282–283.

Cohen, S. (1960). Purification of a nerve-growth promoting protein from the mouse salivary gland and its neuro-cytotoxic antiserum. *Proceedings of the National Academy of Sciences of the United States of America, 46*, 302–311.

Cole, F. J. (1957). Harvey's animals. *Journal of the History of Medicine and Allied Sciences, 12*, 106–113.

Collaborative Cross Consortium. (2012). The genome architecture of the Collaborative Cross mouse genetic reference population. *Genetics, 190*, 389–401.

Collarini, E., Leighton, P., Pedersen, D., Harriman, B., Jacob, R., Mettler-Izquierdo, S., Yi, H., van de Lavoir, M.-C., & Etches, R. J. (2015). Inserting random and site-specific changes into the genome of chickens. *Poultry Science, 94*, 799–803.

Collins, F. S. (2004). What we do and don't know about "race," "ethnicity," genetics and health at the dawn of the genome era. *Nature Genetics, 36*, S13–S15.

Collins, F. S., & Tabak, L. A. (2014). NIH plans to enhance reproducibility. *Nature, 505*, 612–613.

Coltman, B. W., Earley, E. M., Shahar, A., Dudek, F. E., & Ide, C. F. (1995). Factors influencing mossy fiber collateral sprouting in organotypic slice cultures of neonatal mouse hippocampus. *Journal of Comparative Neurology, 362*, 209–222.

Comroe, J. H., & Dripps, R. D. (1974). Ben Franklin and open heart surgery. *Circulation Research, 35*, 361–369.

Conant, J. (2020a). *The Great Secret: The Classified World War II Disaster That Launched the War on Cancer.* Atlantic Books.

Conant, J. (2020b, September). How a chemical weapons disaster in WWII led to a U.S. cover-up—and a new cancer treatment. *Smithsonian Magazine.* https://www.smithsonianmag.com/history/bombing-and -breakthrough-180975505.

Connolly, C. (2005, July 30). Frist breaks with Bush on stem cell research. *The Washington Post.* https://www .washingtonpost.com/archive/politics/2005/07/30/frist-breaks-with-bush-on-stem-cell-research/0abaf147 -bc45-4a73-8b87-1d1639466de8.

Contestabile, A. (2011). The history of the cholinergic hypothesis. *Behavioural Brain Research, 221*, 334–340.

Cook, D., Brown, D., Alexander, R., March, R., Morgan, P., Satterthwaite, G., & Pangalos, M. N. (2014). Lessons learned from the fate of AstraZeneca's drug pipeline: A five-dimensional framework. *Nature Reviews: Drug Discovery, 13*, 419–431.

Cooper, C., Li, R., Lyketsos, C., & Livingston, G. (2013). Treatment for mild cognitive impairment: Systematic review. *British Journal of Psychiatry, 203*, 255–264.

Colquhoun, D. (2014). An investigation of the false discovery rate and the misinterpretation of p-values. *Royal Society Open Science, 1*, Article 140216.

Corbett, K. S., Flynn, B., Foulds, K. E., Francica, J. R., Boyoglu-Barnum, S., Werner, A. P., Flach, B., O'Connell, S., Bock, K. W., Minai, M., et al. (2020). Evaluation of the mRNA-1273 vaccine against SARS-CoV-2 in non-human primates. *New England Journal of Medicine, 383*, 1544–1555.

Cotlar, A. M., Dubose, J. J., & Rose, D. M. (2003). History of surgery for breast cancer: Radical to the sublime. *Current Surgery, 60*, 329–337.

Cottingham, J. (1978). "A brute to the brutes?": Descartes' treatment of animals. *Philosophy, 53*, 551–559.

Court, M. H. (2013). Feline drug metabolism and disposition: Pharmacokinetic evidence for species differences and molecular mechanisms. *Veterinary Clinics of North America: Small Animal Practice, 43*, 1039–1054.

Coutard, H. (1934). Principles of X ray therapy of malignant diseases. *The Lancet, 224*, 1–62.e1–e2. Originally published as Vol. 2(5784).

Cox, L. A. T., Popken, D. A., Kaplan, A. M., Plunkett, L. M., & Becker, R. A. (2016). How well can in vitro data predict in vivo effects of chemicals? Rodent carcinogenicity as a case study. *Regulatory Toxicology and Pharmacology, 77*, 54–64.

Crabbe, J. C., Wahlsten, D., & Dudek, B. C. (1999). Genetics of mouse behavior: Interactions with laboratory environment. *Science, 284*, 1670–1672.

Crawley, J. N. (2007). *What's Wrong with My Mouse? Behavioral Phenotyping of Transgenic and Knockout Mice* (2nd ed.). Wiley.

Creager, A. N. H. (2002). *The Life of a Virus: Tobacco Mosaic Virus as an Experimental Model, 1930–1965*. University of Chicago Press.

Creager, A. N. H., Lunbeck, E., & Wise, M. N. (2007). *Science without Laws: Model Systems, Cases, Exemplary Narratives*. Duke University Press.

Creus-Muncunill, J., & Ehrlich, M. E. (2019). Cell-autonomous and non-cell-autonomous pathogenic mechanisms in Huntington's disease: Insights from in vitro and in vivo models. *Neurotherapeutics, 16*, 957–978.

Crick, F. H. C., Barnett, L., Brenner, S., & Watts-Tobin, R. J. (1961). General nature of the genetic code for proteins. *Nature, 192*, 1227–1232.

Crilly, J. (2007). The history of clozapine and its emergence in the US market: A review and analysis. *History of Psychiatry, 18*, 39–60.

Cross, F. R., Buchler, N. E., & Skotheim, J. M. (2011). Evolution of networks and sequences in eukaryotic cell cycle control. *Philosophical Transactions of the Royal Society of London: Series B, Biological Sciences, 366*, 3532–3544.

Csobonyeiova, M., Polak, S., & Danisovic, L. (2020). Recent overview of the use of iPSCs Huntington's disease modeling and therapy. *International Journal of Molecular Sciences, 21*, Article 2239.

Cully, M. (2019). Zebrafish earn their drug discovery stripes. *Nature Reviews: Drug Discovery, 18*, 811–813.

Cummings, B. J., Uchida, N., Tamaki, S. J., Salazar, D. L., Hooshmand, M., Summers, R., Gage, F. H., & Anderson, A. J. (2005). Human neural stem cells differentiate and promote locomotor recovery in spinal cord-injured mice. *Proceedings of the National Academy of Sciences of the United States of America, 102*, 14069–14074.

Cummings, J. L., Morstorf, T., & Zhong, K. (2014). Alzheimer's disease drug-development pipeline: Few candidates, frequent failures. *Alzheimer's Research and Therapy, 6*, Article 37.

Cunningham, F., Elliott, J., & Lees, P. (Eds.). (2010). *Comparative and Veterinary Pharmacology*. Springer.

Curreri, A. R., Ansfield, F. J., McIver, F. A., Waisman, H. A., & Heidelberger, C. (1958). Clinical studies with 5-fluorouracil. *Cancer Research, 18*, 478–484.

Cushman, D. W., & Ondetti, M. A. (1991). History of the design of captopril and related inhibitors of angiotensin converting enzyme. *Hypertension, 17*, 589–592.

Cuthbert, B. N., & Insel, T. R. (2013). Toward the future of psychiatric diagnosis: The seven pillars of RDoC. *BMC Medicine, 11*, Article 126.

Cutler, G., Marshall, L. A., Chin, N., Baribault, H., & Kassner, P. D. (2007). Significant gene content variation characterizes the genomes of inbred mouse strains. *Genome Research, 17*, 1743–1754.

Cyranoski, D. (2013). Stem cells in Texas: Cowboy culture. *Nature, 494*, 166–168.

Cyranoski, D. (2016). Monkey kingdom. *Nature, 532*, 300–302.

Dagenais, G. R., Leong, D. P., Rangarajan, S., Lanas, F., Lopez-Jaramillo, P., Gupta, R., Diaz, R., Avezum, A., Oliveira, G. B. F., Wielgosz, A., et al. (2020). Variations in common diseases, hospital admissions, and deaths in middle-aged adults in 21 countries from five continents (PURE): A prospective cohort study. *The Lancet, 395*, 785–794.

Dal Canto, M. C., & Gurney, M. E. (1997). A low expressor line of transgenic mice carrying a mutant human Cu, Zn superoxide dismutase (SOD1) gene develops pathological changes that most closely resemble those in human amyotrophic lateral sclerosis. *Acta Neuropathologica, 93*, 537–550.

Darrah, P. A., Zeppa, J. J., Maiello, P., Hackney, J. A., Wadsworth, M. H., Hughes, T. K., Pokkali, S., Swanson, P. A., Grant, N. L., Rodgers, M. A., et al. (2019). Prevention of tuberculosis in macaques after intravenous BCG immunization. *Nature, 577*, 95–102.

Darrow, J. J. (2017). Explaining the absence of surgical procedure regulation. *Cornell Journal of Law and Public Policy, 27*, 189–206.

Davies, S. W., Turmaine, M., Cozens, B. A., DiFiglia, M., Sharp, A. H., Ross, C. A., Scherzinger, E., Wanker, E. E., Mangiarini, L., & Bates, G. P. (1997). Formation of neuronal intranuclear inclusions underlies the neurological dysfunction in mice transgenic for the HD mutation. *Cell, 90*, 537–548.

Davis, E. E., & Katsanis, N. (2017). Zebrafish: A model system to study the architecture of human genetic disease. In P. M. Conn (Ed.), *Animal Models for the Study of Human Disease* (pp. 651–670). Elsevier.

Davis, G. C., Williams, A. C., Markey, S. P., Ebert, M. H., Caine, E. D., Reichert, C. M., & Kopin, I. J. (1979). Chronic parkinsonism secondary to intravenous injection of meperidine analogues. *Psychiatry Research, 1*, 249–254.

Davis, R. H. (2003). *The Microbial Models of Molecular Biology*. Oxford University Press.

Davis, R. H. (2004). The age of model organisms. *Nature Reviews: Genetics, 5*, 69–76.

Day, C.-P., Merlino, G., & Van Dyke, T. (2015). Preclinical mouse cancer models: A maze of opportunities and challenges. *Cell, 163*, 39–53.

Dean, M., & Sung, V. (2018). Review of deutetrabenazine: A novel treatment for chorea associated with Huntington's disease. *Drug Design, Development and Therapy, 12*, 313–319.

De Gregorio, V., Urciuolo, F., Netti, P. A., & Imparato, G. (2020). In vitro organotypic systems to model tumor microenvironment in human papillomavirus (HPV)-related cancers. *Cancers, 12*, Article 1150.

De Kruif, P. (1926). *Microbe Hunters*. Harcourt.

De Lau, L. M., & Breteler, M. M. (2006). Epidemiology of Parkinson's disease. *Lancet Neurology, 5*, 525–535.

Delenclos, M., Burgess, J. D., Lamprokostopoulou, A., Outeiro, T. F., Vekrellis, K., & McLean, P. J. (2019). Cellular models of alpha-synuclein toxicity and aggregation. *Journal of Neurochemistry, 150*, 566–576.

Del Regato, J. A. (1976). Claudius Regaud. *International Journal of Radiation Oncology, Biology, Physics, 1*, 993–1001.

Demin, K. A., Sysoev, M., Chernysh, M. V., Savva, A. K., Koshiba, M., Wappler-Guzzetta, E. A., Song, C., De Abreu, M. S., Leonard, B., Parker, M. O., et al. (2019). Animal models of major depressive disorder and the implications for drug discovery and development. *Expert Opinion on Drug Discovery, 14*, 365–378.

Deroche-Gamonet, V., Belin, D., & Piazza, P. V. (2004). Evidence for addiction-like behavior in the rat. *Science, 305*, 1014–1017.

DeStefano, F., Bodenstab, H. M., & Offit, P. A. (2019). Principal controversies in vaccine safety in the United States. *Clinical Infectious Diseases, 69*, 726–731.

DeVita, V. T., & Chu, E. (2008). A history of cancer chemotherapy. *Cancer Research, 68*, 8643–8653.

De Waal, F. (2016). *Are We Smart Enough to Know How Smart Animals Are?* W. W. Norton.

De Waal, F. (2019). *Mama's Last Hug: Animal Emotions and What They Tell Us about Ourselves.* W. W. Norton.

Dickinson, M. E., Flenniken, A. M., Ji, X., Teboul, L., Wong, M. D., White, J. K., Meehan, T. F., Weninger, W. J., Westerberg, H., Adissu, H., et al. (2016). High-throughput discovery of novel developmental phenotypes. *Nature, 537*, 508–514.

Dickson, D. W., Crystal, H. A., Mattiace, L. A., Masur, D. M., Blau, A. D., Davies, P., Yen, S.-H., & Aronson, M. K. (1992). Identification of normal and pathological aging in prospectively studied nondemented elderly humans. *Neurobiology of Aging, 13*, 179–189.

Dietrich, M. R., Ankeny, R. A., & Chen, P. M. (2014). Publication trends in model organism research. *Genetics, 198*, 787–794.

Dietrich, M. R., Ankeny, R. A., Crowe, N., Green, S., & Leonelli, S. (2020). How to choose your research organism. *Studies in History and Philosophy of Biological and Biomedical Sciences, 80*, Article 101227.

Dietsche, B., Kircher, T., & Falkenberg, I. (2017). Structural brain changes in schizophrenia at different stages of the illness: A selective review of longitudinal magnetic resonance imaging studies. *Australian and New Zealand Journal of Psychiatry, 51*, 500–508.

DiFiglia, M., Sapp, E., Chase, K. O., Davies, S. W., Bates, G. P., Vonsattel, J. P., & Aronin, N. (1997). Aggregation of huntingtin in neuronal intranuclear inclusions and dystrophic neurites in brain. *Science, 277*, 1990–1993.

DiMasi, J. A., Feldman, L., Seckler, A., & Wilson, A. (2010). Trends in risks associated with new drug development: Success rates for investigational drugs. *Clinical Pharmacology and Therapeutics, 87*, 272–277.

Diss, G., Filteau, M., Freschi, L., Leducq, J.-B., Rochette, S., Torres-Quiroz, F., & Landry, C. R. (2013). Integrative avenues for exploring the dynamics and evolution of protein interaction networks. *Current Opinion in Biotechnology, 24*, 775–783.

Djian, P., Hancock, J. M., Chana, H. S. (1996). Codon repeats in genes associated with human diseases: Fewer repeats in the genes of nonhuman primates and nucleotide substitutions concentrated at the sites of reiteration. *Proceedings of the National Academy of Sciences of the United States of America, 93*, 417–421.

Djordjevic, D., Kusumi, K., & Ho, J. W. K. (2016). XGSA: A statistical method for cross-species gene set analysis. *Bioinformatics, 32*, i620–i628.

Djulbegovic, B., & Ioannidis, J. P. A. (2018). Precision medicine for individual patients should use population group averages and larger, not smaller, groups. *European Journal of Clinical Investigation, 348*, Article e13031.

Dobson, G. P. (2014). The August Krogh principle: Seeking unity in diversity. *Shock, 42,* 480–481.

Domagk, G. D. (1947). Further progress in chemotherapy of bacterial infections. *Nobel Lecture.* https://www .nobelprize.org/uploads/2018/06/domagk-lecture.pdf.

Domcke, S., Sinha, R., Levine, D. A., Sander, C., & Schultz, N. (2013). Evaluating cell lines as tumour models by comparison of genomic profiles. *Nature Communications, 4,* Article 2126.

Donaldson, H. H. (1924). *The Rat; Data and Reference Tables for the Albino Rat (Mus Norvegius Albinus) and the Norway Rat (Musnorvegicus).* Wistar Institute of Anatomy and Biology.

Donovan, J., & Kirshblum, S. (2018). Clinical trials in traumatic spinal cord injury. *Neurotherapeutics, 15,* 654–668.

Doorbar, J. (2016). Model systems of human papillomavirus-associated disease. *Journal of Pathology, 238,* 166–179.

Doremalen, N. van, Lambe, T., Spencer, A., Belij-Rammerstorfer, S., Purushotham, J. N., Port, J. R., Avanzato, V., Bushmaker, T., Flaxman, A., Ulaszewska, M., et al. (2020). ChAdOx1 nCoV-19 vaccination prevents SARS-CoV-2 pneumonia in rhesus macaques. *Nature, 586,* 587–582.

Doty, K. R., Guillot-Sestier, M.-V., & Town, T. (2015). The role of the immune system in neurodegenerative disorders: Adaptive or maladaptive? *Brain Research, 1617,* 155–173.

Douglas, W. R. (1972). Of pigs and men and research: A review of applications and analogies of the pig, *Sus scrofa,* in human medical research. *Space Life Sciences, 3,* 226–234.

Draaisma. (2019). The serendipitous story of lithium. *Nature, 572,* 584–585.

Dragatsis, I., Goldowitz, D., Del Mar, N., Deng, Y. P., Meade, C. A., Liu, L., Sun, Z., Dietrich, P., Yue, J., & Reiner, A. (2009). CAG repeat lengths >=335 attenuate the phenotype in the R6/2 Huntington's disease transgenic mouse. *Neurobiology of Disease, 33,* 315–330.

Drago, D., Cossetti, C., Iraci, N., Gaude, E., Musco, G., Bachi, A., & Pluchino, S. (2013). The stem cell secretome and its role in brain repair. *Biochimie, 95,* 2271–2285.

Draper, J. S., Smith, K., Gokhale, P., Moore, H. D., Maltby, E., Johnson, J., Meisner, L., Zwaka, T. P., Thomson, J. A., & Andrews, P. W. (2004). Recurrent gain of chromosomes 17q and 12 in cultured human embryonic stem cells. *Nature Biotechnology, 22,* 53–54.

Drouet, V., Perrin, V., Hassig, R., Dufour, N., Auregan, G., Alves, S., Bonvento, G., Brouillet, E., Luthi-Carter, R., Hantraye, P., & Déglon, N. (2009). Sustained effects of nonallele-specific Huntingtin silencing. *Annals of Neurology, 65,* 276–285.

Drucker, D. J. (2016). Never waste a good crisis: Confronting reproducibility in translational research. *Cell Metabolism, 24,* 348–360.

Druker, B. J., Sawyers, C. L., Kantarjian, H., Resta, D. J., Reese, S. F., Ford, J. M., Capdeville, R., & Talpaz, M. (2001). Activity of a specific inhibitor of the Bcr-Abl tyrosine kinase in the blast crisis of chronic myeloid leukemia and acute lymphoblastic leukemia with the Philadelphia chromosome. *New England Journal of Medicine, 344,* 1038–1042.

Druker, B. J., Tamura, S., Buchdunger, E., Ohno, S., Segal, G. M., Fanning, S., Zimmermann, J., & Lydon, N. B. (1996). Effects of a selective inhibitor of the Abl tyrosine kinase on the growth of Bcr-Abl positive cells. *Nature Medicine, 2,* 561–566.

Drummond, E., & Wisniewski, T. (2017). Alzheimer's disease: Experimental models and reality. *Acta Neuropathologica, 133,* 155–175.

Duan, W. (2013). Targeting sirtuin-1 in Huntington's disease: Rationale and current status. *CNS Drugs, 27*, 345–352.

Ducharme, N. A., Reif, D. M., Gustafsson, J.-A., & Bondesson, M. (2015). Comparison of toxicity values across zebrafish early life stages and mammalian studies: Implications for chemical testing. *Reproductive Toxicology (Elmsford, NY), 55*, 3–10.

Durr, E., Yu, J., Krasinska, K. M., Carver, L. A., Yates, J. R., Testa, J. E., Oh, P., & Schnitzer, J. E. (2004). Direct proteomic mapping of the lung microvascular endothelial cell surface *in vivo* and in cell culture. *Nature Biotechnology, 22*, 985–992.

Dusheck, J. (1985). Protesters prompt halt in animal research. *Science News, 128*, 53.

Dyballa, S., Miñana, R., Rubio-Brotons, M., Cornet, C., Pederzani, T., Escaramis, G., Garcia-Serna, R., Mestres, J., & Terriente, J. (2019). Comparison of zebrafish larvae and hiPSC cardiomyocytes for predicting drug-induced cardiotoxicity in humans. *Toxicological Sciences, 171*, 283–295.

Eastwood, D., Findlay, L., Poole, S., Bird, C., Wadhwa, M., Moore, M., Burns, C., Thorpe, R., & Stebbings, R. (2010). Monoclonal antibody TGN1412 trial failure explained by species differences in CD28 expression on CD4+ effector memory T-cells. *British Journal of Pharmacology, 161*, 512–526.

Eccles, J. C., Schmidt, R. F., & Willis, W. D. (1962). Presynaptic inhibition of the spinal monosynaptic reflex pathway. *Journal of Physiology, 161*, 282–297.

Edler, M. K., Sherwood, C. C., Meindl, R. S., Hopkins, W. D., Ely, J. J., Erwin, J. M., Mufson, E. J., Hof, P. R., & Raghanti, M. A. (2017). Aged chimpanzees exhibit pathologic hallmarks of Alzheimer's disease. *Neurobiology of Aging, 59*, 107–120.

Edsall, G. (1969). A positive approach to the problem of human experimentation. *Daedalus, 98*, 463–479.

Efrat, S. (2021). Epigenetic memory: Lessons from IPS cells derived from human β cells. *Frontiers in Endocrinology, 11*, Article 614234.

Egan, K. J., Vesterinen, H. M., Beglopoulos, V., Sena, E. S., & Macleod, M. R. (2016). From a mouse: Systematic analysis reveals limitations of experiments testing interventions in Alzheimer's disease mouse models. *Evidence-Based Preclinical Medicine, 3*, Article e00015.

Elledge, S. J., Bai, C., & Edwards, M. C. (1993). Cloning mammalian genes using cDNA expression libraries in *Saccharomyces cerevisiae*. *Methods, 5*, 96–101.

Ellenbroek, B., & Youn, J. (2016). Rodent models in neuroscience research: Is it a rat race? *Disease Models and Mechanisms, 9*, 1079–1087.

Eltahla, A., Luciani, F., White, P., Lloyd, A., & Bull, R. (2015). Inhibitors of the hepatitis C virus polymerase; mode of action and resistance. *Viruses, 7*, 5206–5224.

Emborg, M. E. (2017). Nonhuman primate models of neurodegenerative disorders. *ILAR Journal, 58*, 190–201.

Emini Veseli, B., Perrotta, P., De Meyer, G. R. A., Roth, L., Van der Donckt, C., Martinet, W., & De Meyer, G. R. Y. (2017). Animal models of atherosclerosis. *European Journal of Pharmacology, 816*, 3–13.

Enders, J. F., Robbins, F. C., & Weller, T. H. (1980). The cultivation of the poliomyelitis viruses in tissue culture. *Review of Infectious Diseases, 2*, 493–504.

Enders, J. F., Weller, T. H., & Robbins, F. C. (1949). Cultivation of the Lansing strain of poliomyelitis virus in cultures of various human embryonic tissues. *Science, 109*, 85–87.

Endersby, J. (2007). *A Guinea Pig's History of Biology*. Harvard University Press.

Endo, A. (2010). A historical perspective on the discovery of statins. *Proceedings of the Japan Academy: Series B, Physical and Biological Sciences, 86*, 484–493.

Engber, D. (2011, November 17). The trouble with Black-6. *Slate*. http://www.slate.com/articles/health_and_science/the_mouse_trap/2011/11/black_6_lab_mice_and_the_history_of_biomedical_research.html.

Eriksen, J. L., Dawson, T. M., Dickson, D. W., & Petrucelli, L. (2003). Caught in the act: Alpha-synuclein is the culprit in Parkinson's disease. *Neuron, 40*, 453–456.

Esmaeili, M. A., Panahi, M., Yadav, S., Hennings, L., & Kiaei, M. (2013). Premature death of TDP-43 (A315T) transgenic mice due to gastrointestinal complications prior to development of full neurological symptoms of amyotrophic lateral sclerosis. *International Journal of Experimental Pathology, 94*, 56–64.

Espuny-Camacho, I., Arranz, A. M., Fiers, M., Snellinx, A., Ando, K., Munck, S., Bonnefont, J., Lambot, L., Corthout, N., Omodho, L., et al. (2017). Hallmarks of Alzheimer's disease in stem-cell-derived human neurons transplanted into mouse brain. *Neuron, 93*, 1066–1081.

Esteves, P. J., Abrantes, J., Baldauf, H.-M., BenMohamed, L., Chen, Y., Christensen, N., González-Gallego, J., Giacani, L., Hu, J., Kaplan, G., et al. (2018). The wide utility of rabbits as models of human diseases. *Experimental and Molecular Medicine, 50*, 1–10.

Fan, Y., Winanto & Ng, S.-Y. (2020). Replacing what's lost: A new era of stem cell therapy for Parkinson's disease. *Translational Neurodegeneration, 9*, Article 2.

Farber, S., Diamond, L. K., Mercer, R. D., Sylvester, R. F., & Wolff, J. A. (1948). Temporary remissions in acute leukemia in children produced by folic acid antagonist, 4-aminopteroyl-glutamic acid (aminopterin). *New England Journal of Medicine, 238*, 787–793.

Farrar, W. E., Kent, T. H., & Elliott, V. B. (1966). Lethal gram-negative bacterial superinfection in guinea pigs given bacitracin. *Journal of Bacteriology, 92*, 496–501.

Farris, S. M. (2020). The rise to dominance of genetic model organisms and the decline of curiosity-driven organismal research. *PLoS One, 15*, Article e0243088.

Fazio, M., Ablain, J., Chuan, Y., Langenau, D. M., & Zon, L. I. (2020). Zebrafish patient avatars in cancer biology and precision cancer therapy. *Nature Reviews: Cancer, 20*, 263–273.

FDA Commissioner. (2020, September 3). FDA warns about stem cell therapies. *FDA Consumer Updates*. https://www.fda.gov/consumers/consumer-updates/fda-warns-about-stem-cell-therapies.

FDA Commissioner. (2021, June 7). FDA grants accelerated approval for Alzheimer's drug. FDA Press Announcements. https://www.fda.gov/news-events/press-announcements/fda-grants-accelerated-approval-alzheimers-drug.

Feany, M. B., & Bender, W. W. (2000). A *Drosophila* model of Parkinson's disease. *Nature, 404*, 394–398.

Feigin, A., Kieburtz, K., Como, P., Hickey, C., Abwender, D., Zimmerman, C., Steinberg, K., & Shoulson, I. (1996). Assessment of coenzyme Q10 tolerability in Huntington's disease. *Movement Disorders, 11*, 321–323.

Feng, J., & Jing, Z. (2018). Development history of endovascular surgery and devices. In A. Jing, H. Mao, & W. Dai (Eds.), *Endovascular Surgery and Devices* (pp. 3–7). Springer.

Ferens, W. A., & Hovde, C. J. (2011). *Escherichia coli* O157:H7: Animal reservoir and sources of human infection. *Foodborne Pathogens and Disease, 8*, 465–487.

Fernandez-Duque, E., Valeggia, C. R., & Mendoza, S. P. (2009). The biology of paternal care in human and nonhuman primates. *Annual Review of Anthropology, 38*, 115–130.

Ferrante, R. J., Andreassen, O. A., Dedeoglu, A., Ferrante, K. L., Jenkins, B. G., Hersch, S. M., & Beal, M. F. (2002). Therapeutic effects of coenzyme Q10 and remacemide in transgenic mouse models of Huntington's disease. *Journal of Neuroscience, 22*, 1592–1599.

Festing, S., & Wilkinson, R. (2007). The ethics of animal research. Talking point on the use of animals in scientific research. *EMBO Reports, 8,* 526–530.

Fields, S., & Johnston, M. (2005). Whither model organism research? *Science, 307,* 1885–1886.

Filipp, M. E., Travis, B. J., Henry, S. S., Idzikowski, E. C., Magnuson, S. A., Loh, M. Y., Hellenbrand, D. J., & Hanna, A. S. (2019). Differences in neuroplasticity after spinal cord injury in varying animal models and humans. *Neural Regeneration Research, 14,* 7–19.

Finger, S. (2004). *Minds Behind the Brain: A History of the Pioneers and Their Discoveries.* Oxford University Press.

Fink, M. P. (2014). Animal models of sepsis. *Virulence, 5,* 143–153.

Finlay, B. L. (2019). Generic *Homo sapiens* and unique *Mus musculus*: Establishing the typicality of the modeled and the model species. *Brain, Behavior and Evolution, 93,* 122–136.

Finlay, F., & Guiton, S. (2005). Chocolate poisoning. *British Medical Journal, 331,* 633–633.

Fior, R., Póvoa, V., Mendes, R. V., Carvalho, T., Gomes, A., Figueiredo, N., & Ferreira, M. G. (2017). Single-cell functional and chemosensitive profiling of combinatorial colorectal therapy in zebrafish xenografts. *Proceedings of the National Academy of Sciences of the United States of America, 114,* E8234–E8243.

Fischer, A. W., Cannon, B., & Nedergaard, J. (2019). The answer to the question "what is the best housing temperature to translate mouse experiments to humans?" is: Thermoneutrality. *Molecular Metabolism, 26,* 1–3.

Fischman, J. (2020, October 7). Nobel Prize in chemistry goes to discovery of "genetic scissors" called CRISPR/Cas9. *Scientific American.* https://www.scientificamerican.com/article/nobel-prize-in-chemistry-goes-to-discovery-of-genetic-scissors-called-crispr-cas911/.

Fisher, E. M. C., & Bannerman, D. M. (2019). Mouse models of neurodegeneration: Know your question, know your mouse. *Science Translational Medicine, 11,* Article eaaq1818.

Fisher, R. A. (1953). *The Design of Experiments.* Oliver and Boyd.

Flaherty, K. T., Gray, R., Chen, A., Li, S., Patton, D., Hamilton, S. R., Williams, P. M., Mitchell, E. P., Iafrate, A. J., Sklar, J., et al. (2020). The molecular analysis for therapy choice (NCI-MATCH) trial: Lessons for genomic trial design. *Journal of the National Cancer Institute, 112,* Article djz245.

Flajnik, M. F. (2002). Comparative analyses of immunoglobulin genes: Surprises and portents. *Nature Reviews: Immunology, 2,* 688–698.

Fleming, A. (1929). On the antibacterial action of cultures of a *Penicillium*, with special reference to their use in the isolation of *B. influenzae. British Journal of Experimental Pathology, 10,* 226–236.

Fleming, S. M., Fernagut, P.-O., & Chesselet, M.-F. (2005). Genetic mouse models of parkinsonism: Strengths and limitations. *NeuroRx, 2,* 495–503.

Flexner, A. (2017). *The Usefulness of Useless Knowledge (with a Companion Essay by Robbert Dijkgraaf).* Princeton University Press.

Flint, J., & Mackay, T. F. C. (2009). Genetic architecture of quantitative traits in mice, flies, and humans. *Genome Research, 19,* 723–733.

Flood, D. G., Reaume, A. G., Dorfman, K. S., Lin, Y.-G., Lang, D. M., Trusko, S. P., Savage, M. J., Annaert, W. G., De Strooper, B., Siman, R., & Scott, R. W. (2002). FAD mutant PS-1 gene-targeted mice: Increased Aβ42 and Aβ deposition without APP overproduction. *Neurobiology of Aging, 23,* 335–348.

Foidl, B. M., & Humpel, C. (2020). Can mouse models mimic sporadic Alzheimer's disease? *Neural Regeneration Research, 15,* 401–406.

Foley, N. M., Springer, M. S., & Teeling, E. C. (2016). Mammal madness: Is the mammal tree of life not yet resolved? *Philosophical Transactions of the Royal Society of London: Series B, Biological Sciences, 371*, Article 20150140.

Force, A., Lynch, M., Pickett, F. B., Amores, A., Yan, Y. L., & Postlethwait, J. (1999). Preservation of duplicate genes by complementary, degenerative mutations. *Genetics, 151*, 1531–1545.

Formal, S. B., Abrams, G. D., Schneider, H., & Laundy, R. (1963). Penicillin in germ-free guinea pigs. *Nature, 198*, 712–712.

Fornai, F., Longone, P., Cafaro, L., Kastsiuchenka, O., Ferrucci, M., Manca, M. L., Lazzeri, G., Spalloni, A., Bellio, N., Lenzi, P., et al. (2008). Lithium delays progression of amyotrophic lateral sclerosis. *Proceedings of the National Academy of Sciences of the United States of America, 105*, 2052–2057.

Fort, D. G., Herr, T. M., Shaw, P. L., Gutzman, K. E., & Starren, J. B. (2017). Mapping the evolving definitions of translational research. *Journal of Clinical and Translational Science, 1*, 60–66.

Fossat, P., Bacque-Cazenave, J., De Deurwaerdere, P., Delbecque, J.-P., & Cattaert, D. (2014). Anxiety-like behavior in crayfish is controlled by serotonin. *Science, 344*, 1293–1297.

Foury, F. (1997). Human genetic diseases: A cross-talk between man and yeast. *Gene, 195*, 1–10.

Francione, G. L. (1996). *Rain without Thunder: The Ideology of the Animal Rights Movement*. Temple University Press.

Franco, N. H. (2013). Animal experiments in biomedical research: A historical perspective. *Animals, 3*, 238–273.

Franco, N. H., Sandøe, P., & Olsson, I. A. S. (2018). Researchers' attitudes to the 3Rs—an upturned hierarchy? *PLoS One, 13*, Article e0200895.

Freedman, L. P., & Inglese, J. (2014). The increasing urgency for standards in basic biologic research. *Cancer Research, 74*, 4024–4029.

French, R. D. (1975). *Antivivisection and Medical Science in Victorian Society*. Princeton University Press.

Frenette, P. S., & Atweh, G. F. (2007). Sickle cell disease: Old discoveries, new concepts, and future promise. *Journal of Clinical Investigation, 117*, 850–858.

Freund, P., Schmidlin, E., Wannier, T., Bloch, J., Mir, A., Schwab, M. E., & Rouiller, E. M. (2006). Nogo-A–specific antibody treatment enhances sprouting and functional recovery after cervical lesion in adult primates. *Nature Medicine, 12*, 790–792.

Freund, P., Wannier, T., Schmidlin, E., Bloch, J., Mir, A., Schwab, M. E., & Rouiller, E. M. (2007). Anti-Nogo-A antibody treatment enhances sprouting of corticospinal axons rostral to a unilateral cervical spinal cord lesion in adult macaque monkey. *Journal of Comparative Neurology, 502*, 644–659.

Fricker, L. D. (2016). Drug discovery over the past thirty years: Why aren't there more new drugs? *Einstein Journal of Biology and Medicine, 29*, Article 61.

Friedberg, E. C. (2008). Sydney Brenner. *Nature Reviews: Molecular Cell Biology, 9*, 8–9.

Friedmann, H. C. (2004). From "*butyribacterium*" to "*E. coli*": An essay on unity in biochemistry. *Perspectives in Biology and Medicine, 47*, 47–66.

Frith, J. (2012). Syphilis—its early history and treatment until penicillin, and the debate on its origins. *Journal of Military and Veterans' Health, 20*, 49–56.

Fritz, S. L., Butts, R. J., & Wolf, S. L. (2012). Constraint-induced movement therapy: From history to plasticity. *Expert Review of Neurotherapeutics, 12*, 191–198.

Funaki, M., & Janmey, P. A. (2017). Technologies to engineer cell substrate mechanics in hydrogels. In A. Vishwakarma & J. M. Karp (Eds.), *Biology and Engineering of Stem Cell Niches* (pp. 363–373). Elsevier.

Furesz, J. (2006). Developments in the production and quality control of poliovirus vaccines—Historical perspectives. *Biologicals*, *34*, 87–90.

Fux, C. A., Shirtliff, M., Stoodley, P., & Costerton, J. W. (2005). Can laboratory reference strains mirror 'real-world' pathogenesis? *Trends in Microbiology*, *13*, 58–63.

Fye, W. B. (1986). Nitroglycerin: A homeopathic remedy. *Circulation*, *73*, 21–29.

Fye, W. B. (1995). William Murrell. *Clinical Cardiology*, *18*, 426–427.

Gähwiler, B. H. (1981). Organotypic monolayer cultures of nervous tissue. *Journal of Neuroscience Methods*, *4*, 329–342.

Gähwiler, B. H. (1988). Organotypic cultures of neural tissue. *Trends in Neurosciences*, *11*, 484–489.

Gardner, J. (2013). A history of deep brain stimulation: Technological innovation and the role of clinical assessment tools. *Social Studies of Science*, *43*, 707–728.

Garner, J. P. (2014). The significance of meaning: Why do over 90% of behavioral neuroscience results fail to translate to humans, and what can we do to fix it? *ILAR Journal*, *55*, 438–456.

Gartler, S. M. (1968). Apparent HeLa cell contamination of human heteroploid cell lines. *Nature*, *217*, 750–751.

Gay, N., & Pradad, V. (2017, March 8). Few people actually benefit from "breakthrough" cancer immunotherapy. *STAT*. https://www.statnews.com/2017/03/08/immunotherapy-cancer-breakthrough.

Gaztelumendi, N., & Nogués, C. (2014). Chromosome instability in mouse embryonic stem cells. *Scientific Reports*, *4*, Article 5324.

Geerts, H. (2009). Of mice and men: Bridging the translational disconnect in CNS drug discovery. *CNS Drugs*, *23*, 915–926.

Geisler, R., Borel, N., Ferg, M., Maier, J. V., & Strähle, U. (2016). Maintenance of zebrafish lines at the European Zebrafish Resource Center. *Zebrafish*, *13*, S19–S23.

Geison, G. L. (1995). *The Private Science of Louis Pasteur.* Princeton University Press.

Gennari, S. P., MacDonald, M. C., Postle, B. R., & Seidenberg, M. S. (2007). Context-dependent interpretation of words: Evidence for interactive neural processes. *NeuroImage*, *35*, 1278–1286.

Geoffroy, C. G., Lorenzana, A. O., Kwan, J. P., Lin, K., Ghassemi, O., Ma, A., Xu, N., Creger, D., Liu, K., He, Z., & Zheng, B. (2015). Effects of PTEN and Nogo codeletion on corticospinal axon sprouting and regeneration in mice. *Journal of Neuroscience*, *35*, 6413–6428.

Gerlai, R. (1996). Gene-targeting studies of mammalian behavior: Is it the mutation or the background genotype? *Trends in Neurosciences*, *19*, 177–181.

Gest, H. (1995). *Arabidopsis* to Zebrafish: A commentary on "Rosetta Stone" model systems in the biological sciences. *Perspectives in Biology and Medicine*, *39*, 77–85.

Getting, P. A. (1988). Comparative analysis of invertebrate central pattern generators. In A. H. Cohen, S. Rossignol, & S. Grillner (Eds.), *Neural Control of Rhythmic Movements in Vertebrates* (pp. 101–127). Wiley.

Getz, G. S., & Reardon, C. A. (2017). Animal models of atherosclerosis. In P. M. Conneally (Ed.), *Animal Models for the Study of Human Disease* (pp. 205–217). Elsevier.

Geyer, M. A. A., Olivier, B., Joëls, M., & Kahn, R. S. (2012). From antipsychotic to anti-schizophrenia drugs: Role of animal models. *Trends in Pharmacological Sciences*, *33*, 515–521.

Giannakopoulos, P., Herrmann, F. R., Bussiere, T., Bouras, C., Kovari, E., Perl, D. P., Morrison, J. H., Gold, G., & Hof, P. R. (2003). Tangle and neuron numbers, but not amyloid load, predict cognitive status in Alzheimer's disease. *Neurology*, *60*, 1495–1500.

Gibbons, D. L., & Spencer, J. (2011). Mouse and human intestinal immunity: Same ballpark, different players; different rules, same score. *Mucosal Immunology, 4*, 148–157.

Gibbs, R. A., Weinstock, G. M., Metzker, M. L., Muzny, D. M., Sodergren, E. J., Scherer, S., Scott, G., Steffen, D., Burch, P. E., Okwuonu, G., et al. (2004). Genome sequence of the Brown Norway rat yields insights into mammalian evolution. *Nature, 428*, 493–521.

Gieling, E. T., Schuurman, T., Nordquist, R. E., & van der Staay, F. J. (2011). The pig as a model animal for studying cognition and neurobehavioral disorders. In J. J. Hagan (Ed.), *Molecular and Functional Models in Neuropsychiatry* (pp. 359–383). Springer.

Giesy, J. P., Feyk, L. A., Jones, P. D., Kannan, K., & Sanderson, T. (2003). Review of the effects of endocrine-disrupting chemicals in birds. *Pure and Applied Chemistry, 75*, 2287–2303.

Gill, C., Phelan, J. P., Hatzipetros, T., Kidd, J. D., Tassinari, V. R., Levine, B., Wang, M. Z., Moreno, A., Thompson, K., Maier, M., Grimm, J., Gill, A., & Vieira, F. G. (2019). SOD1-positive aggregate accumulation in the CNS predicts slower disease progression and increased longevity in a mutant SOD1 mouse model of ALS. *Scientific Reports, 9*, 6724.

Gillick, K. (2003, April 22). DA asks for more information in chicken chipping case. *United Poultry Concerns.* https://www.upc-online.org/battery_hens/42203woodchipper.htm.

Gilman, A. (1963). The initial clinical trial of nitrogen mustard. *American Journal of Surgery, 105*, 574–578.

Ginsberg, G. L., Pullen Fedinick, K., Solomon, G. M., Elliott, K. C., Vandenberg, J. J., Barone, S., & Bucher, J. R. (2019). New toxicology tools and the emerging paradigm shift in environmental health decision-making. *Environmental Health Perspectives, 127*, Article 125002.

Giovanni, A., Sonsalla, P. K., & Heikkila, R. E. (1994). Studies on species sensitivity to the dopaminergic neurotoxin 1-methyl-4-phenyl-1,2,3,6-tetrahydropyridine. Part 2: Central administration of 1-methyl-4-phenylpyridinium. *Journal of Pharmacology and Experimental Therapeutics, 270*, 1008–1014.

Global Biological Standards Institute. (2013). *The Case for Standards in Life Science Research: Seizing Opportunities at a Time of Critical Need.* Global Biological Standards Institute.

Glossop, N. R. J., & Hardin, P. E. (2002). Central and peripheral circadian oscillator mechanisms in flies and mammals. *Journal of Cell Science, 115*, 3369–3377.

Goldin, A., Venditti, J. M., Humphreys, S. R., Dennis, D., Mantel, N., & Greenhouse, S. W. (1955). A quantitative comparison of the antileukemic effectiveness of two folic acid antagonists in mice. *Journal of the National Cancer Institute, 15*, 1657–1664.

Gonitel, R., Moffitt, H., Sathasivam, K., Woodman, B., Detloff, P. J., Faull, R. L. M., & Bates, G. P. (2008). DNA instability in postmitotic neurons. *Proceedings of the National Academy of Sciences of the United States of America, 105*, 3467–3472.

Gonzalez, F. J., & Yu, A.-M. (2006). Cytochrome p450 and xenobiotic receptor humanized mice. *Annual Review of Pharmacology and Toxicology, 46*, 41–64.

Goodman, J., Chandna, A., & Roe, K. (2015). Trends in animal use at US research facilities. *Journal of Medical Ethics, 41*, 567–569.

Goodman, L. S., & Wintrobe, M. M. (1946). Nitrogen mustard therapy: Use of methyl-bis (beta-chloroethyl) amine hydrochloride and tris (beta-chloroethyl) amine hydrochloride for Hodgkin's disease, lymphosarcoma, leukemia and certain allied and miscellaneous disorders. *JAMA, 132*, 126–132.

Gordon, J. W., Scangos, G. A., Plotkin, D. J., Barbosa, J. A., & Ruddle, F. H. (1980). Genetic transformation of mouse embryos by microinjection of purified DNA. *Proceedings of the National Academy of Sciences of the United States of America, 77*, 7380–7384.

Gosselin, D., Skola, D., Coufal, N. G., Holtman, I. R., Schlachetzki, J. C. M., Sajti, E., Jaeger, B. N., O'Connor, C., Fitzpatrick, C., Pasillas, M. P., et al. (2017). An environment-dependent transcriptional network specifies human microglia identity. *Science, 356*, Article eaal3222.

Gottesman, I. I., & Gould, T. D. (2003). The endophenotype concept in psychiatry: Etymology and strategic intentions. *American Journal of Psychiatry, 160*, 636–645.

Gould, S. E., Junttila, M. R., & de Sauvage, F. J. (2015). Translational value of mouse models in oncology drug development. *Nature Medicine, 21*, 431–439.

Goyal, M., Menon, B. K., Zwam, W. H. van, Dippel, D. W. J., Mitchell, P. J., Demchuk, A. M., Dávalos, A., Majoie, C. B. L. M., Lugt, A. van der, Miquel, M. A. de, et al. (2016). Endovascular thrombectomy after large-vessel ischaemic stroke: A meta-analysis of individual patient data from five randomised trials. *The Lancet, 387*, 1723–1731.

Grade, S., & Götz, M. (2017). Neuronal replacement therapy: Previous achievements and challenges ahead. *NPJ Regenerative Medicine, 2*, Article 29.

Graham, S. M., McCullough, L. D., & Murphy, S. J. (2004). Animal models of ischemic stroke: Balancing experimental aims and animal care. *Comparative Medicine, 54*, 486–496.

Green, S., Dietrich, M. R., Leonelli, S., & Ankeny, R. A. (2018). 'Extreme' organisms and the problem of generalization: Interpreting the Krogh principle. *History and Philosophy of the Life Sciences, 40*, Article 65.

Greene, L. A., Farinelli, S. E., Cunningham, M. E., & Park, D. S. (1998). Culture and experimental use of the PC12 rat pheochromocytoma cell line. In G. Banker & K. Goslin (Eds.), *Culturing Nerve Cells* (pp. 161–188). MIT Press.

Greene, L. A., & Tischler, A. S. (1976). Establishment of a noradrenergic clonal line of rat adrenal pheochromocytoma cells which respond to nerve growth factor. *Proceedings of the National Academy of Sciences of the United States of America, 73*, 2424–2428.

Greenspan, R. J. (2008). The origins of behavioral genetics. *Current Biology, 18*, R192–R198.

Greig, D., & Leu, J.-Y. (2009). Natural history of budding yeast. *Current Biology, 19*, R886–R890.

Griffin, A., Krasniak, C., & Baraban, S. C. (2016). Advancing epilepsy treatment through personalized genetic zebrafish models. *Progress in Brain Research, 226*, 195–207.

Grosberg, R. K., & Strathmann, R. R. (2007). The evolution of multicellularity: A minor major transition? *Annual Review of Ecology, Evolution, and Systematics, 38*, 621–654.

Grothe, B., & Pecka, M. (2014). The natural history of sound localization in mammals—a story of neuronal inhibition. *Frontiers in Neural Circuits, 8*, 116.

Grothe, B., Pecka, M., & McAlpine, D. (2010). Mechanisms of sound localization in mammals. *Physiological Reviews, 90*, 983–1012.

Grundman, K. (2001). Emil von Behring: The founder of serum therapy. *The Nobel Prize.* https://www.nobelprize.org/prizes/medicine/1901/behring/article/.

Grunwald, D. J., & Eisen, J. S. (2002). Headwaters of the zebrafish—emergence of a new model vertebrate. *Nature Reviews: Genetics, 3*, 717–724.

Guan, D., Fan, K., Spence, I., & Matthews, S. (2018). Combining machine learning models of in vitro and in vivo bioassays improves rat carcinogenicity prediction. *Regulatory Toxicology and Pharmacology, 94*, 8–15.

Guenther, M. G., Frampton, G. M., Soldner, F., Hockemeyer, D., Mitalipova, M., Jaenisch, R., & Young, R. A. (2010). Chromatin structure and gene expression programs of human embryonic and induced pluripotent stem cells. *Cell Stem Cell, 7*, 249–257.

Guerrini, A. (2003). *Experimenting with Humans and Animals: From Galen to Animal Rights*. Johns Hopkins University Press.

Guillery, R. W. (2007). Relating the neuron doctrine to the cell theory. Should contemporary knowledge change our view of the neuron doctrine? *Brain Research Reviews, 55*, 411–421.

Gunawardena, J. (2014). Models in biology: "Accurate descriptions of our pathetic thinking." *BMC Biology, 12*, Article 29.

Gunner, C., Gulamhusein, A., & Rosario, D. J. (2016). The modern role of androgen deprivation therapy in the management of localised and locally advanced prostate cancer. *Journal of Clinical Urology, 9*, 24–29.

Gurney, M. E. (1997). Transgenic animal models of familial amyotrophic lateral sclerosis. *Journal of Neurology, 244*, S15–S20.

Gurney, M. E., Pu, H., Chiu, A. Y., Canto, M. D., Polchow, C. Y., Alexander, D. D., Caliendo, J., Hentati, A., Kwon, Y. W., Deng, H. X., & Et, A. (1994). Motor neuron degeneration in mice that express a human Cu, Zn superoxide dismutase mutation. *Science, 264*, 1772–1775.

Gusella, J. F., & MacDonald, M. E. (2009). Huntington's disease: The case for genetic modifiers. *Genome Medicine, 1*, Article 80.

Gutenkunst, R. N., Waterfall, J. J., Casey, F. P., Brown, K. S., Myers, C. R., & Sethna, J. P. (2007). Universally sloppy parameter sensitivities in systems biology models. *PLoS Computational Biology, 3*, Article e189.

Gutierrez, K., Dicks, N., Glanzner, W. G., Agellon, L. B., & Bordignon, V. (2015). Efficacy of the porcine species in biomedical research. *Frontiers in Genetics, 6*, Article 293.

Haddow, A. D., Perez-Sautu, U., Wiley, M. R., Miller, L. J., Kimmel, A. E., Principe, L. M., Wollen-Roberts, S. E., Shamblin, J. D., Valdez, S. M., Cazares, L. H., et al. (2020). Modeling mosquito-borne and sexual transmission of Zika virus in an enzootic host, the African green monkey. *PLoS Neglected Tropical Diseases, 14*, Article e0008107.

Hagell, P., Schrag, A., Piccini, P., Jahanshahi, M., Brown, R., Rehncrona, S., Widner, H., Brundin, P., Rothwell, J. C., Odin, P., et al. (1999). Sequential bilateral transplantation in Parkinson's disease: Effects of the second graft. *Brain, 122*, 1121–1132.

Hahn, W. C., & Weinberg, R. A. (2002). Modelling the molecular circuitry of cancer. *Nature Reviews: Cancer, 2*, 331–341.

Haider, S. (2018). *Early Phase Disease Modification Trials with Selisistat and Optical Coherence Tomography as a Biomarker in Huntington's Disease* [Doctoral dissertation, University College London].

Haigwood, N. L., & Walker, C. M. (2011). Commissioned paper: Comparison of immunity to pathogens in humans, chimpanzees, and macaques. In B. M. Altevogt, D. E. Pankevich, M. K. Shelton-Davenport, & J. P. Kahn (Eds.), *Chimpanzees in Biomedical and Behavioral Research: Assessing the Necessity* (Appendix B). National Academies Press.

Hales, K. G., Korey, C. A., Larracuente, A. M., & Roberts, D. M. (2015). Genetics on the fly: A primer on the *Drosophila* model system. *Genetics, 201*, 815–842.

Haley, P. J. (2003). Species differences in the structure and function of the immune system. *Toxicology, 188*, 49–71.

Hall, E. D., & Springer, J. E. (2004). Neuroprotection and acute spinal cord injury: A reappraisal. *NeuroRx, 1*, 80–100.

Hamre, K., Tharp, R., Poon, K., Xiong, X., & Smeyne, R. J. (1999). Differential strain susceptibility following 1-methyl-4-phenyl-1,2,3,6-tetrahydropyridine (MPTP) administration acts in an autosomal dominant fashion: Quantitative analysis in seven strains of *Mus musculus. Brain Research, 828*, 91–103.

Hamza, A., Tammpere, E., Kofoed, M., Keong, C., Chiang, J., Giaever, G., Nislow, C., & Hieter, P. (2015). Complementation of yeast genes with human genes as an experimental platform for functional testing of human genetic variants. *Genetics, 201*, 1263–1274.

Han, I., You, Y., Kordower, J. H., Brady, S. T., & Morfini, G. A. (2010). Differential vulnerability of neurons in Huntington's disease: The role of cell type-specific features. *Journal of Neurochemistry, 113*, 1037–1091.

Hanahan, D., Wagner, E. F., & Palmiter, R. D. (2007). The origins of oncomice: A history of the first transgenic mice genetically engineered to develop cancer. *Genes and Development, 21*, 2258–2270.

Hanahan, D., & Weinberg, R. A. (2000). The hallmarks of cancer. *Cell, 100*, 57–70.

Hancock, A. M., & Frostig, R. D. (2017). Testing the effects of sensory stimulation as a collateral-based therapeutic for ischemic stroke in C57BL/6J and CD1 mouse strains. *PLoS One, 12*, Article e0183909.

Handler, D. C., & Haynes, P. A. (2019). An experimentally-derived measure of inter-replicate variation in reference samples: The same-same permutation methodology. *bioRxiv*, Article 797217.

Hao, X. (2007). Monkey research in China: Developing a natural resource. *Cell, 129*, 1033–1036.

Hardy, J. A., & Higgins, G. A. (1992). Alzheimer's disease: The amyloid cascade hypothesis. *Science, 256*, 184–185.

Hargis, K. E., & Blalock, E. M. (2017). Transcriptional signatures of brain aging and Alzheimer's disease: What are our rodent models telling us? *Behavioural Brain Research, 322*, 311–328.

Haring, A. P., Sontheimer, H., & Johnson, B. N. (2017). Microphysiological human brain and neural systems-on-a-chip: Potential alternatives to small animal models and emerging platforms for drug discovery and personalized medicine. *Stem Cell Reviews and Reports, 13*, 381–406.

Harnad, S. (2016). Animal sentience: The other-minds problem. *Animal Sentience, 1*, Article 001.

Harrington, R. (2018). *The Angel of Death: The Shocking True Story of Serial Killer Charles Edmund Cullen*. Amazon Digital Services.

Harrison, R. G. (1910). The outgrowth of the nerve fiber as a mode of protoplasmic movement. *Journal of Experimental Zoology, 9*, 787–846.

Harvey, W. (1928). *Exercitatio anatomica de motu cordis et sanguinis in animalibus* (C. D. Leake, Trans.). Thomas. (Original work published 1628.)

Harwerth, I. M., Wels, W., Schlegel, J., Müller, M., & Hynes, N. E. (1993). Monoclonal antibodies directed to the erbB-2 receptor inhibit in vivo tumour cell growth. *British Journal of Cancer, 68*, 1140–1145.

Harwood, P. D. (1963). Therapeutic dosage in small and large mammals. *Science, 139*, 684–685.

Haseman, J. K. (2000). Using the NTP database to assess the value of rodent carcinogenicity studies for determining human cancer risk. *Drug Metabolism Reviews, 32*, 169–186.

Hasselmann, J., & Blurton-Jones, M. (2020). Human iPSC-derived microglia: A growing toolset to study the brain's innate immune cells. *Glia, 68*, 721–739.

Hasselmann, J., Coburn, M. A., England, W., Figueroa Velez, D. X., Kiani Shabestari, S., Tu, C. H., McQuade, A., Kolahdouzan, M., Echeverria, K., Claes, C., et al. (2019). Development of a chimeric model to study and manipulate human microglia in vivo. *Neuron, 103*, 1016–1033.

Hatfield, G. (2007). The passions of the soul and Descartes's machine psychology. *Studies in History and Philosophy of Science, Part A, 38*, 1–35.

Hatziioannou, T., & Evans, D. T. (2012). Animal models for HIV/AIDS research. *Nature Reviews: Microbiology, 10*, 852–867.

Hay, M., Thomas, D. W., Craighead, J. L., Economides, C., & Rosenthal, J. (2014). Clinical development success rates for investigational drugs. *Nature Biotechnology, 32*, 40–51.

Hayflick, L. (1975). Cell biology of aging. *BioScience, 25*, 629–637.

Hayflick, L., & Moorhead, P. S. (1961). The serial cultivation of human diploid cell strains. *Experimental Cell Research, 25*, 585–621.

HD iPSC Consortium. (2012). Induced pluripotent stem cells from patients with Huntington's disease show CAG-repeat-expansion-associated phenotypes. *Cell Stem Cell, 11*, 264–278.

Head, E. (2013). A canine model of human aging and Alzheimer's disease. *Biochimica et Biophysica Acta, 1832*, 1384–1389.

Hebb, A. R., Godwin, T. F., & Gunn, R. W. (1968). A new beta adrenergic blocking agent, propranolol, in the treatment of angina pectoris. *Canadian Medical Association Journal, 98*, 246–251.

Hedges, S. B. (2002). The origin and evolution of model organisms. *Nature Reviews: Genetics, 3*, 838–849.

Heemskerk, J., Tobin, A. J., & Bain, L. J. (2002). Teaching old drugs new tricks. Meeting of the Neurodegeneration Drug Screening Consortium, 7–8 April 2002, Washington, DC, USA. *Trends in Neurosciences, 25*, 494–496.

Heidelberger, C., Chaurhuri, N. K., Danneberg, P., Mooren, D., Griesbach, L., Duschinsky, R., Schnitzer, R. J., Pleven, E., & Scheiner, J. (1957). Fluorinated pyrimidines, a new class of tumour-inhibitory compounds. *Nature, 179*, 663–666.

Hennessy, M. L., & Goldstein, A. M. (2019). Animal models in surgical research. In G. Kennedy, A. Gosain, M. Kibbe, & S. LeMaire (Eds.), *Success in Academic Surgery: Basic Science* (pp. 203–212). Springer.

Herland, A., Maoz, B. M., Das, D., Somayaji, M. R., Prantil-Baun, R., Novak, R., Cronce, M., Huffstater, T., Jeanty, S. S. F., Ingram, M., et al. (2020). Quantitative prediction of human pharmacokinetic responses to drugs via fluidically coupled vascularized organ chips. *Nature Biomedical Engineering, 4*, 421–436.

Hermann, D. M., Popa-Wagner, A., Kleinschnitz, C., & Doeppner, T. R. (2019). Animal models of ischemic stroke and their impact on drug discovery. *Expert Opinion on Drug Discovery, 14*, 315–326.

Herrup, K. (2015). The case for rejecting the amyloid cascade hypothesis. *Nature Neuroscience, 18*, 794–799.

Hersch, S. M., & Ferrante, R. J. (2004). Translating therapies for Huntington's disease from genetic animal models to clinical trials. *NeuroRx, 1*, 298–306.

Hersch, S. M., Schifitto, G., Oakes, D., Bredlau, A.-L., Meyers, C. M., Nahin, R., & Rosas, H. D. (2017). The CREST-E study of creatine for Huntington disease: A randomized controlled trial. *Neurology, 89*, 594–601.

Herschkowitz, J. I., Simin, K., Weigman, V. J., Mikaelian, I., Usary, J., Hu, Z., Rasmussen, K. E., Jones, L. P., Assefnia, S., Chandrasekharan, S., et al. (2007). Identification of conserved gene expression features between murine mammary carcinoma models and human breast tumors. *Genome Biology, 8*, Article R76.

Hesse, M. B. (1963). *Models and Analogies in Science.* Sheed & Ward.

Hikida, T., Jaaro-Peled, H., Seshadri, S., Oishi, K., Hookway, C., Kong, S., Wu, D., Xue, R., Andrade, M., Tankou, S., et al. (2007). Dominant-negative DISC1 transgenic mice display schizophrenia-associated phenotypes detected by measures translatable to humans. *Proceedings of the National Academy of Sciences of the United States of America, 104*, 14501–14506.

Hinnen, A., Hicks, J. B., & Fink, G. R. (1978). Transformation of yeast. *Proceedings of the National Academy of Sciences of the United States of America, 75*, 1929–1933.

Hittner, J. B., Hoogesteijn, A. L., Fair, J. M., van Regenmortel, M. H., & Rivas, A. L. (2019). The third cognitive revolution. *EMBO Reports, 20*, e47647.

Hockly, E., Woodman, B., Mahal, A., Lewis, C. M., & Bates, G. (2003). Standardization and statistical approaches to therapeutic trials in the R6/2 mouse. *Brain Research Bulletin, 61*, 469–479.

Hodgkin, A. L., & Huxley, A. F. (1952). A quantitative description of membrane current and its application to conduction and excitation in nerve. *Journal of Physiology, 117*, 500–544.

Hodos, W., & Campbell, C. B. G. (1969). *Scala naturae*: Why there is no theory in comparative psychology. *Psychological Review, 76*, 337–350.

Hofer, S., Kainz, K., Zimmermann, A., Bauer, M. A., Pendl, T., Poglitsch, M., Madeo, F., & Carmona-Gutierrez, D. (2018). Studying Huntington's disease in yeast: From mechanisms to pharmacological approaches. *Frontiers in Molecular Neuroscience, 11*, 318.

Hoffman, M. (2020). Support, skepticism, and statistics: The aducanumab saga. HCPLive. https://www.hcplive .com/view/support-skepticism-statistics-aducanumab.

Hoffner, G., & Djian, P. (2015). Polyglutamine aggregation in Huntington disease: Does structure determine toxicity? *Molecular Neurobiology, 52*, 1297–1314.

Holliday, R. (1996). Neoplastic transformation: The contrasting stability of human and mouse cells. *Cancer Surveys, 28*, 103–115.

Hollman, A. (1996). Digoxin comes from *Digitalis lanata*. *British Medical Journal, 312*, 912.

Holloway, A. K., Begun, D. J., Siepel, A., & Pollard, K. S. (2008). Accelerated sequence divergence of conserved genomic elements in *Drosophila melanogaster*. *Genome Research, 18*, 1592–1601.

Holst, A., & Frölich, T. (1907). Experimental studies relating to "ship-beri-beri" and scurvy. *Journal of Hygiene (London), 7*, 634–671.

Home Office. (2019). *Annual Statistics of Scientific Procedures on Living Animals, Great Britain: 2018.* https:// www.gov.uk/government/statistics/statistics-of-scientific-procedures-on-living-animals-great-britain-2018.

Hood, L. (2003). Systems biology: Integrating technology, biology, and computation. *Mechanisms of Ageing and Development, 124*, 9–16.

Hornsby, P. J. (2007). Telomerase and the aging process. *Experimental Gerontology, 42*, 575–581.

Horrobin, D. F. (2003). Modern biomedical research: An internally self-consistent universe with little contact with medical reality? *Nature Reviews: Drug Discovery, 2*, 151–154.

Horscroft, N., Lai, V. C., Cheney, W., Yao, N., Wu, J. Z., Hong, Z., & Zhong, W. (2005). Replicon cell culture system as a valuable tool in antiviral drug discovery against hepatitis C virus. *Antiviral Chemistry and Chemotherapy, 16*, 1–12.

Horsley-Silva, J. L., & Vargas, H. E. (2017). New therapies for hepatitis C virus infection. *Gastroenterology Hepatology (NY), 13*, 22–31.

Horstmann, D. M. (1985). The poliomyelitis story: A scientific hegira. *Yale Journal of Biology and Medicine, 58*, 79–90.

Horvitz, H. R. (2002). Worms, life and death. *Nobel Lecture.* https://www.nobelprize.org/prizes/medicine/2002 /horvitz/lecture/.

Horzmann, K. A., & Freeman, J. L. (2018). Making waves: New developments in toxicology with the zebrafish. *Toxicological Sciences, 163*, 5–12.

Howard, R. B., Sayeed, I., & Stein, D. G. (2017). Suboptimal dosing parameters as possible factors in the negative phase III clinical trials of progesterone for traumatic brain injury. *Journal of Neurotrauma, 34*, 1915–1918.

Howe, K., Clark, M. D., Torroja, C. F., Torrance, J., Berthelot, C., Muffato, M., Collins, J. E., Humphray, S., McLaren, K., Matthews, L., et al. (2013). The zebrafish reference genome sequence and its relationship to the human genome. *Nature, 496,* 498–503.

Howells, D. W., Porritt, M. J., Rewell, S. S., O'Collins, V., Sena, E. S., van der Worp, H. B., Traystman, R. J., & Macleod, M. R. (2010). Different strokes for different folks: The rich diversity of animal models of focal cerebral ischemia. *Journal of Cerebral Blood Flow and Metabolism, 30,* 1412–1431.

Howells, D. W., Sena, E. S., & Macleod, M. R. (2014). Bringing rigour to translational medicine. *Nature Reviews: Neurology, 10,* 37–43.

Hsu, P. D., Lander, E. S., & Zhang, F. (2014). Development and applications of CRISPR-Cas9 for genome engineering. *Cell, 157,* 1262–1278.

Huang, Q., & Riviere, J. E. (2014). The application of allometric scaling principles to predict pharmacokinetic parameters across species. *Expert Opinion on Drug Metabolism and Toxicology, 10,* 1241–1253.

Huebner, E. A., & Strittmatter, S. M. (2009). Axon regeneration in the peripheral and central nervous systems. *Results and Problems in Cell Differentiation, 48,* 339–351.

Hugenholtz, H. (2003). Methylprednisolone for acute spinal cord injury: Not a standard of care. *Canadian Medical Association Journal, 168,* 1145–1146.

Huggins, C., Briziarelli, G., & Sutton Jr., H. (1959). Rapid induction of mammary carcinoma in the rat and the influence of hormones on the tumors. *Journal of Experimental Medicine, 109,* 25–42.

Huggins, C., & Hodges, C. V. (1941). Studies on prostatic cancer. I. The effect of castration, of estrogen and of androgen injection on serum phosphatases in metastatic carcinoma of the prostate. *Cancer Research, 1,* 293–297.

Huggins, C., & Stevens, R. A. (1940). The effect of castration on benign hypertrophy of the prostate in man. *Journal of Urology, 43,* 705–714.

Hughes, P., Marshall, D., Reid, Y., Parkes, H., & Gelber, C. (2007). The costs of using unauthenticated, over-passaged cell lines: How much more data do we need? *BioTechniques, 43,* 575–584.

Huh, D., Matthews, B. D., Mammoto, A., Montoya-Zavala, M., Hsin, H. Y., & Ingber, D. E. (2010). Reconstituting organ-level lung functions on a chip. *Science, 328,* 1662–1668.

Humpel, C. (2015). Organotypic brain slice cultures: A review. *Neuroscience, 305,* 86–98.

Hünig, T. (2012). The storm has cleared: Lessons from the CD28 superagonist TGN1412 trial. *Nature Reviews: Immunology, 12,* 317–318.

Hünig, T. (2016). The rise and fall of the CD28 superagonist TGN1412 and its return as TAB08: A personal account. *FEBS Journal, 283,* 3325–3334.

Hunt, P. R. (2017). The *C. elegans* model in toxicity testing. *Journal of Applied Toxicology, 37,* 50–59.

Hunter, R. P. (2010). Interspecies allometric scaling. In F. Cunningham, J. Elliott, & P. Lees (Eds.), *Comparative and Veterinary Pharmacology* (pp. 139–157). Springer.

Huntington Study Group. (2006). Tetrabenazine as antichorea therapy in Huntington disease: A randomized controlled trial. *Neurology, 66,* 366–372.

Hurley, M. J., Deacon, R. M. J., Beyer, K., Ioannou, E., Ibáñez, A., Teeling, J. L., & Cogram, P. (2018). The long-lived *Octodon* degus as a rodent drug discovery model for Alzheimer's and other age-related diseases. *Pharmacology and Therapeutics, 188,* 36–44.

Hutchinson, J. N., & Muller, W. J. (2000). Transgenic mouse models of human breast cancer. *Oncogene, 19,* 6130–6137.

Hutchison, C. A., Chuang, R. Y., Noskov, V. N., Assad-Garcia, N., Deerinck, T. J., Ellisman, M. H., Gill, J., Kannan, K., Karas, B. J., Ma, L., et al. (2016). Design and synthesis of a minimal bacterial genome. *Science, 351*, Article aad6253.

Hyde, C. L., Nagle, M. W., Tian, C., Chen, X., Paciga, S. A., Wendland, J. R., Tung, J. Y., Hinds, D. A., Perlis, R. H., & Winslow, A. R. (2016). Identification of 15 genetic loci associated with risk of major depression in individuals of European descent. *Nature Genetics, 48*, 1031–1036.

Hynds, R. E., Vladimirou, E., & Janes, S. M. (2018). The secret lives of cancer cell lines. *Disease Models and Mechanisms, 11*, Article dmm037366.

Hyun, I., Wilkerson, A., & Johnston, J. (2016). Embryology policy: Revisit the 14-day rule. *Nature, 533*, 169–171.

Ice Bucket Challenge. (2021, February 9). In *Wikipedia*. https://en.wikipedia.org/w/index.php?title=Ice_Bucket_Challenge&oldid=1005734047.

Idalia, V.-M. N., & Bernardo, F. (2017). *Escherichia coli* as a model organism and its application in biotechnology. In A. Samie (Ed.), *Escherichia coli: Recent Advances on Physiology, Pathogenesis and Biotechnological Applications* (pp. 253–274). InTechOpen.

Ideker, T., & Krogan, N. J. (2012). Differential network biology. *Molecular Systems Biology, 8*, 565.

Inman, D. M., & Steward, O. (2003). Physical size does not determine the unique histopathological response seen in the injured mouse spinal cord. *Journal of Neurotrauma, 20*, 33–42.

Inoue, J., Sato, Y., Sinclair, R., Tsukamoto, K., & Nishida, M. (2015). Rapid genome reshaping by multiple-gene loss after whole-genome duplication in teleost fish suggested by mathematical modeling. *Proceedings of the National Academy of Sciences of the United States of America, 112*, 14918–14923.

Insel, T., Cuthbert, B., Garvey, M., Heinssen, R., Pine, D. S., Quinn, K., Sanislow, C., & Wang, P. (2010). Research domain criteria (RDoC): Toward a new classification framework for research on mental disorders. *American Journal of Psychiatry, 167*, 748–751.

Insel, T. R. (2007). From animal models to model animals. *Biological Psychiatry, 62*, 1337–1339.

Insel, T. R., & Cuthbert, B. N. (2009). Endophenotypes: Bridging genomic complexity and disorder heterogeneity. *Biological Psychiatry, 66*, 988–989.

Institute for Health Metrics and Evaluation (IHME), University of Washington. (2018). Global trends in disability. Years lived with disability: 2017 study highlights. Healthdata.org. http://www.healthdata.org/sites/default/files/files/infographics/Infographic_GBD2017-YLDs-Highlights_2018.pdf.

Institute of Medicine & National Research Council. (2011). *Chimpanzees in Biomedical and Behavioral Research: Assessing the Necessity.* National Academies Press.

Institute of Medicine & National Research Council. (2012). *International Animal Research Regulations: Impact on Neuroscience Research: Workshop Summary.* Forum on Neuroscience and Nervous System Disorders. National Academies Press.

Ioannidis, J. P. A. (2005a). Contradicted and initially stronger effects in highly cited clinical research. *JAMA, 294*, 218–228.

Ioannidis, J. P. A. (2005b). Why most published research findings are false. *PLoS Medicine, 2*, Article e124.

Ireson, C. R., Alavijeh, M. S., Palmer, A. M., Fowler, E. R., & Jones, H. J. (2019). The role of mouse tumour models in the discovery and development of anticancer drugs. *British Journal of Cancer, 121*, 101–108.

Iwai, Y., Ishida, M., Tanaka, Y., Okazaki, T., Honjo, T., & Minato, N. (2002). Involvement of PD-L1 on tumor cells in the escape from host immune system and tumor immunotherapy by PD-L1 blockade. *Proceedings of the National Academy of Sciences of the United States of America, 99*, 12293–12297.

Iwanami, A., Kaneko, S., Nakamura, M., Kanemura, Y., Mori, H., Kobayashi, S., Yamasaki, M., Momoshima, S., Ishii, H., Ando, K., et al. (2005). Transplantation of human neural stem cells for spinal cord injury in primates. *Journal of Neuroscience Research*, *80*, 182–190.

Jackson Laboratory. (2021). 3xTG-AD mice. https://www.jax.org/strain/004807.

Jacob, F. (1998). *Of Flies, Mice, and Men* (G. Weiss, Trans.). Harvard University Press.

Jacobsen, C. F., Wolfe, J. B., & Jackson, T. A. (1935). An experimental analysis of the functions of the frontal association areas in primates. *Journal of Nervous and Mental Disease*, *82*, 1–14.

Jagmag, S. A., Tripathi, N., Shukla, S. D., Maiti, S., & Khurana, S. (2016). Evaluation of models of Parkinson's disease. *Frontiers in Neuroscience*, *9*, Article 503.

James, S. L., Abate, D., Abate, K. H., Abay, S. M., Abbafati, C., Abbasi, N., Abbastabar, H., Abd-Allah, F., Abdela, J., Abdelalim, A., et al. (2018). Global, regional, and national incidence, prevalence, and years lived with disability for 354 diseases and injuries for 195 countries and territories, 1990–2017: A systematic analysis for the Global Burden of Disease Study 2017. *The Lancet*, *392*, 1789–1858.

James, S. L., Theadom, A., Ellenbogen, R. G., Bannick, M. S., Montjoy-Venning, W., Lucchesi, L. R., Abbasi, N., Abdulkader, R., Abraha, H. N., Adsuar, J. C., et al. (2019). Global, regional, and national burden of traumatic brain injury and spinal cord injury, 1990–2016: A systematic analysis for the Global Burden of Disease Study 2016. *Lancet Neurology*, *18*, 56–87.

Jankowsky, J. L., & Zheng, H. (2017). Practical considerations for choosing a mouse model of Alzheimer's disease. *Molecular Neurodegeneration*, *12*, Article 89.

Janssen, P. (2009). The "social chemistry" of pharmacological discovery: The haloperidol story. an interview with Dr. Paul Janssen, January 21, 1986. *International Journal of the Addictions*, *27*, 331–346.

Jennings, B. H. (2011). *Drosophila*—a versatile model in biology & medicine. *Materials Today*, *14*, 190–195.

Jensen, E. V., Block, G. E., Ferguson, D. J., & DeSombre, E. R. (1977). Estrogen receptors in breast cancer. *World Journal of Surgery*, *1*, 341–342.

Jeong, J., & Choi, J. (2018). Use of adverse outcome pathways in chemical toxicity testing: Potential advantages and limitations. *Environmental Health and Toxicology*, *33*, Article e2018002.

Jiang, X., Wang, J., Deng, X., Xiong, F., Zhang, S., Gong, Z., Li, X., Cao, K., Deng, H., He, Y., et al. (2020). The role of microenvironment in tumor angiogenesis. *Journal of Experimental and Clinical Cancer Research*, *39*, Article 204.

Joffe, A. R., Bara, M., Anton, N., & Nobis, N. (2016). The ethics of animal research: A survey of the public and scientists in North America. *BMC Medical Ethics*, *17*, Article 17.

John, A., Vinayan, K., & Varghese, J. (2016). Achiasmy: Male fruit flies are not ready to mix. *Frontiers in Cell and Developmental Biology*, *4*, Article 75.

Johnson, R., Teh, C. H., Kunarso, G., Wong, K. Y., Srinivasan, G., Cooper, M. L., Volta, M., Chan, S. S., Lipovich, L., Pollard, S. M., et al. (2008). REST regulates distinct transcriptional networks in embryonic and neural stem cells. *PLoS Biology*, *6*, Article e256.

Johnstone, T. C., Park, G. Y., & Lippard, S. J. (2014). Understanding and improving platinum anticancer drugs—phenanthriplatin. *Anticancer Research*, *34*, 471–476.

Jordan, V. C. (1976). Effect of tamoxifen (ICI 46,474) on initiation and growth of DMBA-induced rat mammary carcinomata. *European Journal of Cancer*, *12*, 419–424.

Jordan, V. C. (2008). Tamoxifen: Catalyst for the change to targeted therapy. *European Journal of Cancer*, *44*, 30–38.

Jørgensen, C. B. (2001). August Krogh and Claude Bernard on basic principles in experimental physiology. *BioScience, 51*, 59–61.

Joyner, M. J. (2011). Giant sucking sound: Can physiology fill the intellectual void left by the reductionists? *Journal of Applied Physiology, 111*, 335–342.

Jucker, M. (2010). The benefits and limitations of animal models for translational research in neurodegenerative diseases. *Nature Medicine, 16*, 1210–1214.

Junod, S. W. (2008). FDA and clinical drug trials: A short history. In M. Davies & F. Kerimani (Eds.), *A Quick Guide to Clinical Trials* (pp. 25–55). Bioplan.

Juntti, S. (2019). The future of gene-guided neuroscience research in non-traditional model organisms. *Brain, Behavior and Evolution, 93*, 108–121.

Kafkafi, N., Agassi, J., Chesler, E. J., Crabbe, J. C., Crusio, W. E., Eilam, D., Gerlai, R., Golani, I., Gomez-Marin, A., Heller, R., et al. (2018). Reproducibility and replicability of rodent phenotyping in preclinical studies. *Neuroscience and Biobehavioral Reviews, 87*, 218–232.

Kaiser, J. (2015, November 18). NIH to end all support for chimpanzee research. *Science*. https://www.sciencemag.org/news/2015/11/nih-end-all-support-chimpanzee-research.

Kaletta, T., & Hengartner, M. O. (2006). Finding function in novel targets: *C. elegans* as a model organism. *Nature Reviews: Drug Discovery, 5*, 387–398.

Kalm, L. M., & Semba, R. D. (2005). They starved so that others be better fed: Remembering Ancel Keys and the Minnesota experiment. *Journal of Nutrition, 135*, 1347–1352.

Kalueff, A. V., Wheaton, M., & Murphy, D. L. (2007). What's wrong with my mouse model? Advances and strategies in animal modeling of anxiety and depression. *Behavioural Brain Research, 179*, 1–18.

Kane, J. M., Honigfeld, G., Singer, J., & Meltzer, H. Y. (1989). Clozapine for the treatment-resistant schizophrenic: Results of a US multicenter trial. *Psychopharmacology (Berlin), 99*, 60–63.

Kaplan, A. (1973). *The Conduct of Inquiry: Methodology for Behavioral Science*. Intertext.

Kas, M. J. H., Fernandes, C., Schalkwyk, L. C., & Collier, D. A. (2007). Genetics of behavioural domains across the neuropsychiatric spectrum; of mice and men. *Molecular Psychiatry, 12*, 324–330.

Kasahara, J., Choudhury, M. E., Yokoyama, H., Kadoguchi, N., & Nomoto, M. (2013). Neurotoxin 1-methyl-4-phenyl-1,2,3,6-tetrahydropyridine-induced animal models for Parkinson's disease. In P. M. Conn (Ed.), *Animal Models for the Study of Human Disease* (pp. 633–650). Elsevier.

Kastrati, A., Mehilli, J., Pache, J., Kaiser, C., Valgimigli, M., Kelbæk, H., Menichelli, M., Sabaté, M., Suttorp, M. J., Baumgart, D., et al. (2007). Analysis of 14 trials comparing sirolimus-eluting stents with bare-metal stents. *New England Journal of Medicine, 356*, 1030–1039.

Katz, P. S. (2016). "Model organisms" in the light of evolution. *Current Biology, 26*, R649–R650.

Kaur, G., & Dufour, J. M. (2012). Cell lines: Valuable tools or useless artifacts. *Spermatogenesis, 2*, 1–5.

Kaye, J. A., & Finkbeiner, S. (2013). Modeling Huntington's disease with induced pluripotent stem cells. *Molecular and Cellular Neurosciences, 56*, 50–64.

Keen, E. C. (2014). A century of phage research: Bacteriophages and the shaping of modern biology. *BioEssays, 37*, 6–9.

Keirstead, H. S., Nistor, G., Bernal, G., Totoiu, M., Cloutier, F., Sharp, K., & Steward, O. (2005). Human embryonic stem cell-derived oligodendrocyte progenitor cell transplants remyelinate and restore locomotion after spinal cord injury. *Journal of Neuroscience, 25*, 4694–4705.

Keller, E. F. (2000). Models of and models for: Theory and practice in contemporary biology. *Philosophy of Science, 67*, S72–S86.

Keller, E. F. (2002). *Making Sense of Life: Explaining Biological Development with Models, Metaphors, and Machines.* Harvard University Press.

Kelly, J. P., Wrynn, A. S., & Leonard, B. E. (1997). The olfactory bulbectomized rat as a model of depression: An update. *Pharmacology and Therapeutics, 74*, 299–316.

Kenyon, C. (1988). The nematode *Caenorhabditis elegans. Science, 240*, 1448–1453.

Keshava, N., Toh, T. S., Yuan, H., Yang, B., Menden, M. P., & Wang, D. (2019). Defining subpopulations of differential drug response to reveal novel target populations. *NPJ Systems Biology and Applications, 5*, Article 36.

Kessler, R. C., Petukhova, M., Sampson, N. A., Zaslavsky, A. M., & Wittchen, H. (2012). Twelve-month and lifetime prevalence and lifetime morbid risk of anxiety and mood disorders in the United States. *International Journal of Methods in Psychiatric Research, 21*, 169–184.

Key, B. (2015). Fish do not feel pain and its implications for understanding phenomenal consciousness. *Biology and Philosophy, 30*, 149–165.

Khan, T., Havey, R. M., Sayers, S. T., Patwardhan, A., & King, W. W. (1999). Animal models of spinal cord contusion injuries. *Laboratory Animal Science, 49*, 161–172.

Kikuchi, T., Morizane, A., Doi, D., Magotani, H., Onoe, H., Hayashi, T., Mizuma, H., Takara, S., Takahashi, R., Inoue, H., et al. (2017). Human iPS cell-derived dopaminergic neurons function in a primate Parkinson's disease model. *Nature, 548*, 592–596.

Kimmel, C. B. (1989). Genetics and early development of zebrafish. *Trends in Genetics, 5*, 283–288.

Kimmelman, J., & Tannock, I. (2018). The paradox of precision medicine. *Nature Reviews: Clinical Oncology, 15*, 341–342.

Kinder, H. A., Clark, E. W., & West, F. D. (2019). The pig as a preclinical traumatic brain injury model: Current models, functional outcome measures, and translational detection strategies. *Neural Regeneration Research, 14*, 413–424.

King, H. D. (1918). Studies on inbreeding. I. The effects in inbreeding on the growth and variability in the body weight of the albino rat. *Journal of Experimental Zoology, 26*, 1–54.

Kirchberger, S., Sturtzel, C., Pascoal, S., & Distel, M. (2017). *Quo natas, Danio?*—Recent progress in modeling cancer in zebrafish. *Frontiers in Oncology, 7*, Article 186.

Kirk, R. G. W. (2012). "Standardization through mechanization": Germ-free life and the engineering of the ideal laboratory animal. *Technology and Culture, 53*, 61–93.

Kirkland, D., Aardema, M., Henderson, L., & Müller, L. (2005). Evaluation of the ability of a battery of three in vitro genotoxicity tests to discriminate rodent carcinogens and non-carcinogens: I. Sensitivity, specificity and relative predictivity. *Mutation Research, 584*, 1–256.

Kirnbauer, R., Booy, F., Cheng, N., Lowy, D. R., & Schiller, J. T. (1992). Papillomavirus L1 major capsid protein self-assembles into virus-like particles that are highly immunogenic. *Proceedings of the National Academy of Sciences of the United States of America, 89*, 12180–12184.

Kirov, S. A., Sorra, K. E., & Harris, K. M. (1999). Slices have more synapses than perfusion-fixed hippocampus from both young and mature rats. *Journal of Neuroscience, 19*, 2876–2886.

Kleim, J. A., & Jones, T. A. (2008). Principles of experience-dependent neural plasticity: Implications for rehabilitation after brain damage. *Journal of Speech, Language, and Hearing Research, 51*, S225–S239.

Klein, C., & Westenberger, A. (2012). Genetics of Parkinson's disease. *Cold Spring Harbor Perspectives in Medicine*, *2*, Article a008888.

Klein, R. A., Vianello, M., Hasselman, F., Adams, B. G., Reginald B. Adams, J., Alper, S., Aveyard, M., Axt, J. R., Babalola, M. T., et al. (2018). Many Labs 2: Investigating variation in replicability across samples and settings: *Advances in Methods and Practices in Psychological Science*, *1*, 443–490.

Kleinjan, D. A., Bancewicz, R. M., Gautier, P., Dahm, R., Schonthaler, H. B., Damante, G., Seawright, A., Hever, A. M., Yeyati, P. L., van Heyningen, V., & Coutinho, P. (2008). Subfunctionalization of duplicated zebrafish pax6 genes by cis-regulatory divergence. *PLoS Genetics*, *4*, Article e29.

Kleinstreuer, N. C., Dix, D. J., Houck, K. A., Kavlock, R. J., Knudsen, T. B., Martin, M. T., Paul, K. B., Reif, D. M., Crofton, K. M., Hamilton, K., et al. (2013). In vitro perturbations of targets in cancer hallmark processes predict rodent chemical carcinogenesis. *Toxicological Sciences*, *131*, 40–55.

Knight, A. (2007). Systematic reviews of animal experiments demonstrate poor human clinical and toxicological utility. *Alternatives to Laboratory Animals*, *35*, 641–659.

Knight, A. W., Little, S., Houck, K., Dix, D., Judson, R., Richard, A., McCarroll, N., Akerman, G., Yang, C., Birrell, L., & Walmsley, R. M. (2009). Evaluation of high-throughput genotoxicity assays used in profiling the US EPA ToxCast™ chemicals. *Regulatory Toxicology and Pharmacology*, *55*, 188–199.

Knight, K. R. G., Kraemer, D. F., & Neuwelt, E. A. (2005). Ototoxicity in children receiving platinum chemotherapy: Underestimating a commonly occurring toxicity that may influence academic and social development. *Journal of Clinical Oncology*, *23*, 8588–8596.

Kochanek, P. M., Dixon, C. E., Mondello, S., Wang, K. K. K., Lafrenaye, A., Bramlett, H. M., Dietrich, W. D., Hayes, R. L., Shear, D. A., Gilsdorf, J. S., et al. (2018). Multi-center pre-clinical consortia to enhance translation of therapies and biomarkers for traumatic brain injury: Operation brain trauma therapy and beyond. *Frontiers in Neurology*, *9*, Article 640.

Kochanek, P. M., Jackson, T. C., Jha, R. M., Clark, R. S. B., Okonkwo, D. O., Bayır, H., Poloyac, S. M., Wagner, A. K., Empey, P. E., Conley, Y. P., et al. (2020). Paths to successful translation of new therapies for severe traumatic brain injury in the golden age of traumatic brain injury research: A Pittsburgh vision. *Journal of Neurotrauma*, *37*, 2353–2371.

Köhler, G., & Milstein, C. (1975). Continuous cultures of fused cells secreting antibody of predefined specificity. *Nature*, *256*, 495–497.

Kohler, R. E. (1994). *Lords of the Fly: Drosophila Genetics and the Experimental Life*. University of Chicago Press.

Koike, H., Arguello, P. A., Kvajo, M., Karayiorgou, M., & Gogos, J. A. (2006). Disc1 is mutated in the 129S6/SvEv strain and modulates working memory in mice. *Proceedings of the National Academy of Sciences of the United States of America*, *103*, 3693–3697.

Kokolus, K. M., Capitano, M. L., Lee, C.-T., Eng, J. W. L., Waight, J. D., Hylander, B. L., Sexton, S., Hong, C.-C., Gordon, C. J., Abrams, S. I., & Repasky, E. A. (2013). Baseline tumor growth and immune control in laboratory mice are significantly influenced by subthermoneutral housing temperature. *Proceedings of the National Academy of Sciences of the United States of America*, *110*, 20176–20181.

Kola, I., & Landis, J. (2004). Can the pharmaceutical industry reduce attrition rates? *Nature Reviews: Drug Discovery*, *3*, 711–715.

Kondrakiewicz, K., Kostecki, M., Szadzińska, W., & Knapska, E. (2019). Ecological validity of social interaction tests in rats and mice. *Genes, Brain, and Behavior*, *18*, Article e12525.

Konstantinov, I. E. (2000). Robert H. Goetz: The surgeon who performed the first successful clinical coronary artery bypass operation. *Annals of Thoracic Surgery, 69,* 1966–1972.

Koonin, E. V., & Novozhilov, A. S. (2017). Origin and evolution of the universal genetic code. *Annual Review of Genetics, 51,* 45–62.

Koppolu, V., & Vasigala, V. K. (2016). Role of *Escherichia coli* in biofuel production. *Microbiology Insights, 9,* 29–35.

Kordasiewicz, H. B., Stanek, L. M., Wancewicz, E. V., Mazur, C., McAlonis, M. M., Pytel, K. A., Artates, J. W., Weiss, A., Cheng, S. H., Shihabuddin, L. S., et al. (2012). Sustained therapeutic reversal of Huntington's disease by transient repression of huntingtin synthesis. *Neuron, 74,* 1031–1044.

Korneev, K. V. (2019). Mouse models of sepsis and septic shock. *Molecular Biology, 53,* 704–717.

Kornum, B. R., & Jennum, P. (2020). The case for narcolepsy as an autoimmune disease. *Expert Review of Clinical Immunology, 16,* 231–233.

Kozak, M. J., & Cuthbert, B. N. (2016). The NIMH research domain criteria initiative: Background, issues, and pragmatics. *Psychophysiology, 53,* 286–297.

Krebs, H. A. (1975). The August Krogh principle: "For many problems there is an animal on which it can be most conveniently studied." *Journal of Experimental Zoology, 194,* 221–226.

Krebs, H. A., & Krebs, J. R. (1980). The "August Krogh principle." *Comparative Biochemistry and Physiology: Part B, Biochemistry and Molecular Biology, 67,* 379–380.

Kreider, J. W., Howett, M. K., Leure-Dupree, A. E., Zaino, R. J., & Weber, J. A. (1987). Laboratory production *in vivo* of infectious human papillomavirus type 11. *Journal of Virology, 61,* 590–593.

Kreider, J. W., Howett, M. K., Wolfe, S. A., Bartlett, G. L., Zaino, R. J., Sedlacek, T. V., & Mortel, R. (1985). Morphological transformation *in vivo* of human uterine cervix with papillomavirus from *condylomata acuminata. Nature, 317,* 639–641.

Kretzschmar, K., & Clevers, H. (2016). Organoids: Modeling development and the stem cell niche in a dish. *Developmental Cell, 38,* 590–600.

Kriks, S., Shim, J.-W., Piao, J., Ganat, Y. M., Wakeman, D. R., Xie, Z., Carrillo-Reid, L., Auyeung, G., Antonacci, C., Buch, A., et al. (2011). Dopamine neurons derived from human ES cells efficiently engraft in animal models of Parkinson's disease. *Nature, 480,* 547–551.

Krishna, A., Biryukov, M., Trefois, C., Antony, P. M., Hussong, R., Lin, J., Heinäniemi, M., Glusman, G., Köglsberger, S., Boyd, O., et al. (2014). Systems genomics evaluation of the SH-SY5Y neuroblastoma cell line as a model for Parkinson's disease. *BMC Genomics, 15,* 1154.

Kritchevsky, D. (1995). Dietary protein, cholesterol and atherosclerosis: A review of the early history. *Journal of Nutrition, 125,* 589S–593S.

Krogh, A. (1929). The progress of physiology. *Science, 70,* 200–204.

Krogh, A. (1941). *The Comparative Physiology of Respiratory Mechanisms.* University of Pennsylvania Press.

Krstic, D., & Knuesel, I. (2013). The airbag problem—a potential culprit for bench-to-bedside translational efforts: Relevance for Alzheimer's disease. *Acta Neuropathologica Communications, 1,* Article 62.

Krummel, M. F., & Allison, J. P. (1995). CD28 and CTLA-4 have opposing effects on the response of T cells to stimulation. *Journal of Experimental Medicine, 182,* 459–465.

Kuhn, A., Goldstein, D. R., Hodges, A., Strand, A. D., Sengstag, T., Kooperberg, C., Becanovic, K., Pouladi, M. A., Sathasivam, K., Cha, J.-H. J., et al. (2007). Mutant huntingtin's effects on striatal gene expression in mice recapitulate changes observed in human Huntington's disease brain and do not differ with mutant huntingtin length or wild-type huntingtin dosage. *Human Molecular Genetics, 16*, 1845–1861.

Kumar, A., Kumar Singh, S., Kumar, V., Kumar, D., Agarwal, S., & Rana, M. K. (2015). Huntington's disease: An update of therapeutic strategies. *Gene, 556*, 91–97.

Kumari, V., & Postma, P. (2005). Nicotine use in schizophrenia: The self medication hypotheses. *Neuroscience and Biobehavioral Reviews, 29*, 1021–1034.

Kvajo, M., McKellar, H., & Gogos, J. A. (2012). Avoiding mouse traps in schizophrenia genetics: Lessons and promises from current and emerging mouse models. *Neuroscience, 211*, 136–164.

Kwon, B. K., Hillyer, J., & Tetzlaff, W. (2010). Translational research in spinal cord injury: A survey of opinion from the SCI community. *Journal of Neurotrauma, 27*, 21–33.

Laborit, H., Huguenard, P., & Alluaume, R. (1952). Un nouveau stabilisateur vegetatif (le 4560 R. P.) [A new vegetative stabilizer; 4560 R. P.]. *Presse Médicale, 60*, 206–208.

LaFollette, H., & Shanks, N. (1993). Animal models in biomedical research: Some epistemological worries. *Public Affairs Quarterly, 7*, 113–130.

LaFollette, H., & Shanks, N. (1995). Two models of models in biomedical research. *Philosophical Quarterly, 45*, 141–160.

Lamoureux, P., Buxbaum, R. E., & Heidemann, S. R. (1989). Direct evidence that growth cones pull. *Nature, 340*, 159–162.

Lancaster, M. A., & Knoblich, J. A. (2014). Organogenesis in a dish: Modeling development and disease using organoid technologies. *Science, 345*, 1247125.

Lancaster, M. A., Renner, M., Martin, C.-A., Wenzel, D., Bicknell, L. S., Hurles, M. E., Homfray, T., Penninger, J. M., Jackson, A. P., & Knoblich, J. A. (2013). Cerebral organoids model human brain development and microcephaly. *Nature, 501*, 373–379.

Landis, S. C., Amara, S. G., Asadullah, K., Austin, C. P., Blumenstein, R., Bradley, E. W., Crystal, R. G., Darnell, R. B., Ferrante, R. J., Fillit, H., et al. (2012). A call for transparent reporting to optimize the predictive value of preclinical research. *Nature, 490*, 187–191.

Langfelder, P., Cantle, J. P., Chatzopoulou, D., Wang, N., Gao, F., Al-Ramahi, I., Lu, X.-H., Ramos, E. M., El-Zein, K., Zhao, Y., et al. (2016). Integrated genomics and proteomics define huntingtin CAG length–dependent networks in mice. *Nature Neuroscience, 19*, 623–633.

Langfelder, P., Mischel, P. S., & Horvath, S. (2013). When is hub gene selection better than standard meta-analysis? *PLoS One, 8*, Article e61505.

Langley, G., Austin, C. P., Balapure, A. K., Birnbaum, L. S., Bucher, J. R., Fentem, J., Fitzpatrick, S. C., Fowle, J. R., Kavlock, R. J., Kitano, H., et al. (2015). Lessons from toxicology: Developing a 21st-century paradigm for medical research. *Environmental Health Perspectives, 123*, A268–A272.

Langston, J. W., Ballard, P., Tetrud, J. W., & Irwin, I. (1983). Chronic parkinsonism in humans due to a product of meperidine-analog synthesis. *Science, 219*, 979–980.

Larsch, J., Ventimiglia, D., Bargmann, C. I., & Albrecht, D. R. (2013). High-throughput imaging of neuronal activity in *Caenorhabditis elegans. Proceedings of the National Academy of Sciences of the United States of America, 110*, E4266–E4273.

Lathrop, D. P., & Forest, C. B. (2011). Magnetic dynamos in the lab. *Physics Today, 64*, 40–45.

Laurencin, C. T., & McClinton, A. (2020). Regenerative cell-based therapies: Cutting edge, bleeding edge, and off the edge. *Regenerative Engineering and Translational Medicine, 6*, 78–89.

LaVail, J. H., & Wolf, M. K. (1973). Postnatal development of the mouse dentate gyrus in organotypic cultures of the hippocampal formation. *American Journal of Anatomy, 137*, 47–65.

Leach, D. R., Krummel, M. F., & Allison, J. P. (1996). Enhancement of antitumor immunity by CTLA-4 blockade. *Science, 271*, 1734–1736.

Leal, P. C., Lins, L. C. R. F., de Gois, A. M., Marchioro, M., & Santos, J. R. (2016). Commentary: Evaluation of models of Parkinson's disease. *Frontiers in Neuroscience, 10*, Article 283.

Leathers, C. W. (1990). Choosing the animal—reasons, excuses, and welfare. In B. E. Rollin (Ed.), *The Experimental Animal in Biomedical Research: A Survey of Scientific and Ethical Issues for Investigators* (pp. 67–80). CRC Press.

Lederberg, J. (1947). Gene recombination and linked segregations in *Escherichia coli*. *Genetics, 32*, 505–525.

Lederberg, J. (1987). Genetic recombination in bacteria: A discovery account. *Annual Review of Genetics, 21*, 23–46.

Lederberg, J. (1998). *Escherichia coli*. In R. Bud & D. J. Warner (Eds.), *Instruments of Science: An Historical Encyclopedia* (pp. 230–232). Garland.

Ledford, H. (2008). Translational research: The full cycle. *Nature, 453*, 843–845.

Lee, J., Cho, Y., Choi, K.-S., Kim, W., Jang, B.-H., Shin, H., Ahn, C., Lim, T. H., & Yi, H.-J. (2019). Efficacy and safety of erythropoietin in patients with traumatic brain injury: A systematic review and meta-analysis. *American Journal of Emergency Medicine, 37*, 1101–1107.

Lee, J.-M., Ramos, E. M., Lee, J.-H., Gillis, T., Mysore, J. S., Hayden, M. R., Warby, S. C., Morrison, P., Nance, M., Ross, C. A., et al. (2012). CAG repeat expansion in Huntington disease determines age at onset in a fully dominant fashion. *Neurology, 78*, 690–695.

Lee, Y. T., Laxton, V., Lin, H. Y., Chan, Y. W. F., Fitzgerald-Smith, S., To, T. L. O., Yan, B. P., Liu, T., & Tse, G. (2017). Animal models of atherosclerosis (Review). *Biomedical Reports, 6*, 259–266.

Lehrman, D. S. (1971). Behavioral science, engineering and poetry. In E. Tobach, L. R. Aronson, & E. Shaw (Eds.), *The Biopsychology of Development* (pp. 459–471). Academic Press.

Leibinger, M., Zeitler, C., Gobrecht, P., Andreadaki, A., Gisselmann, G., & Fischer, D. (2021). Transneuronal delivery of hyper-interleukin-6 enables functional recovery after severe spinal cord injury in mice. *Nature Communications, 12*, 391.

Leonelli, S., & Ankeny, R. A. (2013). What makes a model organism? *Endeavour, 37*, 209–212.

Letai, A. (2017). Functional precision cancer medicine—moving beyond pure genomics. *Nature Medicine, 23*, 1028–1035.

Letten, A. D., & Cornwell, W. K. (2014). Trees, branches and (square) roots: Why evolutionary relatedness is not linearly related to functional distance. *Methods in Ecology and Evolution, 6*, 439–444.

Letvin, N. L., Eaton, K. A., Aldrich, W. R., Sehgal, P. K., Blake, B. J., Schlossman, S. F., King, N. W., & Hunt, R. D. (1983). Acquired immunodeficiency syndrome in a colony of macaque monkeys. *Proceedings of the National Academy of Sciences of the United States of America, 80*, 2718–2722.

Leuchtenberger, R., Leuchtenberger, C., Laszlo, D., & Lewisohn, R. (1945). The influence of "folic acid" on spontaneous breast cancers in mice. *Science, 101*, 46–46.

Levi-Montalcini, R., Meyer, H., & Hamburger, V. (1954). In vitro experiments on the effects of mouse sarcomas 180 and 37 on the spinal and sympathetic ganglia of the chick embryo. *Cancer Research, 14*, 49–57.

Levy, A., & Currie, A. (2015). Model organisms are not (theoretical) models. *British Journal for the Philosophy of Science, 66*, 327–348.

Lewin, R. (1982). Biology is not postage stamp collecting. *Science, 216*, 718–720.

Li, M., Yao, X., Sun, L., Zhao, L., Xu, W., Zhao, H., Zhao, F., Zou, X., Cheng, Z., Li, B., et al. (2020). Effects of electroconvulsive therapy on depression and its potential mechanism. *Frontiers in Psychology, 11*, Article 80.

Li, R., Ma, X., Wang, G., Yang, J., & Wang, C. (2016). Why sex differences in schizophrenia? *Journal of Translational Neurosciences, 1*, 37–42.

Li, S., Hu, N., Zhang, W., Tao, B., Dai, J., Gong, Y., Tan, Y., Cai, D., & Lui, S. (2019). Dysconnectivity of multiple brain networks in schizophrenia: A meta-analysis of resting-state functional connectivity. *Frontiers in Psychiatry, 10*, Article 482.

Li, W., Englund, E., Widner, H., Mattsson, B., van Westen, D., Lätt, J., Rehncrona, S., Brundin, P., Björklund, A., Lindvall, O., & Li, J.-Y. (2016). Extensive graft-derived dopaminergic innervation is maintained 24 years after transplantation in the degenerating parkinsonian brain. *Proceedings of the National Academy of Sciences of the United States of America, 113*, 6544–6549.

Li, X.-J., & Li, S. (2012). Influence of species differences on the neuropathology of transgenic Huntington's disease animal models. *Journal of Genetics and Genomics, 39*, 239–245.

Li, Y.-S. J., Haga, J. H., & Chien, S. (2005). Molecular basis of the effects of shear stress on vascular endothelial cells. *Journal of Biomechanics, 38*, 1949–1971.

Li, Z., Karlovich, C. A., Fish, M. P., Scott, M. P., & Myers, R. M. (1999). A putative *Drosophila* homolog of the Huntington's disease gene. *Human Molecular Genetics, 8*, 1807–1815.

Li, Z., Ridder, B. J., Han, X., Wu, W. W., Sheng, J., Tran, P. N., Wu, M., Randolph, A., Johnstone, R. H., Mirams, G. R., Kuryshev, Y., et al. (2019). Assessment of an *in silico* mechanistic model for proarrhythmia risk prediction under the CIPA initiative. *Clinical Pharmacology and Therapeutics, 105*, 466–475.

Liang, G., & Zhang, Y. (2013). Genetic and epigenetic variations in iPSCs: Potential causes and implications for application. *Cell Stem Cell, 13*, 149–159.

Liao, B.-Y., & Zhang, J. (2008). Null mutations in human and mouse orthologs frequently result in different phenotypes. *Proceedings of the National Academy of Sciences of the United States of America, 105*, 6987–6992.

Libby, P. (2015). Murine "model" monotheism: An iconoclast at the altar of mouse. *Circulation Research, 117*, 921–925.

Lienhard, D. A. (2017, October 11). David H. Hubel and Torsten N. Wiesel's research on optical development in kittens. *The Embryo Project Encyclopedia*. Center for Biology and Society, School of Life Sciences, Arizona State University. https://embryo.asu.edu/handle/10776/12995.

Lienhard, D. A. (2018, February 26). Roger Wolcott Sperry (1913–1994). *The Embryo Project Encyclopedia*. Center for Biology and Society, School of Life Sciences, Arizona State University. https://embryo.asu.edu/handle/10776/13055.

Lilienfeld, S. O. (2014). The Research Domain Criteria (RDoC): An analysis of methodological and conceptual challenges. *Behaviour Research and Therapy, 62*, 129–139.

Lilienfeld, S. O., & Treadway, M. T. (2016). Clashing diagnostic approaches: DSM-ICD versus RDoC. *Annual Review of Clinical Psychology, 12*, 435–463.

Lim, S.-E., Ha, S. J., Jang, W.-H., Jung, K.-M., Jung, M.-S., Yeo, K.-W., Kim, J.-S., Jeong, T.-C., Kang, M.-J., Kim, et al. (2019). Me-too validation study for in vitro eye irritation test with 3D-reconstructed human cornea epithelium, MCTT HCE. *Toxicology in Vitro, 55*, 173–184.

Lin, A., Giuliano, C. J., Palladino, A., John, K. M., Abramowicz, C., Yuan, M. L., Sausville, E. L., Lukow, D. A., Liu, L., Chait, A. R., et al. (2019). Off-target toxicity is a common mechanism of action of cancer drugs undergoing clinical trials. *Science Translational Medicine, 11*, Article eaaw8412.

Lin, L., Faraco, J., Li, R., Kadotani, H., Rogers, W., Lin, X., Qiu, X., Jong, P. J. de, Nishino, S., & Mignot, E. (1999). The sleep disorder canine narcolepsy is caused by a mutation in the hypocretin (orexin) receptor 2 gene. *Cell, 98*, 365–376.

Lin, R., & Lee, J. J. (2020). Novel Bayesian adaptive designs and their applications in cancer clinical trials. In A. Bekker, D. G. Chen, & J. T. Ferreira (Eds.), *Computational and Methodological Statistics and Biostatistics* (pp. 395–426). Springer.

Lindsley, C. W. (2016). Lost in translation: The death of basic science. *ACS Chemical Neuroscience, 7*, Article 1024.

List of people with bipolar disorder. (2021, March 24). In *Wikipedia*. https://en.wikipedia.org/w/index.php ?title=List_of_people_with_bipolar_disorder&oldid=1014007411.

Lister, R., Pelizzola, M., Kida, Y. S., Hawkins, R. D., Nery, J. R., Hon, G., Antosiewicz-Bourget, J., O'Malley, R., Castanon, R., Klugman, S., et al. (2011). Hotspots of aberrant epigenomic reprogramming in human induced pluripotent stem cells. *Nature, 471*, 68–73.

Little, C. C. (1935). A new deal for mice. *Scientific American, 152*, 16–18.

Liu, K., Lu, Y., Lee, J. K., Samara, R., Willenberg, R., Sears-Kraxberger, I., Tedeschi, A., Park, K. K., Jin, D., Cai, B., et al. (2010). PTEN deletion enhances the regenerative ability of adult corticospinal neurons. *Nature Neuroscience, 13*, 1075–1081.

Liu, T., & Khosla, C. (2010). Genetic engineering of *Escherichia coli* for biofuel production. *Annual Review of Genetics, 44*, 53–69.

Liu, W.-C., Wen, L., Xie, T., Wang, H., Gong, J.-B., & Yang, X.-F. (2017). Therapeutic effect of erythropoietin in patients with traumatic brain injury: A meta-analysis of randomized controlled trials. *Journal of Neurosurgery, 127*, 8–15.

Lo, B., & Parham, L. (2009). Ethical issues in stem cell research. *Endocrine Reviews, 30*, 204–213.

Logan, C. A. (2001). "[Are] Norway rats . . . things?": Diversity versus generality in the use of albino rats in experiments on development and sexuality. *Journal of the History of Biology, 34*, 287–314.

Logan, C. A. (2002). Before there were standards: The role of test animals in the production of empirical generality in physiology. *Journal of the History of Biology, 35*, 329–363.

Logan, C. A. (2019). Commercial rodents in America: Standard animals, model animals, and biological diversity. *Brain, Behavior and Evolution, 93*, 70–81.

Logan, C. A., & Brauckmann, S. (2015). Controlling and culturing diversity: Experimental zoology before World War II and Vienna's Biologische Versuchsanstalt. *Journal of Experimental Zoology, Part A, 323A*, 211–226.

Loomer, H. P., Saunders, J. C., & Kline, N. S. (1957). A clinical and pharmacodynamic evaluation of iproniazid as a psychic energizer. *Psychiatric Research Reports, 8*, 129–141.

Louca, S., Mazel, F., Doebeli, M., & Parfrey, L. W. (2019). A census-based estimate of Earth's bacterial and archaeal diversity. *PLoS Biology, 17*, Article e3000106.

Lovett, M. L., Nieland, T. J. F., Dingle, Y.-T. L., & Kaplan, D. L. (2020). Innovations in 3D tissue models of human brain physiology and diseases. *Advanced Functional Materials, 2020*, Article 1909146.

Lowe, D. (2020, February 10). A prospective Alzheimer's trial reports. *Science Translational Medicine: In the Pipeline*. https://blogs.sciencemag.org/pipeline/archives/2020/02/10/a-prospective-alzheimers-trial-reports.

Lu, Y., Huggins, P., & Bar-Joseph, Z. (2009). Cross species analysis of microarray expression data. *Bioinformatics, 25,* 1476–1483.

Luca, S., & Mihaescu, T. (2013). History of BCG vaccine. *Mædica, 8,* 53–58.

Lukjancenko, O., Wassenaar, T. M., & Ussery, D. W. (2010). Comparison of 61 sequenced *Escherichia coli* genomes. *Microbial Ecology, 60,* 708–720.

Lumsden, A. L., Young, R. L., Pezos, N., & Keating, D. J. (2016). Huntingtin-associated protein 1: Eutherian adaptation from a TRAK-like protein, conserved gene promoter elements, and localization in the human intestine. *BMC Evolutionary Biology, 16,* Article 214.

Luo, H.-Q., Gu, W.-W., Huang, L.-W., Wu, L.-H., Tian, Y.-G., Zheng, C.-H., & Yue, M. (2018). Effect of prepregnancy obesity on litter size in primiparous minipigs. *Journal of the American Association for Laboratory Animal Science, 57,* 115–123.

Lutz, C. M., & Osborne, M. A. (2014). Optimizing mouse models of neurodegenerative disorders: Are therapeutics in sight? *Future Neurology, 9,* 67–75.

Lynch, H. T., Shaw, T. G., & Lynch, J. F. (2004). Inherited predisposition to cancer: A historical overview. *American Journal of Medical Genetics: Part C, Seminars in Medical Genetics, 129,* 5–22.

Lynch, T. J., Bell, D. W., Sordella, R., Gurubhagavatula, S., Okimoto, R. A., Brannigan, B. W., Harris, P. L., Haserlat, S. M., Supko, J. G., Haluska, F. G., et al. (2004). Activating mutations in the epidermal growth factor receptor underlying responsiveness of non–small-cell lung cancer to gefitinib. *New England Journal of Medicine, 350,* 2129–2139.

Ma, X., Aravind, A., Pfister, B. J., Chandra, N., & Haorah, J. (2019). Animal models of traumatic brain injury and assessment of injury severity. *Molecular Neurobiology, 56,* 5332–5345.

MacDonald, M. E., Ambrose, C. M., Duyao, M. P., Myers, R. H., Lin, C., Srinidhi, L., Barnes, G., Taylor, S. A., James, M., Groot, N., et al. (1993). A novel gene containing a trinucleotide repeat that is expanded and unstable on Huntington's disease chromosomes. *Cell, 72,* 971–983.

Machado-Vieira, R., Baumann, J., Wheeler-Castillo, C., Latov, D., Henter, I. D., Salvadore, G., & Zarate, C. A. (2010). The timing of antidepressant effects: A comparison of diverse pharmacological and somatic treatments. *Pharmaceuticals, 3,* 19–41.

Macieira-Coelho, A., Diatloff, C., & Malaise, E. (1977). Concept of fibroblast aging in vitro: Implications for cell biology. *Gerontology, 23,* 290–305.

Macy, J., & Horvath, T. L. (2017). Comparative medicine: An inclusive crossover discipline. *Yale Journal of Biology and Medicine, 90,* 493–498.

Mahmood, I. (2007). Application of allometric principles for the prediction of pharmacokinetics in human and veterinary drug development. *Advanced Drug Delivery Reviews, 59,* 1177–1192.

Maienschein, J., Sunderland, M., Ankeny, R. A., & Robert, J. S. (2008). The ethos and ethics of translational research. *American Journal of Bioethics, 8,* 43–51.

Malinovska, L., Palm, S., Gibson, K., Verbavatz, J.-M., & Alberti, S. (2015). *Dictyostelium discoideum* has a highly Q/N-rich proteome and shows an unusual resilience to protein aggregation. *Proceedings of the National Academy of Sciences of the United States of America, 112,* E2620–E2629.

Malkesman, O., Austin, D. R., Chen, G., & Manji, H. K. (2009). Reverse translational strategies for developing animal models of bipolar disorder. *Disease Models and Mechanisms, 2,* 238–245.

Mammis, A., McIntosh, T. K., & Maniker, A. H. (2009). Erythropoietin as a neuroprotective agent in traumatic brain injury: Review. *Surgical Neurology, 71,* 527–531.

Mandai, M., Watanabe, A., Kurimoto, Y., Hirami, Y., Morinaga, C., Daimon, T., Fujihara, M., Akimaru, H., Sakai, N., Shibata, Y., et al. (2017). Autologous induced stem-cell–derived retinal cells for macular degeneration. *New England Journal of Medicine, 376*, 1038–1046.

Manger, P. R., Cort, J., Ebrahim, N., Goodman, A., Henning, J., Karolia, M., Rodrigues, S.-L., & Strkalj, G. (2008). Is 21st Century neuroscience too focussed on the rat/mouse model of brain function and dysfunction? *Frontiers in Neuroanatomy, 2*, Article 5.

Mangiarini, L., Sathasivam, K., Seller, M., Cozens, B., Harper, A., Hetherington, C., Lawton, M., Trottier, Y., Lehrach, H., Davies, S. W., & Bates, G. P. (1996). Exon 1 of the HD gene with an expanded CAG repeat is sufficient to cause a progressive neurological phenotype in transgenic mice. *Cell, 87*, 493–506.

Mansour, A. A., Gonçalves, J. T., Bloyd, C. W., Li, H., Fernandes, S., Quang, D., Johnston, S., Parylak, S. L., Jin, X., & Gage, F. H. (2018). An in vivo model of functional and vascularized human brain organoids. *Nature Biotechnology, 36*, 432–441.

Manzella, F., Maloney, S. E., & Taylor, G. T. (2015). Smoking in schizophrenic patients: A critique of the self-medication hypothesis. *World Journal of Psychiatry, 5*, 35–46.

Marcheque, J., Bussolati, B., Csete, M., & Perin, L. (2019). Stem cells and kidney regeneration: An update. *Stem Cells Translational Medicine, 8*, 82–92.

Marino, L., & Colvin, C. M. (2015). Thinking pigs: A comparative review of cognition, emotion, and personality in *Sus domesticus*. *International Journal of Comparative Psychology, 28*, Article 23859.

Marks, P. W., Witten, C. M., & Califf, R. M. (2017). Clarifying stem-cell therapy's benefits and risks. *New England Journal of Medicine, 376*, 1007–1009.

Maron, D. J., Hochman, J. S., Reynolds, H. R., Bangalore, S., O'Brien, S. M., Boden, W. E., Chaitman, B. R., Senior, R., López-Sendón, J., Alexander, K. P., et al. (2020). Initial invasive or conservative strategy for stable coronary disease. *New England Journal of Medicine, 382*, 1395–1407.

Marsh, J. L., Pallos, J., & Thompson, L. M. (2012). Fly models of Huntington's disease. *Human Molecular Genetics, 12*, R187–R193.

Marsh, N., & Marsh, A. (2000). A short history of nitroglycerine and nitric oxide in pharmacology and physiology. Clinical and Experimental Pharmacology and Physiology, 27, 313–319.

Marsh, S. E., Yeung, S. T., Torres, M., Lau, L., Davis, J. L., Monuki, E. S., Poon, W. W., & Blurton-Jones, M. (2017). HuCNS-SC human NSCs fail to differentiate, form ectopic clusters, and provide no cognitive benefits in a transgenic model of Alzheimer's disease. *Stem Cell Reports, 8*, 235–248.

Martin, B., Ji, S., Maudsley, S., & Mattson, M. P. (2010). "Control" laboratory rodents are metabolically morbid: Why it matters. *Proceedings of the National Academy of Sciences of the United States of America, 107*, 6127–6133.

Martin, U. (2017). Genome stability of programmed stem cell products. *Advanced Drug Delivery Reviews, 120*, 108–117.

Martinez-Coria, H., Green, K. N., Billings, L. M., Kitazawa, M., Albrecht, M., Rammes, G., Parsons, C. G., Gupta, S., Banerjee, P., & LaFerla, F. M. (2010). Memantine improves cognition and reduces Alzheimer's-like neuropathology in transgenic mice. *American Journal of Pathology, 176*, 870–880.

Marx, U., Walles, H., Hoffmann, S., Lindner, G., Horland, R., Sonntag, F., Klotzbach, U., Sakharov, D., Tonevitsky, A., & Lauster, R. (2012). "Human-on-a-chip" developments: A translational cutting-edge alternative to systemic safety assessment and efficiency evaluation of substances in laboratory animals and man? *Alternatives to Laboratory Animals, 40*, 235–257.

Masopust, D., Sivula, C. P., & Jameson, S. C. (2017). Of mice, dirty mice, and men: Using mice to understand human immunology. *Journal of Immunology, 199*, 383–388.

Matarese, G., La Cava, A., & Horvath, T. L. (2012). *In vivo veritas, in vitro artificia. Trends in Molecular Medicine, 18*, 439–442.

Mather, J. A. (2008). Cephalopod consciousness: Behavioural evidence. *Consciousness and Cognition, 17*, 37–48.

Matthews, B. J., & Vosshall, L. B. (2020). How to turn an organism into a model organism in 10 "easy" steps. *Journal of Experimental Biology, 223*, Article 218198.

Matthews, R. A. (2008). Medical progress depends on animal models—doesn't it? *Journal of the Royal Society of Medicine, 101*, 95–98.

Maxwell, E. K., Schnitzler, C. E., Havlak, P., Putnam, N. H., Nguyen, A.-D., Moreland, R. T., & Baxevanis, A. D. (2014). Evolutionary profiling reveals the heterogeneous origins of classes of human disease genes: Implications for modeling disease genetics in animals. *BMC Evolutionary Biology, 14*, Article 212.

Maxwell, R. A., & Eckhardt, S. B. (2012). *Drug Discovery.* Springer Science & Business Media.

Mazure, C. M., & Jones, D. P. (2015). Twenty years and still counting: Including women as participants and studying sex and gender in biomedical research. *BMC Women's Health, 15*, Article 94.

McArthur, R. A., & Borsini, F. (2008). What do you mean by "translational research"? An enquiry through animal and translational models for CNS drug discovery: Neurological disorders. In R. A. McArthur, & F. Borsini (Eds.), *Neurological Disorders: Vol. 2. Animal and Translational Models for CNS Drug Discovery* (pp. xv–xlii). Elsevier.

McCann, J., Choi, E., Yamasaki, E., & Ames, B. N. (1975). Detection of carcinogens as mutagens in the Salmonella/microsome test: Assay of 300 chemicals. *Proceedings of the National Academy of Sciences of the United States of America, 72*, 5135–5139.

McCormick, U., Murray, B., & McNew, B. (2015). Diagnosis and treatment of patients with bipolar disorder: A review for advanced practice nurses. *Journal of the American Association of Nurse Practitioners, 27*, 530–542.

McGarry, A., McDermott, M., Kieburtz, K., de Blieck, E. A., Beal, F., Marder, K., Ross, C., Shoulson, I., Gilbert, P., Mallonee, W. M., et al. (2017). A randomized, double-blind, placebo-controlled trial of coenzyme Q10 in Huntington disease. *Neurology, 88*, 152–159.

McGeer, P. (2003). Is there a future for vaccination as a treatment for Alzheimer's disease? *Neurobiology of Aging, 24*, 391–395.

McKhann, G. M., Knopman, D. S., Chertkow, H., Hyman, B. T., Jack, C. R., Kawas, C. H., Klunk, W. E., Koroshetz, W. J., Manly, J. J., Mayeux, R., et al. (2011). The diagnosis of dementia due to Alzheimer's disease: Recommendations from the National Institute on Aging-Alzheimer's Association workgroups on diagnostic guidelines for Alzheimer's disease. *Alzheimer's and Dementia, 7*, 263–269.

McKinney, W. T., & Bunney, W. E. (1969). Animal model of depression. I. Review of evidence: Implications for research. *Archives of General Psychiatry, 21*, 240–248.

McMaster, P. D. (1922). Do species lacking a gall bladder possess its functional equivalent? *Journal of Experimental Medicine, 35*, 127–140.

McQuade, A., Coburn, M., Tu, C. H., Hasselmann, J., Davtyan, H., & Blurton-Jones, M. (2018). Development and validation of a simplified method to generate human microglia from pluripotent stem cells. *Molecular Neurodegeneration, 13*, Article 67.

Medina, M., Khachaturian, Z. S., Rossor, M., Avila, J., & Cedazo-Minguez, A. (2017). Toward common mechanisms for risk factors in Alzheimer's syndrome. *Alzheimer's and Dementia: Translational Research and Clinical Interventions, 3*, 571–578.

Mehta, P., Kaye, W., Raymond, J., Wu, R., Larson, T., Punjani, R., Heller, D., Cohen, J., Peters, T., Muravov, O., & Horton, K. (2018). Prevalence of amyotrophic lateral sclerosis—United States, 2014. *MMWR, Morbidity and Mortality Weekly Report, 67,* 216–218.

Mekada, K., Abe, K., Murakami, A., Nakamura, S., Nakata, H., Moriwaki, K., Obata, Y., & Yoshiki, A. (2009). Genetic differences among C57BL/6 substrains. *Experimental Animals, 58,* 141–149.

Meltzer, H. Y. (1994). An overview of the mechanism of action of clozapine. *Journal of Clinical Psychiatry, 55*(Suppl. B), 47–52.

Menalled, L., El-Khodor, B. F., Hornberger, M., Park, L., Howland, D., & Brunner, D. (2012). Effect of the rd1 mutation on motor performance in R6/2 and wild type mice. *PLoS Currents, 4,* Article RRN1303.

Menalled, L., Lutz, C., Ramboz, S., Brunner, D., Lager, B., Noble, S., Park, L., & Howland, D. (2014). *A Field Guide to Working with Mouse Models of Huntington's Disease.* The Jackson Laboratory.

Menalled, L. B., Patry, M., Ragland, N., Lowden, P. A. S., Goodman, J., Minnich, J., Zahasky, B., Park, L., Leeds, J., Howland, D., et al. (2010). Comprehensive behavioral testing in the R6/2 mouse model of Huntington's disease shows no benefit from CoQ10 or minocycline. *PLoS One, 5,* Article e9793.

Menalled, L. B., Sison, J. D., Wu, Y., Olivieri, M., Li, X.-J., Li, H., Zeitlin, S., & Chesselet, M.-F. (2002). Early motor dysfunction and striosomal distribution of huntingtin microaggregates in Huntington's disease knock-in mice. *Journal of Neuroscience, 22,* 8266–8276.

Merceron, T. K., & Murphy, S. V. (2015). Hydrogels for 3D bioprinting applications. In A. Atala & J. J. Yoo (Eds.), *Essentials of 3D Biofabrication and Translation* (pp. 249–270). Elsevier.

Merkle, F. T., Ghosh, S., Kamitaki, N., Mitchell, J., Avior, Y., Mello, C., Kashin, S., Mekhoubad, S., Ilic, D., Charlton, M., et al. (2017). Human pluripotent stem cells recurrently acquire and expand dominant negative P53 mutations. *Nature, 545,* 229–233.

Merton, R. K. (1968). The Matthew effect in science. The reward and communication systems of science are considered. *Science, 159,* 56–63.

Messaoudi, I., Estep, R., Robinson, B., & Wong, S. W. (2011). Nonhuman primate models of human immunology. *Antioxidants and Redox Signaling, 14,* 261–273.

Mestas, J., & Hughes, C. C. W. (2004). Of mice and not men: Differences between mouse and human immunology. *Journal of Immunology, 172,* 2731–2738.

Meyer, J. M., & Simpson, G. M. (1997). From chlorpromazine to olanzapine: A brief history of antipsychotics. *Psychiatric Services, 48,* 1137–1139.

Meyers, J. R. (2018). Zebrafish: Development of a vertebrate model organism. *Current Protocols: Essential Laboratory Techniques,* Article e19. https://doi.org/10.1002/cpet.19.

Miczek, K. A., & de Wit, H. (2008). Challenges for translational psychopharmacology research—some basic principles. *Psychopharmacology (Berlin), 199,* 291–301.

Milan, D. J., Peterson, T. A., Ruskin, J. N., Peterson, R. T., & MacRae, C. A. (2003). Drugs that induce repolarization abnormalities cause bradycardia in zebrafish. *Circulation, 107,* 1355–1358.

Miller, C. T., Hale, M. E., Okano, H., Okabe, S., & Mitra, P. (2019). Comparative principles for next-generation neuroscience. *Frontiers in Behavioral Neuroscience, 13,* Article 12.

Millet, L. J., & Gillette, M. U. (2012). Over a century of neuron culture: From the hanging drop to microfluidic devices. *Yale Journal of Biology and Medicine, 85,* 501–521.

Mitsuya, H., Weinhold, K. J., Furman, P. A., St Clair, M. H., Lehrman, S. N., Gallo, R. C., Bolognesi, D., Barry, D. W., & Broder, S. (1985). 3′-Azido-3′-deoxythymidine (BW A509U): An antiviral agent that inhibits the infectivity and cytopathic effect of human T-lymphotropic virus type III/lymphadenopathy-associated virus in vitro. *Proceedings of the National Academy of Sciences of the United States of America, 82*, 7096–7100.

Mitsuya, H., Yarchoan, R., & Broder, S. (1990). Molecular targets for AIDS therapy. *Science, 249*, 1533–1544.

Mittelman, D., & Wilson, J. H. (2013). The fractured genome of HeLa cells. *Genome Biology, 14*, 111.

Mogil, J. S., & Macleod, M. R. (2017). No publication without confirmation. *Nature, 542*, 409–411.

Mohammed, R., Opara, K., Lall, R., Ojha, U., & Xiang, J. (2020). Evaluating the effectiveness of anti-Nogo treatment in spinal cord injuries. *Neural Development, 15*, Article 1.

Mokbel, K., & Hassanally, D. (2001). From HER2 to Herceptin. *Current Medical Research and Opinion, 17*, 51–59.

Monamy, V. (2000). *Animal Experimentation: A Guide to the Issues.* Cambridge University Press.

Monson, C. A., & Sadler, K. C. (2010). Inbreeding depression and outbreeding depression are evident in wild-type zebrafish lines. *Zebrafish, 7*, 189–197.

Monteys, A. M., Ebanks, S. A., Keiser, M. S., & Davidson, B. L. (2017). CRISPR/Cas9 editing of the mutant huntingtin allele in vitro and in vivo. *Molecular Therapy, 25*, 12–23.

Monteys, A. M., Spengler, R. M., Dufour, B. D., Wilson, M. S., Oakley, C. K., Sowada, M. J., McBride, J. L., & Davidson, B. L. (2014). Single nucleotide seed modification restores in vivo tolerability of a toxic artificial miRNA sequence in the mouse brain. *Nucleic Acids Research, 42*, 13315–13327.

Monticello, T. M., Jones, T. W., Dambach, D. M., Potter, D. M., Bolt, M. W., Liu, M., Keller, D. A., Hart, T. K., & Kadambi, V. J. (2017). Current nonclinical testing paradigm enables safe entry to first-in-human clinical trials: The IQ consortium nonclinical to clinical translational database. *Toxicology and Applied Pharmacology, 334*, 100–109.

Morgan, T. H. (1915). *The Mechanism of Mendelian Heredity.* Holt.

Morice, M.-C., Serruys, P. W., Barragan, P., Bode, C., Van Es, G.-A., Stoll, H.-P., Snead, D., Mauri, L., Cutlip, D. E., & Sousa, E. (2007). Long-term clinical outcomes with sirolimus-eluting coronary stents: Five-year results of the RAVEL trial. *Journal of the American College of Cardiology, 50*, 1299–1304.

Morlacchi, P., & Nelson, R. R. (2011). How medical practice evolves: Learning to treat failing hearts with an implantable device. *Research Policy, 40*, 511–525.

Morrison, M., & Morgan, M. S. (1999). Models as mediating instruments. In M. Morrison, M. S. Morgan, & Q. Skinner (Eds.), *Models as Mediators: Perspectives on Natural and Social Science* (pp. 10–37). Cambridge University Press.

Mozhui, K., Karlsson, R. M., Kash, T. L., Ihne, J., Norcross, M., Patel, S., Farrell, M. R., Hill, E. E., Graybeal, C., Martin, K. P., et al. (2010). Strain differences in stress responsivity are associated with divergent amygdala gene expression and glutamate-mediated neuronal excitability. *Journal of Neuroscience, 30*, 5357–5367.

Mullane, K., & Williams, M. (2013). Alzheimer's therapeutics: Continued clinical failures question the validity of the amyloid hypothesis—but what lies beyond? *Biochemical Pharmacology, 85*, 289–305.

Mullane, K., & Williams, M. (2019). Preclinical models of Alzheimer's disease: Relevance and translational validity. *Current Protocols in Pharmacology, 84*, Article e57.

Müller, B., & Grossniklaus, U. (2010). Model organisms—a historical perspective. *Journal of Proteomics, 73*, 2054–2063.

Murray, C. J. L., Vos, T., Lozano, R., Naghavi, M., Flaxman, A. D., Michaud, C., Ezzati, M., Shibuya, K., Salomon, J. A., Abdalla, S., et al. (2012). Disability-adjusted life years (DALYs) for 291 diseases and injuries in 21 regions, 1990–2010: A systematic analysis for the Global Burden of Disease Study 2010. *The Lancet, 380*, 2197–2223.

Murray, E., Wise, S., & Graham, K. (2016). *The Evolution of Memory Systems.* Oxford University Press.

Nadeau, J. H., & Auwerx, J. (2019). The virtuous cycle of human genetics and mouse models in drug discovery. *Nature Reviews: Drug Discovery, 18,* 255–272.

Nam, S. A., Seo, E., Kim, J. W., Kim, H. W., Kim, H. L., Kim, K., Kim, T.-M., Ju, J. H., Gomez, I. G., Uchimura, K., et al. (2019). Graft immaturity and safety concerns in transplanted human kidney organoids. *Experimental and Molecular Medicine, 51,* Article 145.

Namikawa, R., Kaneshima, H., Lieberman, M., Weissman, I. L., & McCune, J. M. (1988). Infection of the SCID-hu mouse by HIV-1. *Science, 242,* 1684–1686.

Narayan, R. K., Michel, M. E., Ansell, B., Baethmann, A., Biegon, A., Bracken, M. B., Bullock, M. R., Choi, S. C., Clifton, G. L., Contant, C. F., et al. (2002). Clinical trials in head injury. *Journal of Neurotrauma, 19,* 503–557.

Nardone, R., Florea, C., Höller, Y., Brigo, F., Versace, V., Lochner, P., Golaszewski, S., & Trinka, E. (2017). Rodent, large animal and non-human primate models of spinal cord injury. *Zoology, 123,* 101–114.

Narver, H. L., Hoogstraten-Miller, S., Linkenhoker, J., & Weichbrod, R. H. (2017). Tributes for animals and the dedicated people entrusted with their care: A practical how-to guide. *Laboratory Animals, 46,* 369–372.

Nashold, B. S., & Slaughter, D. G. (1969). Some observation on tremor. In F. J. Gillingham, & I. M. L. Donaldson (Eds.), *Third Symposium on Parkinson's Disease* (pp. 241–246). Livingstone.

Nasir, A., & Caetano-Anollés, G. (2015). A phylogenomic data-driven exploration of viral origins and evolution. *Science Advances, 1,* Article e1500527.

Nathanson, N., & Kew, O. M. (2010). From emergence to eradication: The epidemiology of poliomyelitis deconstructed. *American Journal of Epidemiology, 172,* 1213–1229.

National Cancer Institute. (2007). What is cancer? https://www.cancer.gov/about-cancer/understanding/what-is-cancer.

National Cancer Institute. (2018). The Cancer Genome Atlas Program. https://www.cancer.gov/about-nci/organization/ccg/research/structural-genomics/tcga.

National Center for Biotechnology Information (NCBI). (2020). Genome List—Genome—NCBI. https://www.ncbi.nlm.nih.gov/genome/browse/#!/overview/.

National Research Council. (1985). *Models for Biomedical Research: A New Perspective.* National Academies Press.

National Research Council. (2006). *Toxicity Testing for Assessment of Environmental Agents: Interim Report.* National Academies Press.

National Research Council. (2007). *Toxicity Testing in the 21st Century: A Vision and a Strategy.* National Academies Press.

National Research Council. (2011a). *Toward Precision Medicine: Building a Knowledge Network for Biomedical Research and a New Taxonomy of Disease.* National Academies Press.

National Research Council. (2011b). *Guide for the Care and Use of Laboratory Animals* (8th ed.). National Academies Press.

Nazha, B., Mishra, M., Pentz, R., & Owonikoko, T. K. (2019). Enrollment of racial minorities in clinical trials: Old problem assumes new urgency in the age of immunotherapy. *American Society of Clinical Oncology Educational Book, 39,* 3–10.

Neigh, G., & Mitzelfelt, M. (2016). *Sex Differences in Physiology.* Academic Press.

Nelson, N. C. (2018). *Model Behavior: Animal Experiments, Complexity, and the Genetics of Psychiatric Disorders.* University of Chicago Press.

Nelson, P. T., Greenberg, S. G., & Saper, C. B. (1994). Neurofibrillary tangles in the cerebral cortex of sheep. *Neuroscience Letters, 170,* 187–190.

Nestler, E. J., & Hyman, S. E. (2010). Animal models of neuropsychiatric disorders. *Nature Neuroscience, 13,* 1161–1169.

Ng, K. K. F., & Vane, J. R. (1968). Fate of angiotensin 1 in the circulation. *Nature, 218,* 144–150.

Nicoll, J. A. R., Wilkinson, D., Holmes, C., Steart, P., Markham, H., & Weller, R. O. (2003). Neuropathology of human Alzheimer disease after immunization with amyloid-beta peptide: A case report. *Nature Medicine, 9,* 448–452.

Noble, D. (2011). Differential and integral views of genetics in computational systems biology. *Interface Focus, 1,* 7–15.

Nosek, B. A., Ebersole, C. R., DeHaven, A. C., & Mellor, D. T. (2018). The preregistration revolution. *Proceedings of the National Academy of Sciences of the United States of America, 115,* 2600–2606.

Nunes, L. M., Robles-Escajeda, E., Santíago-Vazquez, Y., Ortega, N. M., Lema, C., Muro, A., Almodovar, G., Das, U., Das, S., Dimmock, J. R., Aguilera, R. J., & Varela-Ramirez, A. (2014). The gender of cell lines matters when screening for novel anti-cancer drugs. *AAPS Journal, 16,* 872–874.

Nunney, L., Maley, C. C., Breen, M., Hochberg, M. E., & Schiffman, J. D. (2015). Peto's paradox and the promise of comparative oncology. *Philosophical Transactions of the Royal Society of London: Series B, Biological Sciences, 370,* Article 20140177.

Oakley, H., Cole, S. L., Logan, S., Maus, E., Shao, P., Craft, J., Guillozet-Bongaarts, A., Ohno, M., Disterhoft, J., Eldik, L. V., et al. (2006). Intraneuronal β-amyloid aggregates, neurodegeneration, and neuron loss in transgenic mice with five familial Alzheimer's disease mutations: Potential factors in amyloid plaque formation. *Journal of Neuroscience, 26,* 10129–10140.

Oberg, A. (2016). All too human? Speciesism, racism, and sexism. *Think, 15,* 39–50.

O'Brien, K. P., Westerlund, I., & Sonnhammer, E. L. L. (2004). OrthoDisease: A database of human disease orthologs. *Human Mutation, 24,* 112–119.

Ocana, A., Pandiella, A., Siu, L. L., & Tannock, I. F. (2011). Preclinical development of molecular-targeted agents for cancer. *Nature Reviews: Clinical Oncology, 8,* 200–209.

O'Collins, V. E., Donnan, G. A., Macleod, M. R., & Howells, D. W. (2017). Animal models of ischemic stroke versus clinical stroke: Comparison of infarct size, cause, location, study design, and efficacy of experimental therapies. In P. M. Conn (Ed.), *Animal Models for the Study of Human Disease* (pp. 481–523). Academic Press.

Oddo, S., Billings, L., Kesslak, J. P., Cribbs, D. H., & LaFerla, F. M. (2004). Aβ immunotherapy leads to clearance of early, but not late, hyperphosphorylated tau aggregates via the proteasome. *Neuron, 43,* 321–332.

Oddo, S., Caccamo, A., Shepherd, J. D., Murphy, M. P., Golde, T. E., Kayed, R., Metherate, R., Mattson, M. P., Akbari, Y., & LaFerla, F. M. (2003). Triple-transgenic model of Alzheimer's disease with plaques and tangles: Intracellular Aβ and synaptic dysfunction. *Neuron, 39,* 409–421.

Office of Minority Health. (2020). Profile: Black/African American. US Department of Health and Human Services. https://www.minorityhealth.hhs.gov/omh/browse.aspx?lvl=3&lvlid=61.

Okano, H., Sasaki, E., Yamamori, T., Iriki, A., Shimogori, T., Yamaguchi, Y., Kasai, K., & Miyawaki, A. (2016). Brain/MINDS: A Japanese national brain project for marmoset neuroscience. *Neuron, 92,* 582–590.

O'Keefe, J., & Nadel, L. (1978). *The Hippocampus as a Cognitive Map*. Oxford University Press.

Olanow, C. W., Goetz, C. G., Kordower, J. H., Stoessl, A. J., Sossi, V., Brin, M. F., Shannon, K. M., Nauert, G. M., Perl, D. P., Godbold, J., & Freeman, T. B. (2003). A double-blind controlled trial of bilateral fetal nigral transplantation in Parkinson's disease. *Annals of Neurology, 54*, 403–414.

Oleaga, C., Lavado, A., Riu, A., Rothemund, S., Carmona-Moran, C. A., Persaud, K., Yurko, A., Lear, J., Narasimhan, N. S., Long, C. J., et al. (2019). Long-term electrical and mechanical function monitoring of a human-on-a-chip system. *Advanced Functional Materials, 29*, Article 1805792.

Oliveria, S. F. (2018). The dark history of early deep brain stimulation. *Lancet Neurology, 17*, Article 748.

Olson, H., Betton, G., Robinson, D., Thomas, K., Monro, A., Kolaja, G., Lilly, P., Sanders, J., Sipes, G., Bracken, W., et al. (2000). Concordance of the toxicity of pharmaceuticals in humans and in animals. *Regulatory Toxicology and Pharmacology, 32*, 56–67.

Olsson, L., & Kaplan, H. S. (1980). Human-human hybridomas producing monoclonal antibodies of predefined antigenic specificity. *Proceedings of the National Academy of Sciences of the United States of America, 77*, 5429–5431.

Online Mendelian Inheritance in Man. (2020). OMIM gene map statistics. https://omim.org/statistics/geneMap.

Onstad, S., Skre, I., Torgersen, S., & Kringlen, E. (1991). Twin concordance for DSM-III-R schizophrenia. *Acta Psychiatrica Scandinavica, 83*, 395–401.

Ooi, J., Langley, S. R., Xu, X., Utami, K. H., Sim, B., Huang, Y., Harmston, N. P., Tay, Y. L., Ziaei, A., Zeng, R., et al. (2019). Unbiased profiling of isogenic Huntington disease hPSC-derived CNS and peripheral cells reveals strong cell-type specificity of CAG length effects. *Cell Reports, 26*, 2494–2508.

Open Science Collaboration. (2015). Estimating the reproducibility of psychological science. *Science, 349*, Article aac4716.

Oreskes, N. (2007). From scaling to simulation: Changing meanings and ambitions of models in geology. In A. N. H. Creager, E. Lunbeck, & M. N. Wise (Eds.), *Science without Laws: Model Systems, Cases, Exemplary Narratives* (pp. 93–124). Duke University Press.

Organisation for Economic Co-operation and Development (OECD). (2019). Test Guideline No. 492: Reconstructed human cornea-like epithelium (RhCE) test method for identifying chemicals not requiring classification and labelling for eye irritation or serious eye damage. https://www.oecd-ilibrary.org/docserver/9789264242548-en.pdf.

Orrù, G., & Carta, M. G. (2018). Genetic variants involved in bipolar disorder, a rough road ahead. *Clinical Practice and Epidemiology in Mental Health, 14*, 37–45.

Ortmann, D., & Vallier, L. (2017). Variability of human pluripotent stem cell lines. *Current Opinion in Genetics and Development, 46*, 179–185.

Orzack, S. H. (2012). The philosophy of modelling or does the philosophy of biology have any use? *Philosophical Transactions of the Royal Society of London: Series B, Biological Sciences, 367*, 170–180.

Orzack, S. H., & McLoone, B. (2019). Modeling in biology: Looking backward and looking forward. *Studia Metodologiczne, 39*, 73–98.

Osuchowski, M. F., Remick, D. G., Lederer, J. A., Lang, C. H., Aasen, A. O., Aibiki, M., Azevedo, L. C., Bahrami, S., Boros, M., Cooney, R., et al. (2014). Abandon the mouse research ship? Not just yet! *Shock, 41*, 463–475.

Osuna, C. E., & Whitney, J. B. (2017). Nonhuman primate models of Zika virus infection, immunity, and therapeutic development. *Journal of Infectious Diseases, 216*, S928–S934.

Paez, J. G., Jänne, P. A., Lee, J. C., Tracy, S., Greulich, H., Gabriel, S., Herman, P., Kaye, F. J., Lindeman, N., Boggon, T. J., et al. (2004). EGFR mutations in lung cancer: Correlation with clinical response to gefitinib therapy. *Science, 304*, 1497–1500.

Paganoni, S., Macklin, E. A., Hendrix, S., Berry, J. D., Elliott, M. A., Maiser, S., Karam, C., Caress, J. B., Owegi, M. A., Quick, A., et al. (2020). Trial of sodium phenylbutyrate–taurursodiol for amyotrophic lateral sclerosis. *New England Journal of Medicine, 383*, 919–930.

Paleacu, D. (2007). Tetrabenazine in the treatment of Huntington's disease. *Neuropsychiatric Disease and Treatment, 3*, 545–551.

Palfi, S., Brouillet, E., Jarraya, B., Bloch, J., Jan, C., Shin, M., Condé, F., Li, X.-J., Aebischer, P., Hantraye, P., & Déglon, N. (2007). Expression of mutated huntingtin fragment in the putamen is sufficient to produce abnormal movement in non-human primates. *Molecular Therapy, 15*, 1444–1451.

Palumbo, A., Facon, T., Sonneveld, P., Bladè, J., Offidani, M., Gay, F., Moreau, P., Waage, A., Spencer, A., Ludwig, H., et al. (2008). Thalidomide for treatment of multiple myeloma: 10 years later. *Blood, 111*, 3968–3977.

Parashar, A., & Udayabanu, M. (2017). Gut microbiota: Implications in Parkinson's disease. *Parkinsonism and Related Disorders, 38*, 1–7.

Pardi, N., Hogan, M. J., Porter, F. W., & Weissman, D. (2018). mRNA vaccines—a new era in vaccinology. *Nature Reviews: Drug Discovery, 17*, 261–279.

Pardoll, D. M. (2012). The blockade of immune checkpoints in cancer immunotherapy. *Nature Reviews: Cancer, 12*, 252–264.

Park, J. E., & Silva, A. C. (2019). Generation of genetically engineered non-human primate models of brain function and neurological disorders. *American Journal of Primatology, 81*, Article e22931.

Park, J. E., Zhang, X. F., Choi, S.-H., Okahara, J., Sasaki, E., & Silva, A. C. (2016). Generation of transgenic marmosets expressing genetically encoded calcium indicators. *Scientific Reports, 6*, Article 34931.

Park, M.-N., Park, J. H., Paik, H. Y., & Lee, S. K. (2015). Insufficient sex description of cells supplied by commercial vendors. *American Journal of Physiology: Cell Physiology, 308*, C578–C580.

Parker, J. D., & Parker, J. O. (1998). Nitrate therapy for stable angina pectoris. *New England Journal of Medicine, 338*, 520–531.

Parkkinen, V.-P. (2017). Are model organisms theoretical models? *Disputatio, 9*, 471–498.

Parsons, C. G., Danysz, W., & Quack, G. (1999). Memantine is a clinically well tolerated N-methyl-d-aspartate (NMDA) receptor antagonist—a review of preclinical data. *Neuropharmacology, 38*, 735–767.

Paşca, S. P. (2018). The rise of three-dimensional human brain cultures. *Nature, 553*, 437–445.

Paspalas, C. D., Carlyle, B. C., Leslie, S., Preuss, T. M., Crimins, J. L., Huttner, A. J., van Dyck, C. H., Rosene, D. L., Nairn, A. C., & Arnsten, A. F. T. (2018). The aged rhesus macaque manifests Braak stage III/IV Alzheimer's-like pathology. *Alzheimer's and Dementia, 14*, 680–691.

Patten, S. A., Parker, J. A., Wen, X.-Y., & Drapeau, P. (2016). Simple animal models for amyotrophic lateral sclerosis drug discovery. *Expert Opinion on Drug Discovery, 11*, 797–804.

Pawar, S. (2003). Taxonomic chauvinism and the methodologically challenged. *BioScience, 53*, 861–864.

Payne, M. M. (2001). Charles Theodore Dotter: The father of intervention. *Texas Heart Institute Journal, 28*, 28–38.

Pearce, R. M. (1912). Chance and the prepared mind. *Science, 35*, 941–956.

Pease, C. M., & Bull, J. J. (1992). Is science logical? *BioScience*, *42*, 293–298.

Pegram, M., & Ngo, D. (2006). Application and potential limitations of animal models utilized in the development of trastuzumab (Herceptin®): A case study. *Advanced Drug Delivery Reviews*, *58*, 723–734.

Peirson, B. R. E., Kropp, H., Damerow, J., & Laubichler, M. D. (2017). The diversity of experimental organisms in biomedical research may be influenced by biomedical funding. *BioEssays*, *39*, Article 1600258.

Perel, P., Roberts, I., Sena, E., Wheble, P., Briscoe, C., Sandercock, P., Macleod, M., Mignini, L. E., Jayaram, P., & Khan, K. S. (2007). Comparison of treatment effects between animal experiments and clinical trials: Systematic review. *BMJ*, *334*, Article 197.

Perlman, R. L. (2016). Mouse models of human disease: An evolutionary perspective. *Evolution, Medicine, and Public Health*, *2016*, 170–176.

Perlow, M. J., Freed, W. J., Hoffer, B. J., Seiger, A., Olson, L., & Wyatt, R. J. (1979). Brain grafts reduce motor abnormalities produced by destruction of nigrostriatal dopamine system. *Science*, *204*, 643–647.

Perrin, S. (2014). Preclinical research: Make mouse studies work. *Nature*, *507*, 423–425.

Personalized Med Coalition. (2019). *Personalized Medicine at FDA: The Scope and Significance of Progress in 2019*. https://personalizedmedicinecoalition.org/Userfiles/PMC-Corporate/file/PM_at_FDA_The_Scope_and _Significance_of_Progress_in_2019.pdf.

Péterfi, O., Boda, F., Szabó, Z., Ferencz, E., & Bába, L. (2019). Hypotensive snake venom components—a mini-review. *Molecules*, *24*, Article 2778.

Peto, R. (1977). Epidemiology, multistage models, and short-term mutagenicity tests. In H. H. Hyatt, J. D. Watson, J. A. Winsten (Eds.), *Origins of Human Cancer* (pp. 1403–1428). Cold Spring Harbor Press.

Peto, R. (2016). Epidemiology, multistage models, and short-term mutagenicity tests. *International Journal of Epidemiology*, *45*, 621–637.

Petrov, D., Mansfield, C., Moussy, A., & Hermine, O. (2017). ALS clinical trials review: 20 years of failure. Are we any closer to registering a new treatment? *Frontiers in Aging Neuroscience*, *9*, Article 68.

Philip, M., Aldrich, M., Fisher, M., & Savitz, S. I. (2009). Methodological quality of animal studies of neuroprotective agents currently in phase II/III acute ischemic stroke trials. *Stroke*, *40*, 577–581.

Phillips, K. A., Bales, K. L., Capitanio, J. P., Conley, A., Czoty, P. W., 't Hart, B. A., Hopkins, W. D., Hu, S.-L., Miller, L. A., Nader, M. A., et al. (2014). Why primate models matter. *American Journal of Primatology*, *76*, 801–827.

Phillips, W. S., Herly, M., Del Negro, C. A., & Rekling, J. C. (2015). Organotypic slice cultures containing the preBötzinger complex generate respiratory-like rhythms. *Journal of Neurophysiology*, *115*, 1063–1070.

Pickl, M., & Ries, C. H. (2009). Comparison of 3D and 2D tumor models reveals enhanced HER2 activation in 3D associated with an increased response to trastuzumab. *Oncogene*, *28*, 461–468.

Pienaar, I. S., Götz, J., & Feany, M. B. (2010). Parkinson's disease: Insights from non-traditional model organisms. *Progress in Neurobiology*, *92*, 558–571.

Pierre-Victor Galtier. (2020, October 11). In *Wikipedia*. https://en.wikipedia.org/w/index.php?title=Pierre -Victor_Galtier&oldid=982988855.

Pifferi, F., Epelbaum, J., & Aujard, F. (2019). Strengths and weaknesses of the gray mouse lemur (*Microcebus murinus*) as a model for the behavioral and psychological symptoms and neuropsychiatric symptoms of dementia. *Frontiers in Pharmacology*, *10*, Article 1291.

Pinheiro, A., Neves, F., Lemos de Matos, A., Abrantes, J., van der Loo, W., Mage, R., & Esteves, P. J. (2016). An overview of the lagomorph immune system and its genetic diversity. *Immunogenetics*, *68*, 83–107.

Plaisier, S. B., Taschereau, R., Wong, J. A., & Graeber, T. G. (2010). Rank–rank hypergeometric overlap: Identification of statistically significant overlap between gene-expression signatures. *Nucleic Acids Research, 38*, Article e169.

Platt, A., David, B. T., & Fessler, R. G. (2020). Stem cell clinical trials in spinal cord injury: A brief review of studies in the United States. *Medicines, 7*, Article 27.

Pollak, G. D. (2014). A changing view of the auditory system obtained from the ears of bats. In A. N. Popper & R. R. Fay (Eds.), *Perspectives on Auditory Research* (pp. 441–466). Springer.

Pomerantz, M. M., & Freedman, M. L. (2011). The genetics of cancer risk. *Cancer Journal (Sudbury Mass.), 17*, 416–422.

Pons, T. P., Garraghty, P. E., Ommaya, A. K., Kaas, J. H., Taub, E., & Mishkin, M. (1991). Massive cortical reorganization after sensory deafferentation in adult macaques. *Science, 252*, 1857–1860.

Poo, M., Du, J., Ip, N. Y., Xiong, Z.-Q., Xu, B., & Tan, T. (2016). China brain project: Basic neuroscience, brain diseases, and brain-inspired computing. *Neuron, 92*, 591–596.

Potashkin, J. A., Blume, S. R., & Runkle, N. K. (2010). Limitations of animal models of Parkinson's disease. *Parkinson's Disease, 2011*, Article 658083.

Pound, P., & Bracken, M. (2014). Is animal research sufficiently evidence based to be a cornerstone of biomedical research? *BMJ, 348*, Article g3387.

Povinelli, D. J., & Bering, J. M. (2002). The mentality of apes revisited. *Current Directions in Psychological Science, 11*, 115–119.

Powell, S. B. (2010). Models of neurodevelopmental abnormalities in schizophrenia. *Current Topics in Behavioral Neurosciences, 4*, 435–481.

Pradhan, B. S., & Majumdar, S. S. (2016). An efficient method for generation of transgenic rats avoiding embryo manipulation. *Molecular Therapy: Nucleic Acids, 2165*, Article e293.

Pratt, J., Winchester, C., Dawson, N., & Morris, B. (2012). Advancing schizophrenia drug discovery: Optimizing rodent models to bridge the translational gap. *Nature Reviews: Drug Discovery, 11*, 560–579.

Preer, J. R. (1997). Whatever happened to paramecium genetics? *Genetics, 145*, 217–225.

Preuss, T. M. (2019). Critique of pure marmoset. *Brain, Behavior and Evolution, 93*, 92–107.

Preuss, T. M. (2000). Taking the measure of diversity: Comparative alternatives to the model-animal paradigm in cortical neuroscience. *Brain, Behavior and Evolution, 55*, 287–299.

Preuss, T. M., & Robert, J. S. (2014). Animal models of the human brain: Repairing the paradigm. In M. S. Gazzaniga & G. R. Mangun (Eds.), *The Cognitive Neurosciences* (pp. 59–66). MIT Press.

Preza, E., Hardy, J., Warner, T., & Wray, S. (2016). Review: Induced pluripotent stem cell models of frontotemporal dementia. *Neuropathology and Applied Neurobiology, 42*, 497–520.

Pringsheim, T., Wiltshire, K., Day, L., Dykeman, J., Steeves, T., & Jette, N. (2012). The incidence and prevalence of Huntington's disease: A systematic review and meta-analysis. *Movement Disorders, 27*, 1083–1091.

Prinz, F., Schlange, T., & Asadullah, K. (2011). Believe it or not: How much can we rely on published data on potential drug targets? *Nature Reviews: Drug Discovery, 10*, Article 712.

Prpar Mihevc, S., & Majdič, G. (2019). Canine cognitive dysfunction and Alzheimer's disease—two facets of the same disease? *Frontiers in Neuroscience, 13*, Article 604.

Pugh, P. L., Ahmed, S. F., Smith, M. I., Upton, N., & Hunter, A. J. (2004). A behavioural characterisation of the FVB/N mouse strain. *Behavioural Brain Research, 155*, 283–289.

Quirke, V. M. (2017). Tamoxifen from failed contraceptive pill to best-selling breast cancer medicine: A case-study in pharmaceutical innovation. *Frontiers in Pharmacology, 8,* Article 620.

Rabchevsky, A. G., Fugaccia, I., Sullivan, P. G., Blades, D. A., & Scheff, S. W. (2002). Efficacy of methylprednisolone therapy for the injured rat spinal cord. *Journal of Neuroscience Research, 68,* 7–18.

Racker, E. (Ed.). (1954). *Cellular Metabolism and Infection.* Academic Press.

Rader, K. A. (2004). *Making Mice.* Princeton University Press.

Rahimpour, A., Ahani, R., Najaei, A., Adeli, A., Barkhordari, F., & Mahboudi, F. (2016). Development of genetically modified Chinese Hamster Ovary host cells for the enhancement of recombinant tissue plasminogen activator expression. *Malaysian Journal of Medical Sciences, 23,* 6–13.

Rajagopalan, R., Pan, L., Schaefer, C., Nicholas, J., Lim, S., Misialek, S., Stevens, S., Hooi, L., Aleskovski, N., Ruhrmund, D., et al. (2016). Preclinical characterization and human microdose pharmacokinetics of ITMN-8187, a nonmacrocyclic inhibitor of the hepatitis C virus NS3 protease. *Antimicrobial Agents and Chemotherapy, 61,* Article e01569.

Rangarajan, A., & Weinberg, R. A. (2003). Comparative biology of mouse versus human cells: Modelling human cancer in mice. *Nature Reviews: Cancer, 3,* 952–959.

Rappuoli, R. (2014). 1885, the first rabies vaccination in humans. *Proceedings of the National Academy of Sciences of the United States of America, 111,* Article 12273.

Rasko, J., & Power, C. (2017, September 1). Dr Con Man: The rise and fall of a celebrity scientist who fooled almost everyone. *The Guardian.* https://www.theguardian.com/science/2017/sep/01/paolo-macchiarini-scientist-surgeon-rise-and-fall.

Raymond, L. A., André, V. M., Cepeda, C., Gladding, C. M., Milnerwood, A. J., & Levine, M. S. (2011). Pathophysiology of Huntington's disease: Time-dependent alterations in synaptic and receptor function. *Neuroscience, 198,* 252–273.

Reardon, S. (2018). Lab-grown 'mini brains' produce electrical patterns that resemble those of premature babies. *Nature, 563,* 453–453.

Reaume, A. G., Howland, D. S., Trusko, S. P., Savage, M. J., Lang, D. M., Greenberg, B. D., Siman, R., & Scott, R. W. (1996). Enhanced amyloidogenic processing of the beta-amyloid precursor protein in gene-targeted mice bearing the Swedish familial Alzheimer's disease mutations and a "humanized" Aβ sequence. *Journal of Biological Chemistry, 271,* 23380–23388.

Reddy, A. S., & Zhang, S. (2013). Polypharmacology: Drug discovery for the future. *Expert Review of Clinical Pharmacology, 6,* 41–47.

Rehman, W., Arfons, L. M., & Lazarus, H. M. (2011). The rise, fall and subsequent triumph of thalidomide: Lessons learned in drug development. *Therapeutic Advances in Hematology, 2,* 291–308.

Reis, L. O., Zani, E. L., & García-Perdomo, H. A. (2018). Estrogen therapy in patients with prostate cancer: A contemporary systematic review. *International Urology and Nephrology, 50,* 993–1003.

Richter, C. P. (1968). Experiences of a reluctant rat-catcher the common Norway rat—friend or enemy? *Proceedings of the American Philosophical Society, 112,* 403–415.

Richter, S. H., Garner, J. P., Auer, C., Kunert, J., & Würbel, H. (2010). Systematic variation improves reproducibility of animal experiments. *Nature Methods, 7,* 167–168.

Riedel, S. (2005). Edward Jenner and the history of smallpox and vaccination. *Proceedings (Baylor University Medical Center), 18,* 21–25.

Rietdijk, C. D., Perez-Pardo, P., Garssen, J., van Wezel, R. J. A., & Kraneveld, A. D. (2017). Exploring Braak's hypothesis of Parkinson's disease. *Frontiers in Neurology*, *8*, Article 37.

Rivera-Hernandez, T., Carnathan, D. G., Moyle, P. M., Toth, I., West, N. P., Young, P. L., Silvestri, G., & Walker, M. J. (2014). The contribution of non-human primate models to the development of human vaccines. *Discovery Medicine*, *18*, 313–322.

Robanus-Maandag, E., Dekker, M., van der Valk, M., Carrozza, M. L., Jeanny, J. C., Dannenberg, J. H., Berns, A., & te Riele, H. (1998). p107 is a suppressor of retinoblastoma development in pRb-deficient mice. *Genes and Development*, *12*, 1599–1609.

Roberts, R. C. (2007). Disrupted in schizophrenia (DISC1): Integrating clinical and basic findings. *Schizophrenia Bulletin*, *33*, 11–15.

Roguev, A., Bandyopadhyay, S., Zofall, M., Zhang, K., Fischer, T., Collins, S. R., Qu, H., Shales, M., Park, H.-O., Hayles, J., et al. (2008). Conservation and rewiring of functional modules revealed by an epistasis map in fission yeast. *Science*, *322*, 405–410.

Roizin, L., Stellar, S., & Liu, J. C. (1979). Neuronal nuclear-cytoplasmic changes in Huntington's chorea: Electron microscope investigations. In T. N. Chase, N. S. Wexler, & A. Barbeau (Eds.), *Huntington's Disease* (pp. 95–122). Raven Press.

Roldão, A., Mellado, M. C. M., Castilho, L. R., Carrondo, M. J. T., & Alves, P. M. (2010). Virus-like particles in vaccine development. *Expert Review of Vaccines*, *9*, 1149–1176.

Rosen, R. F., Tomidokoro, Y., Farberg, A. S., Dooyema, J., Ciliax, B., Preuss, T. M., Neubert, T. A., Ghiso, J. A., LeVine, H., & Walker, L. C. (2016). Comparative pathobiology of β-amyloid and the unique susceptibility of humans to Alzheimer's disease. *Neurobiology of Aging*, *44*, 185–196.

Rosenberg, B., Vancamp, L., Trosko, J. E., & Mansour, V. H. (1969). Platinum compounds: A new class of potent antitumour agents. *Nature*, *222*, 385–386.

Rosenblueth, A., & Wiener, N. (1945). The role of models in science. *Philosophy of Science*, *12*, 316–321.

Rosenfield, L. C. (1940). *From Beast-machine to Man-machine: The Theme of Animal Souls in French Letters from Descartes to La Mettrie*. Oxford University Press.

Rosenquist, Å., Samuelsson, B., Johansson, P.-O., Cummings, M. D., Lenz, O., Raboisson, P., Simmen, K., Vendeville, S., de Kock, H., Nilsson, M., et al. (2014). Discovery and development of simeprevir (TMC435), a HCV NS3/4A protease inhibitor. *Journal of Medical Chemistry*, *57*, 1673–1693.

Rosenthal, M. F., Gertler, M., Hamilton, A. D., Prasad, S., & Andrade, M. C. B. (2017). Taxonomic bias in animal behaviour publications. *Animal Behavior*, *127*, 83–89.

Roth, B. L., Sheffler, D. J., & Kroeze, W. K. (2004). Magic shotguns versus magic bullets: Selectively non-selective drugs for mood disorders and schizophrenia. *Nature Reviews: Drug Discovery*, *3*, 353–359.

Rous, P. (1959). Surmise and fact on the nature of cancer. *Nature*, *183*, 1357–1361.

Rous, P. (1967). The challenge to man of the neoplastic cell. *Science*, *157*, 24–28.

Ru, J., Huo, Y., & Yang, Y. (2020). Microbial degradation and valorization of plastic wastes. *Frontier in Microbiology*, *11*, Article 442.

Rüb, U., Hentschel, M., Stratmann, K., Brunt, E., Heinsen, H., Seidel, K., Bouzrou, M., Auburger, G., Paulson, H., Vonsattel, J.-P., et al. (2014). Huntington's disease (HD): Degeneration of select nuclei, widespread occurrence of neuronal nuclear and axonal inclusions in the brainstem. *Brain Pathology*, *24*, 247–260.

Rubin, E. B., Shemesh, Y., Cohen, M., Elgavish, S., Robertson, H. M., & Bloch, G. (2006). Molecular and phylogenetic analyses reveal mammalian-like clockwork in the honey bee (*Apis mellifera*) and shed new light on the molecular evolution of the circadian clock. *Genome Research, 16*, 1352–1365.

Rubin, G. M., & Spradling, A. C. (1982). Genetic transformation of *Drosophila* with transposable element vectors. *Science, 218*, 348–353.

Rubin, G. M., Yandell, M. D., Wortman, J. R., Gabor Miklos, G. L., Nelson, C. R., Hariharan, I. K., Fortini, M. E., Li, P. W., Apweiler, R., Fleischmann, W., et al. (2000). Comparative genomics of the eukaryotes. *Science, 287*, 2204–2215.

Rubio, D. M., Schoenbaum, E. E., Lee, L. S., Schteingart, D. E., Marantz, P. R., Anderson, K. E., Platt, L. D., Baez, A., & Esposito, K. (2010). Defining translational research: Implications for training. *Academic Medicine, 85*, 470–475.

Rubinsztein, D. C., Amos, W., Leggo, J., Goodburn, S., Ramesar, R. S., Old, J., Bontrop, R., McMahon, R., Barton, D. E., & Ferguson-Smith, M. A. (1994). Mutational bias provides a model for the evolution of Huntington's disease and predicts a general increase in disease prevalence. *Nature Genetics, 7*, 525–530.

Ruoß, M., Vosough, M., Königsrainer, A., Nadalin, S., Wagner, S., Sajadian, S., Huber, D., Heydari, Z., Ehnert, S., Hengstler, J. G., & Nussler, A. K. (2020). Towards improved hepatocyte cultures: Progress and limitations. *Food and Chemical Toxicology, 138*, Article 111188.

Ruprecht, R. M., Bernard, L. D., Chou, T.-C., Sosa, M. A. G., Fazely, F., Koch, J., Sharma, P. L., & Mullaney, S. (1990). Murine models for evaluating antiretroviral therapy. *Cancer Research, 50*, 5618s–5627s.

Russell, L. B. (2013). The Mouse House: A brief history of the ORNL mouse-genetics program, 1947–2009. *Mutation Research, 753*, 69–90.

Russell, W. M. S., & Burch, R. L. (1959). *The Principles of Humane Experimental Technique.* Methuen.

Sabin, A. B. (1956). Present status of attenuated live-virus poliomyelitis vaccine. *JAMA, 162*, 1589–1596.

Sabin, A. B. (1965). Oral poliovirus vaccine: History of its development and prospects for eradication of poliomyelitis. *JAMA, 194*, 872–876.

Saito, T., Matsuba, Y., Mihira, N., Takano, J., Nilsson, P., Itohara, S., Iwata, N., & Saido, T. C. (2014). Single APP knock-in mouse models of Alzheimer's disease. *Nature Neuroscience, 17*, 661–663.

Sánchez, N. S., Mills, G. B., & Mills Shaw, K. R. (2017). Precision oncology: Neither a silver bullet nor a dream. *Pharmacogenomics, 18*, 1525–1539.

Sandler, M. (1990). Monoamine oxidase inhibitors in depression: History and mythology. *Journal of Psychopharmacology (Oxford), 4*, 136–139.

Sargent, J. (Director). (2004). *Something the Lord Made* [Film]. HBO Films.

Sasaki, E., Suemizu, H., Shimada, A., Hanazawa, K., Oiwa, R., Kamioka, M., Tomioka, I., Sotomaru, Y., Hirakawa, R., Eto, T., et al. (2009). Generation of transgenic non-human primates with germline transmission. *Nature, 459*, 523–527.

Sato, K., & Sasaki, E. (2017). Genetic engineering in nonhuman primates for human disease modeling. *Journal of Human Genetics, 63*, 125–131.

Saudou, F., Finkbeiner, S., Devys, D., & Greenberg, M. E. (1998). Huntingtin acts in the nucleus to induce apoptosis but death does not correlate with the formation of intranuclear inclusions. *Cell, 95*, 55–66.

Savageau, M. A. (1983). *Escherichia coli* habitats, cell types, and molecular mechanisms of gene control. *American Naturalist, 122*, 732–744.

Savoji, H., Mohammadi, M. H., Rafatian, N., Toroghi, M. K., Wang, E. Y., Zhao, Y., Korolj, A., Ahadian, S., & Radisic, M. (2019). Cardiovascular disease models: A game changing paradigm in drug discovery and screening. *Biomaterials*, *198*, 3–26.

Sawada, H., Ishiguro, H., Nishii, K., Yamada, K., Tsuchida, K., Takahashi, H., Goto, J., Kanazawa, I., & Nagatsu, T. (2007). Characterization of neuron-specific huntingtin aggregates in human huntingtin knock-in mice. *Neuroscience Research*, *57*, 559–573.

Scannell, J. W., Blanckley, A., Boldon, H., & Warrington, B. (2012). Diagnosing the decline in pharmaceutical R&D efficiency. *Nature Reviews: Drug Discovery*, *11*, 191–200.

Schaffner, K. F. (1998). Model organisms and behavioral genetics: A rejoinder. *Philosophy of Science*, *65*, 276–288.

Schellinck, H. M., Cyr, D. P., & Brown, R. E. (2010). *How Many Ways Can Mouse Behavioral Experiments Go Wrong? Confounding Variables in Mouse Models of Neurodegenerative Diseases and How to Control Them.* Elsevier.

Schenk, D., Barbour, R., Dunn, W., Gordon, G., Grajeda, H., Guido, T., Hu, K., Huang, J., Johnson-Wood, K., Khan, K., et al. (1999). Immunization with amyloid-beta attenuates Alzheimer-disease-like pathology in the PDAPP mouse. *Nature*, *400*, 173–177.

Schizophrenia Working Group of the Psychiatric Genomics Consortium. (2014). Biological insights from 108 schizophrenia-associated genetic loci. *Nature*, *511*, 421–427.

Schnabel, J. (2008). Standard model. *Nature*, *454*, 682–685.

Schneider, L. (2020, September 30). Paolo Macchiarini indicted for aggravated assault in Sweden. *For Better Science.* https://forbetterscience.com/2020/09/30/paolo-macchiarini-indicted-for-aggravated-assault-in-sweden/.

Schnell, L., & Schwab, M. E. (1990). Axonal regeneration in the rat spinal cord produced by an antibody against myelin-associated neurite growth inhibitors. *Nature*, *343*, 269–272.

Schou, M., Juel-Nielsen, N., Stromgren, E., & Voldby, H. (1954). The treatment of manic psychoses by the administration of lithium salts. *Journal of Neurology, Neurosurgery, and Psychiatry*, *17*, 250–260.

Schumann, G., Binder, E. B., Holte, A., de Kloet, E. R., Oedegaard, K. J., Robbins, T. W., Walker-Tilley, T. R., Bitter, I., Brown, V. J., Buitelaar, J., et al. (2014). Stratified medicine for mental disorders. *European Neuropsychopharmacology*, *24*, 5–50.

Schwab, C., Hosokawa, M., & McGeer, P. L. (2004). Transgenic mice overexpressing amyloid beta protein are an incomplete model of Alzheimer disease. *Experimental Neurology*, *188*, 52–64.

Schwartzkroin, P. A. (1975). Characteristics of CA1 neurons recorded intracellularly in the hippocampal in vitro slice preparation. *Brain Research*, *85*, 423–436.

Science History Institute. (2016). Howard Walter Florey and Ernst Boris Chain. https://www.sciencehistory.org/historical-profile/howard-walter-florey-and-ernst-boris-chain.

Scotland, R. W. (2010). Deep homology: A view from systematics. *BioEssays*, *32*, 438–449.

Scott, S., Kranz, J. E., Cole, J., Lincecum, J. M., Thompson, K., Kelly, N., Bostrom, A., Theodoss, J., Al Nakhala, B. M., Vieira, F. G., Ramasubbu, J., & Heywood, J. A. (2008). Design, power, and interpretation of studies in the standard murine model of ALS. *Amyotrophic Lateral Sclerosis*, *9*, 4–15.

Sedaghat, A. R., Sherman, A., & Quon, M. J. (2002). A mathematical model of metabolic insulin signaling pathways. *American Journal of Physiology: Endocrinology and Metabolism*, *283*, E1084–E1101.

Seluanov, A., Gladyshev, V. N., Vijg, J., & Gorbunova, V. (2018). Mechanisms of cancer resistance in long-lived mammals. *Nature Reviews: Cancer*, *18*, 433–441.

Seluanov, A., Hine, C., Azpurua, J., Feigenson, M., Bozzella, M., Mao, Z., Catania, K. C., & Gorbunova, V. (2009). Hypersensitivity to contact inhibition provides a clue to cancer resistance of naked mole-rat. *Proceedings of the National Academy of Sciences of the United States of America, 106,* 19352–19357.

Sengupta, S., & Higgs, P. G. (2015). Pathways of genetic code evolution in ancient and modern organisms. *Journal of Molecular Evolution, 80,* 229–243.

Seok, J., Warren, H. S., Cuenca, A. G., Mindrinos, M. N., Baker, H. V., Xu, W., Richards, D. R., McDonald-Smith, G. P., Gao, H., Hennessy, L., et al. (2013). Genomic responses in mouse models poorly mimic human inflammatory diseases. *Proceedings of the National Academy of Sciences of the United States of America, 110,* 3507–3512.

Sezonov, G., Joseleau-Petit, D., & D'Ari, R. (2007). *Escherichia coli* physiology in Luria-Bertani broth. *Journal of Bacteriology, 189,* 8746–8749.

Shah, E. J., Gurdziel, K., & Ruden, D. M. (2019). Mammalian models of traumatic brain injury and a place for *Drosophila* in TBI research. *Frontiers in Neuroscience, 13,* 409.

Shah, K., McCormack, C. E., & Bradbury, N. A. (2014). Do you know the sex of your cells? *American Journal of Physiology: Cell Physiology, 306,* C3–C18.

Shakir, R. (2018). The struggle for stroke reclassification. *Nature Reviews: Neurology, 14,* 447–448.

Shanks, N., Greek, R., & Greek, J. (2009). Are animal models predictive for humans? *Philosophy, Ethics, and Humanities in Medicine, 4,* Article 2.

Shapiro, K. J. (2004). Animal model research: The apples and oranges quandary. *Alternatives to Laboratory Animals, 32,* S405–S409.

Sharif-Alhoseini, M., Khormali, M., Rezaei, M., Safdarian, M., Hajighadery, A., Khalatbari, M. M., Safdarian, M., Meknatkhah, S., Rezvan, M., Chalangari, M., Derakhshan, P., & Rahimi-Movaghar, V. (2017). Animal models of spinal cord injury: A systematic review. *Spinal Cord, 55,* 714–721.

Sharma, R., Khristov, V., Rising, A., Jha, B. S., Dejene, R., Hotaling, N., Li, Y., Stoddard, J., Stankewicz, C., Wan, Q., Zhang, C., et al. (2019). Clinical-grade stem cell–derived retinal pigment epithelium patch rescues retinal degeneration in rodents and pigs. *Science Translational Medicine, 11,* Article eaat5580.

Sharma, V., & McNeill, J. H. (2009). To scale or not to scale: The principles of dose extrapolation. *British Journal of Pharmacology, 157,* 907–921.

Shaw, I. C., & Jones, H. B. (1994). Mechanisms of non-genotoxic carcinogenesis. *Trends in Pharmacological Sciences, 15,* 89–93.

Shay, T., Lederer, J. A., & Benoist, C. (2015). Genomic responses to inflammation in mouse models mimic humans: We concur, apples to oranges comparisons won't do. *Proceedings of the National Academy of Sciences of the United States of America, 112,* Article E346.

Shaye, D. D., & Greenwald, I. (2011). Ortholist: A compendium of *C. elegans* genes with human orthologs. *PLoS One, 6,* Article e20085.

Shelbourne, P. F., Keller-McGandy, C., Bi, W. L., Yoon, S.-R., Dubeau, L., Veitch, N. J., Vonsattel, J.-P., Wexler, N. S., Arnheim, N., & Augood, S. J. (2007). Triplet repeat mutation length gains correlate with cell-type specific vulnerability in Huntington disease brain. *Human Molecular Genetics, 16,* 1133–1142.

Shen, W. W. (1999). A history of antipsychotic drug development. *Comprehensive Psychiatry, 40,* 407–414.

Shen, Y.-T., Chen, L., Testani, J. M., Regan, C. P., & Shannon, R. P. (2017). Models for cardiovascular research. In P. M. Conn (Ed.), *Animal Models for the Study of Human Disease* (pp. 147–174). Elsevier.

Sheth, S., Mukherjea, D., Rybak, L. P., & Ramkumar, V. (2017). Mechanisms of cisplatin-induced ototoxicity and otoprotection. *Frontiers in Cellular Neuroscience, 11,* Article 338.

Shorter, E. (2009). The history of lithium therapy. *Bipolar Disorders, 11*, 4–9.

Shubin, N., Tabin, C., & Carroll, S. (1997). Fossils, genes and the evolution of animal limbs. *Nature, 388*, 639–648.

Shubin, N., Tabin, C., & Carroll, S. (2009). Deep homology and the origins of evolutionary novelty. *Nature, 457*, 818–823.

Shultz, L. D., Ishikawa, F., & Greiner, D. L. (2007). Humanized mice in translational biomedical research. *Nature Reviews: Immunology, 7*, 118–130.

Shultz, S. R., McDonald, S. J., Haar, C. V., Meconi, A., Vink, R., van Donkelaar, P., Taneja, C., Iverson, G. L., & Christie, B. R. (2017). The potential for animal models to provide insight into mild traumatic brain injury: Translational challenges and strategies. *Neuroscience and Biobehavioral Reviews, 76*, 396–414.

Siegel, J. M. (2008). Do all animals sleep? *Trends in Neurosciences, 31*, 208–213.

Silverman, M. E. (1989). William Withering and an account of the foxglove. *Clinical Cardiology, 12*, 415–418.

Siman, R., Reaume, A. G., Savage, M. J., Trusko, S., Lin, Y.-G., Scott, R. W., & Flood, D. G. (2000). Presenilin-1 P264L knock-in mutation: Differential effects on Aβ production, amyloid deposition, and neuronal vulnerability. *Journal of Neuroscience, 20*, 8717–8726.

Simmons, J. P., Nelson, L. D., & Simonsohn, U. (2011). False-positive psychology: Undisclosed flexibility in data collection and analysis allows presenting anything as significant. *Psychological Science, 22*, 1359–1366.

Simon, M. M., Greenaway, S., White, J. K., Fuchs, H., Gailus-Durner, V., Wells, S., Sorg, T., Wong, K., Bedu, E., Cartwright, E. J., et al. (2013). A comparative phenotypic and genomic analysis of C57BL/6J and C57BL/6N mouse strains. *Genome Biology, 14*, Article R82.

Simons, D. J., Shoda, Y., & Lindsay, D. S. (2017). Constraints on generality (COG): A proposed addition to all empirical papers. *Perspectives on Psychological Science, 12*, 1123–1128.

Singer, E., Walter, C., Weber, J. J., Krahl, A.-C., Mau-Holzmann, U. A., Rischert, N., Riess, O., Clemensson, L. E., & Nguyen, H. P. (2017). Reduced cell size, chromosomal aberration and altered proliferation rates are characteristics and confounding factors in the STHdh cell model of Huntington disease. *Scientific Reports, 7*, Article 16880.

Singer, P. (1990). *Animal Liberation*. Avon Books.

Singh, M. S., Park, S. S., Albini, T. A., Canto-Soler, M. V., Klassen, H., MacLaren, R. E., Takahashi, M., Nagiel, A., Schwartz, S. D., & Bharti, K. (2020). Retinal stem cell transplantation: Balancing safety and potential. *Progress in Retinal Eye Research, 75*, Article 100779.

Sinha, V. K., De Buck, S. S., Fenu, L. A., Smit, J. W., Nijsen, M., Gilissen, R. A. H. J., Van Peer, A., Lavrijsen, K., & Mackie, C. E. (2008). Predicting oral clearance in humans: How close can we get with allometry? *Clinical Pharmacokinetics, 47*, 35–45.

Sipes, N. S., Padilla, S., & Knudsen, T. B. (2011). Zebrafish—as an integrative model for twenty-first century toxicity testing. *Birth Defects Research: Part C, Embryo Today, 93*, 256–267.

Skinner, B. F. (1938). *The Behavior of Organisms: An Experimental Analysis*. Appleton-Century.

Skloot, R. (2010). *The Immortal Life of Henrietta Lacks*. Crown.

Sladek, J. R., Collier, T. J., Haber, S. N., Roth, R. H., & Eugene Redmond, D. (1986). Survival and growth of fetal catecholamine neurons transplanted into primate brain. *Brain Research Bulletin, 17*, 809–818.

Sleigh, J., Harvey, M., Voss, L., & Denny, B. (2014). Ketamine—more mechanisms of action than just NMDA blockade. *Trends in Anaesthesia and Critical Care, 4*, 76–81.

Small, W. S. (1900). An experimental study of the mental processes of the rat. *American Journal of Psychology*, *11*, 133–165.

Smeyne, M., Jiao, Y., Shepherd, K. R., & Smeyne, R. J. (2005). Glia cell number modulates sensitivity to MPTP in mice. *Glia*, *52*, 144–152.

Smith, A. M., & Dragunow, M. (2014). The human side of microglia. *Trends in Neurosciences*, *37*, 125–135.

Smith, C. J., & Anderson, S. P. (2017). High discordance in development and organ site distribution of tumors in rats and mice in NTP two-year inhalation studies. *Toxicology Research and Application*, *1*, 1–22.

Smith, C. J., & Perfetti, T. A. (2018). The "false-positive" conundrum in the NTP 2-year rodent cancer study database. *Toxicology Research and Application*, *2*, 1–13.

Smith, C. U. M. (2010). The triune brain in antiquity: Plato, Aristotle, Erasistratus. *Journal of the History of the Neurosciences*, *19*, 1–14.

Smith, D. F. (2013, May 23). Vivien Thomas and the role of dogs in experimental surgery. *Perspectives in Veterinary Medicine*. https://ecommons.cornell.edu/handle/1813/46023.

Smith, M. R., Syed, A., Lukacsovich, T., Purcell, J., Barbaro, B. A., Worthge, S. A., Wei, S. R., Pollio, G., Magnoni, L., Scali, C., et al. (2014). A potent and selective Sirtuin 1 inhibitor alleviates pathology in multiple animal and cell models of Huntington's disease. *Human Molecular Genetics*, *23*, 2995–3007.

Sneader, W. (1998). The discovery of heroin. *The Lancet*, *352*, 1697–1699.

Sneader, W. (2005). *Drug Discovery: A History*. Wiley.

Sneddon, L. U. (2009). Pain perception in fish: Indicators and endpoints. *ILAR Journal*, *50*, 338–342.

Sommer, C. J. (2017). Ischemic stroke: Experimental models and reality. *Acta Neuropathologica*, *133*, 245–261.

Sorby-Adams, A. J., Vink, R., & Turner, R. J. (2018). Large animal models of stroke and traumatic brain injury as translational tools. *American Journal of Physiology: Regulatory. Integrative and Comparative Physiology*, *315*, R165–R190.

Sorge, R. E., Martin, L. J., Isbester, K. A., Sotocinal, S. G., Rosen, S., Tuttle, A. H., Wieskopf, J. S., Acland, E. L., Dokova, A., Kadoura, B., et al. (2014). Olfactory exposure to males, including men, causes stress and related analgesia in rodents. *Nature Methods*, *11*, 629–632.

Sorrells, S. F., Paredes, M. F., Cebrian-Silla, A., Sandoval, K., Qi, D., Kelley, K. W., James, D., Mayer, S., Chang, J., Auguste, K. I., et al. (2018). Human hippocampal neurogenesis drops sharply in children to undetectable levels in adults. *Nature*, *555*, 377–381.

Soto, P. C., Stein, L. L., Hurtado-Ziola, N., Hedrick, S. M., & Varki, A. (2010). Relative over-reactivity of human versus chimpanzee lymphocytes: Implications for the human diseases associated with immune activation. *Journal of Immunology*, *184*, 4185–4195.

Southwell, A. L., Smith-Dijak, A., Kay, C., Sepers, M., Villanueva, E. B., Parsons, M. P., Xie, Y., Anderson, L., Felczak, B., Waltl, S., et al. (2017). An enhanced Q175 knock-in mouse model of Huntington disease with higher mutant huntingtin levels and accelerated disease phenotypes. *Human Molecular Genetics*, *25*, 3654–3675.

Southwell, A. L., Warby, S. C., Carroll, J. B., Doty, C. N., Skotte, N. H., Zhang, W., Villanueva, E. B., Kovalik, V., Xie, Y., Pouladi, M. A., et al. (2013). A fully humanized transgenic mouse model of Huntington disease. *Human Molecular Genetics*, *22*, 18–34.

Souza, S. P., Splitt, S. D., Alvarez, J. A., Wizzard, S., Wilson, J. N., Luo, Z., Baumgarth, N., & Jensen, K. D. (2020). Nfkbid is required for immunity and antibody responses to *Toxoplasma gondii*. *bioRxiv*, Article 2020.06.26.174151.

Speaking of Research Editor. (2011, February 1). Albert Sabin and the monkeys who gave summer back to the children. *Speaking of Research*. https://speakingofresearch.com/2011/02/01/the-monkeys-who-gave-summer-back-to-the-children/.

Speakman, J. R. (2005). Body size, energy metabolism and lifespan. *Journal of Experimental Biology, 208*, 1717–1730.

Spector, J. M., Harrison, R. S., & Fishman, M. C. (2018). Fundamental science behind today's important medicines. *Science Translational Medicine, 10*, Article eaaq1787.

Spera, A. M., Eldin, T. K., Tosone, G., & Orlando, R. (2016). Antiviral therapy for hepatitis C: Has anything changed for pregnant/lactating women? *World Journal of Hepatology, 8*, 557–565.

Spillantini, M. G., Schmidt, M. L., Lee, V. M.-Y., Trojanowski, J. Q., Jakes, R., & Goedert, M. (1997). α-Synuclein in Lewy bodies. *Nature, 388*, 839–840.

Stadnicki, S. W., Fleischman, R. W., Schaeppi, U., & Merriam, P. (1975). Cis-dichlorodiammineplatinum (II) (NSC-119875): Hearing loss and other toxic effects in rhesus monkeys. *Cancer Chemotherapy Reports, 59*, 467–480.

Stanek, L. M., Sardi, S. P., Mastis, B., Richards, A. R., Treleaven, C. M., Taksir, T., Misra, K., Cheng, S. H., & Shihabuddin, L. S. (2014). Silencing mutant huntingtin by adeno-associated virus-mediated RNA interference ameliorates disease manifestations in the YAC128 mouse model of Huntington's disease. *Human Gene Therapy, 25*, 461–474.

Starr, P. A., Kang, G. A., Heath, S., Shimamoto, S., Turner, R. S. (2008). Pallidal neuronal discharge in Huntington's disease: Support for selective loss of striatal cells originating the indirect pathway. *Experimental Neurology, 211*, 227–233.

Statista Research Department. (2016, April 1). Animals used in research experiments worldwide 2016. *Statista*. https://www.statista.com/statistics/639954/animals-used-in-research-experiments-worldwide/.

Stebbings, R., Findlay, L., Edwards, C., Eastwood, D., Bird, C., North, D., Mistry, Y., Dilger, P., Liefooghe, E., Cludts, I., et al. (2007). "Cytokine storm" in the phase I trial of monoclonal antibody TGN1412: Better understanding the causes to improve preclinical testing of immunotherapeutics. *Journal of Immunology, 179*, 3325–3331.

Stebbings, R., Poole, S., & Thorpe, R. (2009). Safety of biologics, lessons learnt from TGN1412. *Current Opinion in Biotechnology, 20*, 673–677.

Steeds, H., Carhart-Harris, R. L., & Stone, J. M. (2015). Drug models of schizophrenia. *Therapeutic Advances in Psychopharmacology, 5*, 43–58.

Steel, D. (2007). *Across the Boundaries: Extrapolation in Biology and Social Science*. Oxford University Press.

Stein, D. G. (2015). Embracing failure: What the phase III progesterone studies can teach about TBI clinical trials. *Brain Injury, 29*, 1259–1272.

Steinbrook, R. (2021). The accelerated approval of aducanumab for treatment of patients with Alzheimer Disease. JAMA Internal Medicine. https://doi.org/10.1001/jamainternmed.2021.4622.

Stent, G. S. (1963). *Molecular Biology of Bacterial Viruses*. Freeman.

Sterling, P., & Laughlin, S. (2015). *Principles of Neural Design*. MIT Press.

Stern, A., & Sorek, R. (2011). The phage-host arms-race: Shaping the evolution of microbes. *BioEssays, 33*, 43–51.

Stern, C. D. (2005). The chick: A great model system becomes even greater. *Developmental Cell, 8*, 9–17.

Steward, O. (2016). A rhumba of "r's": Replication, reproducibility, rigor, robustness: What does a failure to replicate mean? *eNeuro, 3*, e0072–16.2016.

Steward, O., Popovich, P. G., Dietrich, W. D., & Kleitman, N. (2012). Replication and reproducibility in spinal cord injury research. *Experimental Neurology*, *233*, 597–605.

Stockwell, S. (1983). George Thomas Beatson M. D.(1848–1933). *CA: A Cancer Journal for Clinicians*, *33*, 105–107.

Stoker, T., Blair, N., & Barker, R. (2017). Neural grafting for Parkinson's disease: Challenges and prospects. *Neural Regeneration Research*, *12*, 389–392.

Stoppini, L., Buchs, P.-A., & Muller, D. (1991). A simple method for organotypic cultures of nervous tissue. *Journal of Neuroscience Methods*, *37*, 173–182.

Strähle, U., Scholz, S., Geisler, R., Greiner, P., Hollert, H., Rastegar, S., Schumacher, A., Selderslaghs, I., Weiss, C., Witters, H., & Braunbeck, T. (2012). Zebrafish embryos as an alternative to animal experiments—a commentary on the definition of the onset of protected life stages in animal welfare regulations. *Reproductive Toxicology (Elmsford, NY)*, *33*, 128–132.

Strausfeld, N. J., & Hirth, F. (2013). Deep homology of arthropod central complex and vertebrate basal ganglia. *Science*, *340*, 157–161.

Strebhardt, K., & Ullrich, A. (2008). Paul Ehrlich's magic bullet concept: 100 years of progress. *Nature Reviews: Cancer*, *8*, 473–480.

Striedter, G. F. (1998). Stepping into the same river twice: Homologues as recurring attractors in epigenetic landscapes. *Brain, Behavior and Evolution*, *52*, 218–231.

Striedter, G. F. (2005). *Principles of Brain Evolution.* Sinauer Associates.

Striedter, G. F. (2016). Lack of neocortex does not imply fish cannot feel pain. *Animal Sentience*, *1*, Article 021.

Striedter, G. F. (2019). Variation across species and levels: Implications for model species research. *Brain, Behavior and Evolution*, *93*, 1–13.

Striedter, G. F., Belgard, T. G., Chen, C.-C., Davis, F. P., Finlay, B. L., Güntürkün, O., Hale, M. E., Harris, J. A., Hecht, E. E., Hof, P. R., et al. (2014). NSF workshop report: Discovering general principles of nervous system organization by comparing brain maps across species. *Brain, Behavior and Evolution*, *83*, 1–8.

Striedter, G. F., & Northcutt, R. G. (1991). Biological hierarchies and the concept of homology. *Brain, Behavior and Evolution*, *38*, 177–189.

Strong, M. J., Kesavapany, S., & Pant, H. C. (2005). The pathobiology of amyotrophic lateral sclerosis: A proteinopathy? *Journal of Neuropathology and Experimental Neurology*, *64*, 649–664.

Strong, M. K., Southwell, A. L., Yonan, J. M., Hayden, M. R., Macgregor, G. R., Thompson, L. M., & Steward, O. (2012). Age-dependent resistance to excitotoxicity in Htt CAG140 mice and the effect of strain background. *Journal of Huntington's Disease*, *1*, 221–241.

Sturtevant, A. H. (1954). Social implications of the genetics of man. *Science, 120*, 405–407.

Sullivan, J. C., & Finnerty, J. R. (2007). A surprising abundance of human disease genes in a simple "basal" animal, the starlet sea anemone (*Nematostella vectensis*). *Genome*, *50*, 689–692.

Sullivan, P. F. (2005). The genetics of schizophrenia. *PLoS Medicine*, *2*, Article e212.

Sullivan, P. F., Daly, M. J., & O'Donovan, M. (2012). Genetic architectures of psychiatric disorders: The emerging picture and its implications. *Nature Reviews: Genetics*, *13*, 537–551.

Sulston, J. E. (2002). *C. elegans*: The cell lineage and beyond. *Nobel Prize Lecture.* https://www.nobelprize.org/prizes/medicine/2002/sulston/lecture/.

Sumi, T., Tsuneyoshi, N., Nakatsuji, N., & Suemori, H. (2007). Apoptosis and differentiation of human embryonic stem cells induced by sustained activation of c-Myc. *Oncogene, 26*, 5564–5576.

Summers, W. C. (2004). Bacteriophage research: Early history. In E. Kutter & A. Sulakvelidze (Eds.), *Bacteriophages: Biology and Applications* (pp. 5–27). CRC Press.

Summers, W. K., Majovski, L. V., Marsh, G. M., Tachiki, K., & Kling, A. (1986). Oral tetrahydroaminoacridine in long-term treatment of senile dementia, Alzheimer type. *New England Journal of Medicine, 315*, 1241–1245.

Sun, L., Ferreira, J. C. B., & Mochly-Rosen, D. (2011). ALDH2 activator inhibits increased myocardial infarction injury by nitroglycerin tolerance. *Science Translational Medicine, 3*, Article 107ra111.

Sung, N. S., Crowley, W. F., Genel, M., Salber, P., Sandy, L., Sherwood, L. M., Johnson, S. B., Catanese, V., Tilson, H., Getz, K., et al. (2003). Central challenges facing the national clinical research enterprise. *JAMA, 289*, 1278–1287.

Süssmuth, S. D., Haider, S., Landwehrmeyer, G. B., Farmer, R., Frost, C., Tripepi, G., Andersen, C. A., Di Bacco, M., Lamanna, C., Diodato, E., et al. (2015). An exploratory double-blind, randomized clinical trial with selisistat, a SirT1 inhibitor, in patients with Huntington's disease. *British Journal of Clinical Pharmacology, 79*, 465–476.

Suurväli, J., Whiteley, A. R., Zheng, Y., Gharbi, K., Leptin, M., & Wiehe, T. (2020). The laboratory domestication of zebrafish: From diverse populations to inbred substrains. *Molecular Biology and Evolution, 37*, 1056–1069.

Suzich, J. A., Ghim, S. J., Palmer-Hill, F. J., White, W. I., Tamura, J. K., Bell, J. A., Newsome, J. A., Jenson, A. B., & Schlegel, R. (1995). Systemic immunization with papillomavirus L1 protein completely prevents the development of viral mucosal papillomas. *Proceedings of the National Academy of Sciences of the United States of America, 92*, 11553–11557.

Swarup, V., Hinz, F. I., Rexach, J. E., Noguchi, K., Toyoshiba, H., Oda, A., Hirai, K., Sarkar, A., Seyfried, N. T., Cheng, C., et al. (2019). Identification of evolutionarily conserved gene networks mediating neurodegenerative dementia. *Nature Medicine, 25*, 152–164.

Swayze, V. W. (1995). Frontal leukotomy and related psychosurgical procedures in the era before antipsychotics (1935–1954): A historical overview. *American Journal of Psychiatry, 152*, 505–515.

Swindle, M. M., Makin, A., Herron, A. J., Clubb, F. J., Jr., & Frazier, K. S. (2012). Swine as models in biomedical research and toxicology testing. *Veterinary Pathology, 49*, 344–356.

Swinney, D. C., & Anthony, J. (2011). How were new medicines discovered? *Nature Reviews: Drug Discovery, 10*, 507–519.

Tabrizi, S. J., Leavitt, B. R., Landwehrmeyer, G. B., Wild, E. J., Saft, C., Barker, R. A., Blair, N. F., Craufurd, D., Priller, J., Rickards, H., et al. (2019). Targeting huntingtin expression in patients with Huntington's disease. *New England Journal of Medicine, 380*, 2307–2316.

Takahashi, K., Tanabe, K., Ohnuki, M., Narita, M., Ichisaka, T., Tomoda, K., & Yamanaka, S. (2007). Induction of pluripotent stem cells from adult human fibroblasts by defined factors. *Cell, 131*, 861–872.

Takao, K., & Miyakawa, T. (2015). Genomic responses in mouse models greatly mimic human inflammatory diseases. *Proceedings of the National Academy of Sciences of the United States of America, 112*, 1167–1172.

Tallaksen-Greene, S. J., Crouse, A. B., Hunter, J. M., Detloff, P. J., & Albin, R. L. (2005). Neuronal intranuclear inclusions and neuropil aggregates in Hdh CAG(150) knockin mice. *Neuroscience, 131*, 843–852.

Tan, R. H., Ke, Y. D., Ittner, L. M., & Halliday, G. M. (2017). ALS/FTLD: Experimental models and reality. *Acta Neuropathologica, 133*, 177–196.

Tan, S. Y., & Yip, A. (2014). António Egas Moniz (1874–1955): Lobotomy pioneer and Nobel laureate. *Singapore Medical Journal, 55*, 175–176.

Tang, H., & Mayersohn, M. (2006). A global examination of allometric scaling for predicting human drug clearance and the prediction of large vertical allometry. *Journal of Pharmaceutical Sciences, 95*, 1783–1799.

Tannenbaum, J., & Bennet, B. T. (2015). Russell and Burch's 3Rs then and now: The need for clarity in definition and purpose. *Journal of the American Association for Laboratory Animal Science, 54*, 120–132.

Tao, L., & Reese, T. A. (2017). Making mouse models that reflect human immune responses. *Trends in Immunology, 38*, 181–193.

Tarantino, L. (2000). Dissection of behavior and psychiatric disorders using the mouse as a model. *Human Molecular Genetics, 9*, 953–965.

Tartari, M., Gissi, C., Lo Sardo, V., Zuccato, C., Picardi, E., Pesole, G., & Cattaneo, E. (2008). Phylogenetic comparison of huntingtin homologues reveals the appearance of a primitive polyQ in sea urchin. *Molecular Biology and Evolution, 25*, 330–338.

Taub, E. (1980). Somatosensory deafferention research with monkeys: Implications for rehabilitation medicine. In L. P. Ince (Ed.), *Behavioral Psychology in Rehabilitation Medicine; Clinical Applications* (pp. 371–401). Williams & Wilkins.

Taub, E., Crago, J. E., Burgio, L. D., Groomes, T. E., Cook, E. W., DeLuca, S. C., & Miller, N. E. (1994). An operant approach to rehabilitation medicine: Overcoming learned nonuse by shaping. *Journal of the Experimental Analysis of Behavior, 61*, 281–293.

Taubenfeld, S. M., Stevens, K. A., Pollonini, G., Ruggiero, J., & Alberini, C. M. (2002). Profound molecular changes following hippocampal slice preparation: Loss of AMPA receptor subunits and uncoupled mRNA/protein expression. *Journal of Neurochemistry, 81*, 1348–1360.

Telenius, H., Kremer, B., Goldberg, Y. P., Theilmann, J., Andrew, S. E., Zeisler, J., Adam, S., Greenberg, C., Ives, E. J., & Clarke, L. A. (1994). Somatic and gonadal mosaicism of the Huntington disease gene CAG repeat in brain and sperm. *Nature Genetics, 6*, 409–414.

Temple, R., & Stockbridge, N. L. (2007). BiDil for heart failure in black patients: The U.S. food and drug administration perspective. *Annals of Internal Medicine, 146*, 57–62.

Tenaillon, O., Rodriguez-Verdugo, A., Gaut, R. L., McDonald, P., Bennett, A. F., Long, A. D., & Gaut, B. S. (2012). The molecular diversity of adaptive convergence. *Science, 335*, 457–461.

Tennant, R. W., Margolin, B. H., Shelby, M. D., Zeiger, E., Haseman, J. K., Spalding, J., Caspary, W., Resnick, M., Stasiewicz, S., Anderson, B., & Et, A. (1987). Prediction of chemical carcinogenicity in rodents from in vitro genetic toxicity assays. *Science, 236*, 933–941.

Thadani, U., & Rodgers, T. (2006). Side effects of using nitrates to treat angina. *Expert Opinion on Drug Safety, 5*, 667–674.

Thames, H. D. (1988). Early fractionation methods and the origins of the NSD concept. *Acta Oncologica (Stockholm, Sweden), 27*, 89–103.

Thannickal, T. C., Moore, R. Y., Nienhuis, R., Ramanathan, L., Gulyani, S., Aldrich, M., Cornford, M., & Siegel, J. M. (2000). Reduced number of hypocretin neurons in human narcolepsy. *Neuron, 27*, 469–474.

Thariat, J., Hannoun-Levi, J.-M., Sun Myint, A., Vuong, T., & Gérard, J.-P. (2013). Past, present, and future of radiotherapy for the benefit of patients. *Nature Reviews: Clinical Oncology, 10*, 52–60.

Thippeshappa, R., Kimata, J. T., & Kaushal, D. (2020). Toward a macaque model of HIV-1 infection: Roadblocks, progress, and future strategies. *Frontier in Microbiology, 11*, Article 882.

Thomas, J. A., Welch, J. J., Lanfear, R., & Bromham, L. (2010). A generation time effect on the rate of molecular evolution in invertebrates. *Molecular Biology and Evolution, 27*, 1173–1180.

Thomas, J. A., Welch, J. J., Woolfit, M., & Bromham, L. (2006). There is no universal molecular clock for invertebrates, but rate variation does not scale with body size. *Proceedings of the National Academy of Sciences of the United States of America, 103*, 7366–7371.

Thomas, J. W., Touchman, J. W., Blakesley, R. W., Bouffard, G. G., Beckstrom-Sternberg, S. M., Margulies, E. H., Blanchette, M., Siepel, A. C., Thomas, P. J., et al. (2003). Comparative analyses of multi-species sequences from targeted genomic regions. *Nature, 424*, 788–793.

Thomas, K. R., Folger, K. R., & Capecchi, M. R. (1986). High frequency targeting of genes to specific sites in the mammalian genome. *Cell, 44*, 419–428.

Thomas, R. M., Dyke, T. V., Merlino, G., & Day, C.-P. (2016). Concepts in cancer modeling: A brief history. *Cancer Research, 76*, 5921–5925.

Thomas, S. J., & Grossberg, G. T. (2009). Memantine: A review of studies into its safety and efficacy in treating Alzheimer's disease and other dementias. *Clinical Interventions in Aging, 4*, 367–377.

Thompson, D. W. (1917). *On Growth and Form.* Cambridge University Press.

Thomson, J. A., Itskovitz-Eldor, J., Shapiro, S. S., Waknitz, M. A., Swiergiel, J. J., Marshall, V. S., & Jones, J. M. (1998). Embryonic stem cell lines derived from human blastocysts. *Science, 282*, 1145–1147.

Thornton, A. (2019, February 8). This is how many animals we eat each year. *World Economic Forum.* https://www.weforum.org/agenda/2019/02/chart-of-the-day-this-is-how-many-animals-we-eat-each-year/.

Threadgill, D. W., Miller, D. R., Churchill, G. A., & de Villena, F. P.-M. (2011). The Collaborative Cross: A recombinant inbred mouse population for the systems genetic era. *ILAR Journal, 52*, 24–31.

Tian, X., Azpurua, J., Hine, C., Vaidya, A., Myakishev-Rempel, M., Ablaeva, J., Mao, Z., Nevo, E., Gorbunova, V., & Seluanov, A. (2013). High-molecular-mass hyaluronan mediates the cancer resistance of the naked mole rat. *Nature, 499*, 346–349.

Tierney, A. J. (2000). Egas Moniz and the origins of psychosurgery: A review commemorating the 50th anniversary of Moniz's Nobel prize. *Journal of the History of the Neurosciences, 9*, 22–36.

Tolman, E. C. (1948). Cognitive maps in rats and men. *Psychological Review, 55*, 189–208.

Tolosa, E., Vila, M., Klein, C., & Rascol, O. (2020). LRRK2 in Parkinson disease: Challenges of clinical trials. *Nature Reviews: Neurology, 16*, 97–107.

Tomasetti, C., & Vogelstein, B. (2015). Variation in cancer risk among tissues can be explained by the number of stem cell divisions. *Science, 347*, 78–81.

Tomioka, K., & Matsumoto, A. (2010). A comparative view of insect circadian clock systems. *Cellular and Molecular Life Sciences, 67*, 1397–1406.

Trias, E., Ibarburu, S., Barreto-Núñez, R., Babdor, J., Maciel, T. T., Guillo, M., Gros, L., Dubreuil, P., Díaz-Amarilla, P., Cassina, P., et al. (2016). Post-paralysis tyrosine kinase inhibition with masitinib abrogates neuroinflammation and slows disease progression in inherited amyotrophic lateral sclerosis. *Journal of Neuroinflammation, 13*, Article 177.

Troudet, J., Grandcolas, P., Blin, A., Vignes-Lebbe, R., & Legendre, F. (2017). Taxonomic bias in biodiversity data and societal preferences. *Scientific Reports, 7*, Article 9132.

Trubody, B. (2016, June/July). Richard Feynman's philosophy of science. *Philosophy Now, 114.* https://philosophynow.org/issues/114/Richard_Feynmans_Philosophy_of_Science.

Tsilidis, K. K., Panagiotou, O. A., Sena, E. S., Aretouli, E., Evangelou, E., Howells, D. W., Salman, R. A.-S., Macleod, M. R., & Ioannidis, J. P. A. (2013). Evaluation of excess significance bias in animal studies of neurological diseases. *PLoS Biology, 11*, Article e1001609.

Tsuji, O., Sugai, K., Yamaguchi, R., Tashiro, S., Nagoshi, N., Kohyama, J., Iida, T., Ohkubo, T., Itakura, G., Isoda, M., et al. (2019). Concise review: Laying the groundwork for a first-in-human study of an induced pluripotent stem cell-based intervention for spinal cord injury. *Stem Cells, 37*, 6–13.

Tugendreich, S., Bassett, D. E., McKusick, V. A., Boguski, M. S., & Hieter, P. (1994). Genes conserved in yeast and humans. *Human Molecular Genetics, 3*, 1509–1517.

Turner, E. H., Matthews, A. M., Linardatos, E., Tell, R. A., & Rosenthal, R. (2008). Selective publication of antidepressant trials and its influence on apparent efficacy. *New England Journal of Medicine, 358*, 252–260.

Turner, L., & Knoepfler, P. (2016). Selling stem cells in the USA: Assessing the direct-to-consumer industry. *Cell Stem Cell, 19*, 154–157.

Ulndreaj, A., Badner, A., & Fehlings, M. G. (2017). Promising neuroprotective strategies for traumatic spinal cord injury with a focus on the differential effects among anatomical levels of injury. *F1000Research, 6*, Article 1907.

Umeda, T., Maekawa, S., Kimura, T., Takashima, A., Tomiyama, T., & Mori, H. (2014). Neurofibrillary tangle formation by introducing wild-type human tau into APP transgenic mice. *Acta Neuropathologica, 127*, 685–698.

Unethical human experimentation. (2020, October 21). In *Wikipedia*. https://en.wikipedia.org/w/index.php ?title=Unethical_human_experimentation&oldid=984641547.

US Department of Agriculture. (2020, July). *USDA Animal Care: Animal Welfare Act and Animal Welfare Regulations.* (Blue Book.) APHIS 41-35-076. Animal and Plant Health Inspection Service, US Department of Agriculture.

Vahlne, A. (2009). A historical reflection on the discovery of human retroviruses. *Retrovirology, 6*, Article 40.

Valenza, M., Marullo, M., Di Paolo, E., Cesana, E., Zuccato, C., Biella, G., & Cattaneo, E. (2015). Disruption of astrocyte-neuron cholesterol cross talk affects neuronal function in Huntington's disease. *Cell Death and Differentiation, 22*, 690–702.

Valeriani, G., Corazza, O., Bersani, F. S., Melcore, C., Metastasio, A., Bersani, G., & Schifano, F. (2015). Olanzapine as the ideal "trip terminator"? Analysis of online reports relating to antipsychotics' use and misuse following occurrence of novel psychoactive substance-related psychotic symptoms. *Human Psychopharmacology, 30*, 249–254.

Van Bogaert, M. J. V., Groenink, L., Oosting, R. S., Westphal, K. G. C., van der Gugten, J., & Olivier, B. (2006). Mouse strain differences in autonomic responses to stress. *Genes, Brain, and Behavior, 5*, 139–149.

VandeBerg, J. L., & Zola, S. M. (2005). A unique biomedical resource at risk. *Nature, 437*, 30–32.

Van de Lavoir, M.-C., Diamond, J. H., Leighton, P. A., Mather-Love, C., Heyer, B. S., Bradshaw, R., Kerchner, A., Hooi, L. T., Gessaro, T. M., Swanberg, S. E., et al. (2006). Germline transmission of genetically modified primordial germ cells. *Nature, 441*, 766–769.

Van der Graaf, P. H., Benson, N., & Peletier, L. A. (2016). Topics in mathematical pharmacology. *Journal of Dynamics and Differential Equations, 28*, 1337–1356.

Van der Laan, A. L., & Boenink, M. (2015). Beyond bench and bedside: Disentangling the concept of translational research. *Health Care Analysis, 23*, 32–49.

Van der Staay, F. J. (2006). Animal models of behavioral dysfunctions: Basic concepts and classifications, and an evaluation strategy. *Brain Research Reviews, 52*, 131–159.

Van der Staay, F. J., Nordquist, R. E., & Arndt, S. S. (2017). Large farm animal models of human neurobehavioral and psychiatric disorders: Methodological and practical considerations. In P. M. Conn (Ed.), *Animal Models for the Study of Human Disease* (pp. 71–100). Elsevier.

Van der Worp, H. B., Howells, D. W., Sena, E. S., Porritt, M. J., Rewell, S., O'Collins, V., & Macleod, M. R. (2010). Can animal models of disease reliably inform human studies? *PLoS Medicine, 7*, Article e1000245.

Vanes, L. D., Mouchlianitis, E., Patel, K., Barry, E., Wong, K., Thomas, M., Szentgyorgyi, T., Joyce, D., & Shergill, S. (2019). Neural correlates of positive and negative symptoms through the illness course: An fMRI study in early psychosis and chronic schizophrenia. *Scientific Reports, 9*, Article 14444.

Vanhooren, V., & Libert, C. (2013). The mouse as a model organism in aging research: Usefulness, pitfalls and possibilities. *Ageing Research Reviews, 12*, 8–21.

Van Riel, D., & de Wit, E. (2020). Next-generation vaccine platforms for COVID-19. *Nature Materials, 19*, 810–812.

Van Valen, L. M., & Maiorana, V. (1991). HeLa, a new microbial species. *Evolutionary Theory, 10*, 71–74.

Varki, A., & Altheide, T. K. (2005). Comparing the human and chimpanzee genomes: Searching for needles in a haystack. *Genome Research, 15*, 1746–1758.

Varki, N., Anderson, D., Herndon, J. G., Pham, T., Gregg, C. J., Cheriyan, M., Murphy, J., Strobert, E., Fritz, J., Else, J. G., & Varki, A. (2009). Heart disease is common in humans and chimpanzees, but is caused by different pathological processes. *Evolutionary Applications, 2*, 101–112.

Varki, N. M., Strobert, E., Dick Jr., E. J., Benirschke, K., & Varki, A. (2011). Biomedical differences between human and nonhuman hominids: Potential roles for uniquely human aspects of sialic acid biology. *Annual Review of Pathology: Mechanisms of Disease, 6*, 365–393.

Varma, H., Voisine, C., DeMarco, C. T., Cattaneo, E., Lo, D. C., Hart, A. C., & Stockwell, B. R. (2007). Selective inhibitors of death in mutant huntingtin cells. *Nature Chemical Biology, 3*, 99–100.

Verhaegen, F., Dubois, L., Gianolini, S., Hill, M. A., Karger, C. P., Lauber, K., Prise, K. M., Sarrut, D., Thorwarth, D., Vanhove, C., et al. (2018). ESTRO ACROP: Technology for precision small animal radiotherapy research: Optimal use and challenges. *Radiotherapy and Oncology, 126*, 471–478.

Verstraelen, S., & Van Rompay, A. R. (2018). CON4EI: Development of serious eye damage and eye irritation testing strategies with respect to the requirements of the UN GHS/EU CLP hazard categories. *Toxicology in Vitro, 49*, 2–5.

Vidinská, D., Vochozková, P., Šmatlíková, P., Ardan, T., Klíma, J., Juhás, Š., Juhásová, J., Bohuslavová, B., Baxa, M., Valeková, I., Motlík, J., & Ellederová, Z. (2018). Gradual phenotype development in Huntington disease transgenic minipig model at 24 months of age. *Neuro-degenerative Diseases, 18*, 107–119.

Villeneuve, D. L., Crump, D., Garcia-Reyero, N., Hecker, M., Hutchinson, T. H., LaLone, C. A., Landesmann, B., Lettieri, T., Munn, S., Nepelska, M., et al. (2014). Adverse Outcome Pathway (AOP) development I: Strategies and principles. *Toxicological Sciences, 142*, 312–320.

Vinken, M., Vanhaecke, T., & Rogiers, V. (2012). Primary hepatocyte cultures as *in vitro* tools for toxicity testing: Quo vadis? *Toxicology in Vitro, 26*, 541–544.

Vogel, G. (2020). Monkey facility in China lures neuroscientist. *Science, 367*, 496–497.

Vogt, P. K. (2012). Retroviral oncogenes: A historical primer. *Nature Reviews: Cancer, 12*, 639–648.

Volpato, V., & Webber, C. (2020). Addressing variability in iPSC-derived models of human disease: Guidelines to promote reproducibility. *Disease Models and Mechanisms, 13*, Article dmm042317.

Wagner, A. (2015). Causal drift, robust signaling, and complex disease. *PLoS One, 10*, Article e0118413.

Wagner, G. P. (2007). The developmental genetics of homology. *Nature Reviews: Genetics, 8*, 473–479.

Wahlsten, D., Metten, P., Phillips, T. J., Boehm, S. L., Burkhart-Kasch, S., Dorow, J., Doerksen, S., Downing, C., Fogarty, J., Rodd-Henricks, K., et al. (2003). Different data from different labs: Lessons from studies of gene-environment interaction. *Journal of Neurobiology, 54*, 283–311.

Walker, L. C., & Jucker, M. (2017). The exceptional vulnerability of humans to Alzheimer's disease. *Trends in Molecular Medicine, 23*, 534–545.

Walmsley, R. M., & Billinton, N. (2011). How accurate is in vitro prediction of carcinogenicity? *British Journal of Pharmacology, 162*, 1250–1258.

Wang, H., Chen, X., Li, Y., Tang, T.-S., & Bezprozvanny, I. (2010). Tetrabenazine is neuroprotective in Huntington's disease mice. *Molecular Neurodegeneration, 5*, Article 18.

Wang, M., Cao, R., Zhang, L., Yang, X., Liu, J., Xu, M., Shi, Z., Hu, Z., Zhong, W., & Xiao, G. (2020). Remdesivir and chloroquine effectively inhibit the recently emerged novel coronavirus (2019-nCoV) in vitro. *Cell Research, 30*, 269–271.

Wang, R. N., Green, J., Wang, Z., Deng, Y., Qiao, M., Peabody, M., Zhang, Q., Ye, J., Yan, Z., Denduluri, S., et al. (2014). Bone morphogenetic protein (BMP) signaling in development and human diseases. *Genes and Diseases, 1*, 87–105.

Wang, T., Ma, J., Hogan, A. N., Fong, S., Licon, K., Tsui, B., Kreisberg, J. F., Adams, P. D., Carvunis, A.-R., Bannasch, D. L., et al. (2020). Quantitative translation of dog-to-human aging by conserved remodeling of the DNA methylome. *Cell Systems, 11*, 176–185.

Warren, T. K., Jordan, R., Lo, M. K., Ray, A. S., Mackman, R. L., Soloveva, V., Siegel, D., Perron, M., Bannister, R., Hui, et al. (2016). Therapeutic efficacy of the small molecule GS-5734 against Ebola virus in rhesus monkeys. *Nature, 531*, 381–385.

Warrick, J. M., Paulson, H. L., Gray-Board, G. L., Bui, Q. T., Fischbeck, K. H., Pittman, R. N., & Bonini, N. M. (1998). Expanded polyglutamine protein forms nuclear inclusions and causes neural degeneration in *Drosophila. Cell, 93*, 939–949.

Watnick, R. S. (2012). The role of the tumor microenvironment in regulating angiogenesis. *Cold Spring Harbor Perspectives in Medicine, 2*, Article a006676.

Watson, J. B. (1903). *Animal Education: An Experimental Study on the Psychical Development of the White Rat, Correlated with the Growth of Its Nervous System.* University of Chicago Press.

Watson, J. B. (1913). Psychology as the behaviorist views it. *Psychological Review, 20*, 158–177.

Watson, J. D., & Crick, F. H. C. (1953). Molecular structure of nucleic acids: A structure for deoxyribose nucleic acid. *Nature, 171*, 737–738.

Watt, F. M., Frye, M., & Benitah, S. A. (2008). MYC in mammalian epidermis: How can an oncogene stimulate differentiation? *Nature Reviews: Cancer, 8*, 234–242.

Wayman, S. (1966, Feb 4) Concentration camps for dogs. Life Magazine.

Weatherall, D. (2006). *The Use of Non-human Primates in Research—The Weatherall Report* The Academy of Medical Sciences. https://mrc.ukri.org/documents/pdf/the-use-of-non-human-primates-in-research/.

Wege, A. K. (2018). Humanized mouse models for the preclinical assessment of cancer immunotherapy. *BioDrugs, 32*, 245–266.

Wegorzewska, I., Bell, S., Cairns, N. J., Miller, T. M., & Baloh, R. H. (2009). TDP-43 mutant transgenic mice develop features of ALS and frontotemporal lobar degeneration. *Proceedings of the National Academy of Sciences of the United States of America, 106*, 18809–18814.

Weiner, J. (1999). *Time, Love, Memory: A Great Biologist and His Quest for the Origins of Behavior.* Vintage.

Wellbourne-Wood, J., & Chatton, J.-Y. (2018). From cultured rodent neurons to human brain tissue: Model systems for pharmacological and translational neuroscience. *ACS Chemical Neuroscience, 9*, 1975–1985.

Wellems, T. E. (2010). Optimism, persistence, and our collective crystal ball. *American Journal of Tropical Medicine and Hygiene, 83*, 1–6.

Wernersson, R., Schierup, M. H., Jørgensen, F. G., Gorodkin, J., Panitz, F., Stærfeldt, H.-H., Christensen, O. F., Mailund, T., Hornshøj, H., Klein, A., et al. (2005). Pigs in sequence space: A 0.66X coverage pig genome survey based on shotgun sequencing. *BMC Genomics, 6*, Article 70.

West, L. J., Pierce, C. M., & Thomas, W. D. (1962). Lysergic acid diethylamide: Its effects on a male Asiatic elephant. *Science, 138*, 1100–1103.

Westergaard, D., Moseley, P., Sørup, F. K. H., Baldi, P., & Brunak, S. (2019). Population-wide analysis of differences in disease progression patterns in men and women. *Nature Communications, 10*, Article 666.

Westerink, R. H. S., & Ewing, A. G. (2008). The PC12 cell as model for neurosecretion. *Acta Physiologica (Oxford), 192*, 273–285.

Whishaw, I. Q. (1999). The laboratory rat, the Pied Piper of twentieth century neuroscience. *Brain Research Bulletin, 50*, Article 411.

Whishaw, I. Q., & Tomie, J. A. (1996). Of mice and mazes: Similarities between mice and rats on dry land but not water mazes. *Physiology and Behavior, 60*, 1191–1197.

Whiteford, H. A., Ferrari, A. J., Degenhardt, L., Feigin, V., & Vos, T. (2015). The global burden of mental, neurological and substance use disorders: An analysis from the Global Burden of Disease study 2010. *PLoS One, 10*, Article e0116820.

Whitesides, G. M. (2006). The origins and the future of microfluidics. *Nature, 442*, 368–373.

Wild, D. E. J., & Tabrizi, S. J. (2017). Therapies targeting DNA and RNA in Huntington's disease. *Lancet Neurology, 16*, 837–847.

Wild, E. J. (2016). Huntington's disease: The most curable incurable brain disorder? *EBioMedicine, 8*, 3–4.

Wiley, D. S., Redfield, S. E., & Zon, L. I. (2017). Chemical screening in zebrafish for novel biological and therapeutic discovery. *Methods in Cell Biology, 138*, 651–679.

Wilhelmus, K. R. (2001). The Draize eye test. *Survey of Ophthalmology, 45*, 493–515.

Williams, K. J. (2009). The introduction of "chemotherapy" using arsphenamine—the first magic bullet. *Journal of the Royal Society of Medicine, 102*, 343–348.

Willner, P. (1984). The validity of animal models of depression. *Psychopharmacology (Berlin), 83*, 1–16.

Willner, P. (2005). Chronic mild stress (CMS) revisited: Consistency and behavioural-neurobiological concordance in the effects of CMS. *Neuropsychobiology, 52*, 90–110.

Wiseman, F. K., Al-Janabi, T., Hardy, J., Karmiloff-Smith, A., Nizetic, D., Tybulewicz, V. L. J., Fisher, E. M. C., & Strydom, A. (2015). A genetic cause of Alzheimer disease: Mechanistic insights from Down syndrome. *Nature Reviews: Neuroscience, 16*, 564–574.

Withering, W. (1785). *An Account of the Foxglove, and Some of Its Medical Uses: With Practical Remarks on Dropsy, and Other Diseases*. M. Swinney.

Witkowski, J. A. (1980). Dr. Carrel's immortal cells. *Medical History, 24*, 129–142.

Woerz, I., Lohmann, V., & Bartenschlager, R. (2009). Hepatitis C virus replicons: Dinosaurs still in business? *Journal of Viral Hepatitis, 16*, 1–9.

Wojnarowicz, M. W., Fisher, A. M., Minaeva, O., & Goldstein, L. E. (2017). Considerations for experimental animal models of concussion, traumatic brain injury, and chronic traumatic encephalopathy—these matters matter. *Frontiers in Neurology, 8*, Article 240.

Wolf, E., Kemter, E., Klymiuk, N., & Reichart, B. (2019). Genetically modified pigs as donors of cells, tissues, and organs for xenotransplantation. *Animal Frontiers, 9,* 13–20.

Wolf, S. L., Winstein, C. J., Miller, J. P., Thompson, P. A., Taub, E., Uswatte, G., Morris, D., Blanton, S., Nichols-Larsen, D., & Clark, P. C. (2008). Retention of upper limb function in stroke survivors who have received constraint-induced movement therapy: The EXCITE randomised trial. *Lancet Neurology, 7,* 33–40.

Wong, D. T., Horng, J. S., Bymaster, F. P., Hauser, K. L., & Molloy, B. B. (1974). A selective inhibitor of serotonin uptake: Lilly 110140, 3-(p-trifluoromethylphenoxy)-n-methyl-3-phenylpropylamine. *Life Science, 15,* 471–479.

Wong, D. T., Perry, K. W., & Bymaster, F. P. (2005). The discovery of fluoxetine hydrochloride (Prozac). *Nature Reviews: Drug Discovery, 4,* 764–774.

Wong, H., Grossman, S. J., Bai, S. A., Diamond, S., Wright, M. R., Grace, J. E., Qian, M., He, K., Yeleswaram, K., & Christ, D. D. (2004). The chimpanzee (*Pan troglodytes*) as a pharmacokinetic model for selection of drug candidates: Model characterization and application. *Drug Metabolism and Disposition, 32,* 1359–1369.

Wood Library Museum of Anesthesiology. (2021). History of anesthesia—interactive timeline. https://www.woodlibrarymuseum.org/history-of-anesthesia/.

Wood, V., Gwilliam, R., Rajandream, M.-A., Lyne, M., Lyne, R., Stewart, A., Sgouros, J., Peat, N., Hayles, J., Baker, S., et al. (2002). The genome sequence of *Schizosaccharomyces pombe. Nature, 415,* 871–880.

Woodcock, M. E., Idoko-Akoh, A., & McGrew, M. J. (2017). Gene editing in birds takes flight. *Mammalian Genome, 28,* 315–323.

Woodman, B., Butler, R., Landles, C., Lupton, M. K., Tse, J., Hockly, E., Moffitt, H., Sathasivam, K., & Bates, G. P. (2007). The HdhQ150/Q150 knock-in mouse model of HD and the R6/2 exon 1 model develop comparable and widespread molecular phenotypes. *Brain Research Bulletin, 72,* 83–97.

Workman, A. D., Charvet, C. J., Clancy, B., Darlington, R. B., & Finlay, B. L. (2013). Modeling transformations of neurodevelopmental sequences across mammalian species. *Journal of Neuroscience, 33,* 7368–7383.

World Health Organization. (2018a). The top 10 causes of death. https://www.who.int/news-room/fact-sheets/detail/the-top-10-causes-of-death.

World Health Organization. (2018b). Global Health Estimates 2016: Deaths by cause, age, sex, by country and by region, 2000–2016. http://www.who.int/healthinfo/global_burden_disease/estimates/en/.

World Health Organization. (2020). WHO Model Lists of Essential Medicines. https://www.who.int/groups/expert-committee-on-selection-and-use-of-essential-medicines/essential-medicines-lists.

World Medical Association. (2013). World Medical Association Declaration of Helsinki: Ethical principles for medical research involving human subjects. *JAMA, 310,* 2191.

Wu, L., Wang, D., & Evans, J. A. (2019). Large teams have developed science and technology; small teams have disrupted it. *Nature, 566,* 378–382.

Würbel, H. (2000). Behaviour and the standardization fallacy. *Nature Genetics, 26,* 263–263.

Würbel, H. (2007). Refinement of rodent research through environmental enrichment and systematic randomization. *NC3Rs, 9,* 1–9.

Wurm, F. M. (2004). Production of recombinant protein therapeutics in cultivated mammalian cells. *Nature Biotechnology, 22,* 1393–1398.

Wyand, M. S. (1992). The use of SIV-infected Rhesus monkeys for the preclinical evaluation of AIDS drugs and vaccines. *AIDS Research and Human Retroviruses, 8,* 349–356.

Xie, R., Zheng, W., Guan, L., Ai, Y., & Liang, Q. (2020). Engineering of hydrogel materials with perfusable microchannels for building vascularized tissues. *Small, 16*, 1902838.

Xiong, Y., Mahmood, A., & Chopp, M. (2013). Animal models of traumatic brain injury. *Nature Reviews: Neuroscience, 14*, 128–142.

Xu, E. G., Lin, N., Cheong, R. S., Ridsdale, C., Tahara, R., Du, T. Y., Das, D., Zhu, J., Silva, L. P., Azimzada, et al. (2019). Artificial turf infill associated with systematic toxicity in an amniote vertebrate. *Proceedings of the National Academy of Sciences of the United States of America, 116*, 25156–25161.

Xu, Y.-P., Qiu, Y., Zhang, B., Chen, G., Chen, Q., Wang, M., Mo, F., Xu, J., Wu, J., Zhang, R.-R., et al. (2019). Zika virus infection induces RNAi-mediated antiviral immunity in human neural progenitors and brain organoids. *Cell Research, 29*, 265–273.

Yang, N., Ng, Y. H., Pang, Z. P., Südhof, T. C., & Wernig, M. (2011). Induced neuronal (iN) cells: How to make and define a neuron. *Cell Stem Cell, 9*, 517–525.

Yang, S.-H., Cheng, P.-H., Banta, H., Piotrowska-Nitsche, K., Yang, J.-J., Cheng, E. C. H., Snyder, B., Larkin, K., Liu, J., Orkin, J., et al. (2008). Towards a transgenic model of Huntington's disease in a non-human primate. *Nature, 453*, 921–924.

Yanofsky, C. (2007). Establishing the triplet nature of the genetic code. *Cell, 128*, 815–818.

Yarchoan, R., & Broder, S. (1987). Development of antiretroviral therapy for the acquired immunodeficiency syndrome and related disorders. *New England Journal of Medicine, 316*, 557–564.

Yartsev, M. M. (2017). The emperor's new wardrobe: Rebalancing diversity of animal models in neuroscience research. *Science, 358*, 466–469.

Yeger-Lotem, E., Riva, L., Su, L. J., Gitler, A. D., Cashikar, A. G., King, O. D., Auluck, P. K., Geddie, M. L., Valastyan, J. S., Karger, D. R., et al. (2009). Bridging high-throughput genetic and transcriptional data reveals cellular responses to alpha-synuclein toxicity. *Nature Genetics, 41*, 316–323.

Yeung, J. (2020, March 29). The US keeps millions of chickens in secret farms to make flu vaccines. But their eggs won't work for coronavirus. *CNN Health.* https://www.cnn.com/2020/03/27/health/chicken-egg-flu-vaccine-intl-hnk-scli/index.html.

Yirmiya, R. (1996). Endotoxin produces a depressive-like episode in rats. *Brain Research, 711,* 163–174.

Yong, K. S. M., Her, Z., & Chen, Q. (2018). Humanized mice as unique tools for human-specific studies. *Archivum Immunologiae et Therapiae Experimentalis, 66*, 245–266.

Young, R. M. (1990). *Mind, Brain, and Adaptation in the Nineteenth Century: Cerebral Localization and Its Biological Context from Gall to Ferrier.* Oxford University Press.

Yu, J., & Thomson, J. A. (2008). Pluripotent stem cell lines. *Genes and Development, 22*, 1987–1997.

Yudell, M., Roberts, D., DeSalle, R., & Tishkoff, S. (2016). Taking race out of human genetics. *Science, 351*, 564–565.

Yu-Taeger, L., Bonin, M., Stricker-Shaver, J., Riess, O., & Nguyen, H. H. P. (2017). Dysregulation of gene expression in the striatum of BACHD rats expressing full-length mutant huntingtin and associated abnormalities on molecular and protein levels. *Neuropharmacology, 117*, 260–272.

Yutzy, S. H., Woofter, C. R., Abbott, C. C., Melhem, I. M., & Parish, B. S. (2012). The increasing frequency of mania and bipolar disorder. *Journal of Nervous and Mental Disease, 200*, 380–387.

Zahs, K. R., & Ashe, K. H. (2010). "Too much good news"—are Alzheimer mouse models trying to tell us how to prevent, not cure, Alzheimer's disease? *Trends in Neurosciences, 33*, 381–389.

Zarate, C. A., Singh, J. B., Carlson, P. J., Brutsche, N. E., Ameli, R., Luckenbaugh, D. A., Charney, D. S., & Manji, H. K. (2006). A randomized trial of an N-methyl-D-aspartate antagonist in treatment-resistant major depression. *Archives of General Psychiatry, 63*, 856–864.

Zdobnov, E. M., von Mering, C., Letunic, I., & Bork, P. (2005). Consistency of genome-based methods in measuring Metazoan evolution. *FEBS Letters, 579*, 3355–3361.

Zeiss, C. J. (2015). Improving the predictive value of interventional animal models data. *Drug Discovery Today, 20*, 475–482.

Zeiss, C. J., Allore, H. G., & Beck, A. P. (2017). Established patterns of animal study design undermine translation of disease-modifying therapies for Parkinson's disease. *PLoS One, 12*, Article e0171790.

Zeller, E. A., & Barsky, J. (1952). In vivo inhibition of liver and brain monoamine oxidase by 1-Isonicotinyl-2 -isopropyl hydrazine. *Proceedings of the Society for Experimental Biology and Medicine, 81*, 459–461.

Zerhouni, E. (2003). The NIH Roadmap. *Science, 302*, 63–72.

Zhao, H., Sun, Q.-L., Duan, L.-J., Yang, Y.-D., Gao, Y.-S., Zhao, D.-Y., Xiong, Y., Wang, H.-J., Song, J.-W., Yang, K.-T., et al. (2019). Is cell transplantation a reliable therapeutic strategy for spinal cord injury in clinical practice? A systematic review and meta-analysis from 22 clinical controlled trials. *European Spine Journal, 28*, 1092–1112.

Zheng, B., Atwal, J., Ho, C., Case, L., He, X. -l., Garcia, K. C., Steward, O., & Tessier-Lavigne, M. (2005). Genetic deletion of the Nogo receptor does not reduce neurite inhibition *in vitro* or promote corticospinal tract regeneration in vivo. *Proceedings of the National Academy of Sciences of the United States of America, 102*, 1205–1210.

Zheng, B., Ho, C., Li, S., Keirstead, H., Steward, O., & Tessier-Lavigne, M. (2003). Lack of enhanced spinal regeneration in Nogo-deficient mice. *Neuron, 38*, 213–224.

Zheng, J. Q., Felder, M., Connor, J. A., & Poo, M. (1994). Turning of nerve growth cones induced by neurotransmitters. *Nature, 368*, 140–144.

Zinder, N. D., & Lederberg, J. (1952). Genetic exchange in *Salmonella. Journal of Bacteriology, 64*, 679–699.

Zoghbi, H. Y. (2013). The basics of translation. *Science, 339*, 250.

Zoll, P. M. (1973). Development of electric control of cardiac rhythm. *JAMA, 226*, 881–886.

Zuccato, C., Ciammola, A., Rigamonti, D., Leavitt, B. R., Goffredo, D., Conti, L., MacDonald, M. E., Friedlander, R. M., Silani, V., Hayden, M. R., et al. (2001). Loss of huntingtin-mediated BDNF gene transcription in Huntington's disease. *Science, 293*, 493–498.

Zwiegers, P., Lee, G., & Shaw, C. A. (2014). Reduction in hSOD1 copy number significantly impacts ALS phenotype presentation in G37R (line 29) mice: Implications for the assessment of putative therapeutic agents. *Journal of Negative Results in Biomedicine, 13*, Article 14.

INDEX

Note: Page numbers in italics indicate a figure and page numbers in bold indicate a table.